Contributions to Palaeobotany

A retirement tribute to
Professor W. S. Lacey

Edited by M. W. Dick and
D. Edwards

Reprinted from the Botanical Journal of the Linnean Society
Volume 86, Numbers 1 & 2, 1983

Published for the Linnean Society of London

ACADEMIC PRESS

London   New York   Toronto   Sydney   San Francisco

ACADEMIC PRESS INC. (LONDON) LIMITED
24/28 Oval Road
London NW1
(Registered Office)

US edition published by
ACADEMIC PRESS INC.
111 Fifth Avenue
New York
New York 10003

ISBN 0 12 215120 8

Printed in Great Britain by
The Whitefriars Press Ltd., Tonbridge

Contents

Cover: *Archaeosigillaria kidstonii* Kräusel & Weyland (1949)
(syn.: *Clwydia decussata* Lacey (1962), Chaloner & Boureau (1967))

Professor W. S. Lacey, B.Sc., Ph.D., D.Sc., F.L.S., F.G.S.

Botanical Journal of the Linnean Society (1983), *86:* iii–ix

Professor W. S. Lacey, B.Sc., Ph.D., D.Sc., F.L.S., F.G.S.

Bill Lacey is known internationally for his contributions to the study of fossil plants—an interest that, although it played a major role in his academic life, never completely submerged his interest in living plants, their ecology and latterly their protection.

As a schoolboy in Leicestershire, the pull of Charnwood Forest was to prove a strong influence. Involving the patronage of a landowner in the area, he developed his interest in local plants and their biology. This was to form a firm basis that influenced his career and that has stayed with him throughout his working life.

In 1936 he went to Reading to study first a B.Sc. General Hons. then a Special Hons. degree in Botany. Here, under the influence of the teaching of Professor Tom Harris, his interest in Palaeobotany was awakened; for his final year he did a special project with Tom Harris. Thus began a friendship that remains to the present day. This friendship was to prove a strong influence on Bill Lacey throughout a number of lean years, both during and after the war, when opportunities were few and far between.

Following his graduation, and a year obtaining a Diploma in Education, Bill Lacey took up employment as an analytical chemist at the Royal Ordnance Factory at Chorley in Lancashire. There he was responsible for the quality control of explosives. I imagine that he looked at many a quarry face in later years and wished for a pound, or two, of the stuff he had analysed to make the quarry reveal its secrets! Work of this kind failed to deflect him from his first love—botany—and soon he published a little book entitled the *Flora of Chorley and District*, undoubtedly the result of many weekend and evening walks diligently spent recording many observations which later were used as a basis for the publication. Suffice it to say that when Travis' *Flora of West Lancashire* was published, some 30 years later, the *Flora of Chorley and District* remained the principal citation of work from that area.

From 1943 he became associated with Education and held a part-time lecturership at the Wigan and District Mining and Technical College. In 1944 he left the Ministry of Supply and became a full-time lecturer at the College. Soon after, he was instrumental in establishing the Wigan and District Field Club and led its first field trip in 1945. This was a Fungus Foray and Bill, as thorough as ever, produced a key to the *Larger Fungi found in the Wigan District* which was distributed to members on this trip and no doubt contributed to the success of the occasion.

Palaeobotany was not forgotten—the Millstone Grit quarries attracted his interest, as did coal balls from the waste heaps of Collieries in the Burnley area. Here he had the help of a member of his family. His wife's 'Uncle Vernon' (Vernon Dean of Accrington) was a respected amateur geologist who was able to help Bill to become familiar with the area in a relatively short space of time. Then there was the famous cycling trip to North Wales in 1945 when, in the company of

0024–4074/83/010iii+07 $03.00/0

the now Reader in Geology at Birmingham University, E. D. Lacy, who was also at the R.O.F. Chorley at the time, he was introduced to the great Carboniferous Limestone deposits of the area. The third member of that group was Leslie Marchant, who was at Reading with Bill, later also at the R.O.F., and who, to this day, remains a great personal friend. All the material collected on that trip was sent back to Wigan by rail, where it was first analysed.

This interest, once awakened, received its greatest possible stimulus when, in 1946, he was appointed Assistant Lecturer in Botany at the University College of North Wales, Bangor, where he was to stay for the remainder of his working life, being made Lecturer in 1949, Senior Lecturer in 1956, Reader in 1968 and finally being awarded a personal chair in 1976. Not that leaving Chorley produced a sudden loss of interest in the area—for the *Chorley Citizen* in 1947 was still carrying 'Nature Notes' written by him—no doubt still visiting the area to see his wife's family. For another of his 'achievements' whilst at the R.O.F. was to meet Edna and make her his wife; she quickly established herself as a constant support of his many personal ambitions.

One of these ambitions was to read for a higher degree and so, once established in Bangor, he began his extra-mural Ph.D. study, by part-time research, under the direction of Tom Harris. The subject of his thesis was the flora of the Lower Carboniferous of North Wales. Field work was, in those days, entirely by train, bus or bicycle; this no doubt led to a thorough examination of hand specimens before they became part of the load to take home! In 1955 he gained his Ph.D. and began to publish accounts of his work which made a major contribution to a previously little known strata in an area of the country that had received little attention. Little attention, at least, from palaeobotanists that is, for Professor T. Neville George and Dr Ernest Neaverson had published many papers concerning its geology. Preliminary work by Professor John Walton had been carried out in the 1920's and there was much fruitful co-operation between Professor Walton and Bill Lacey that greatly influenced the work of the latter.

Initially Professor Walton provided some of his own plant fossil material from the Kilpatrick Hills which although clearly not North Wales nevertheless was of a similar geological age to the deposits found there. There were many visits by Bill to Glasgow and these formed a bond of respect and co-operation that was maintained up until the time of Professor Walton's death. Dr Neaverson, who was at the University of Liverpool, generously assisted Bill Lacey with collecting trips, contributing in no mean way from his almost encyclopaedic knowledge of Carboniferous stratigraphical palaeontology.

In 1958 Bill Lacey spent several months at the University College of Rhodesia and Nyasaland, where with the assistance of Professor Geoffrey Bond an investigation into the Karroo Floras began. This was to flourish and on subsequent trips to the southern Hemisphere, in particular to India and back to the African continent, established him as a leading authority in Gondwana palaeobotany and contributed greatly to our understanding of the nature of the fructifications of *Glossopteris, Gangamopteris* and related genera. In later years he was able to assist the British Antarctic Survey in the analysis of plant remains of Permian age from the Central Transantarctic mountains.

In 1963 he went on sabbatical leave as visiting Professor to the University of Southern Illinois, to discover that coal balls 'grow' to huge sizes in America— where else!

Early in his researches into the three-dimensional nature of fossils, along with Dr A. J. Willis and Dr Ken Joy in 1956, he developed a technique that enabled the speed of palaeobotanical enquiry to take off at an almost exponential rate. Whereas the previous technique, incidentally following a paper by Walton in *Nature* in 1926, had enabled researchers to produce one thin section a day by making a 'cellulose peel' of the fossil material, the Joy, Willis & Lacey 'rapid cellulose peel technique' produced one thin section about every 20 min. Many palaeobotanists, therefore, found their productivity increased overnight!

Many papers followed in the 1960's and 1970's that concerned his interests in Carboniferous and Permian floras. Latterly he showed an interest in the Devonian of Eire. The early quality of this work was recognized when, in 1968, he was awarded a D.Sc. degree of the University of Wales, to be followed shortly by his appointment as Reader of that University.

In January 1963, in a rather sombre lecture room in the University College at Bangor, a meeting was held out of which the North Wales Naturalists was born. Bill Lacey was a founder member and became its first Secretary. Thus began an almost tireless involvement with the conservation of wildlife in North Wales. At the outset, in addition to the rigours of being secretary to a new venture, he also edited the news letter; subsequently, perhaps even inevitably, in 1973 he became Chairman of the Trust, a post that he holds to the present day. There is adequate evidence of his efforts in the many interesting and varied wildlife sites in North Wales that, thanks to his own drive and enthusiasm, are now protected against the advancing technological age.

This involvement with the Trust resulted in his being invited to serve on the Nature Conservancy Council for Wales, where for 4 years he served the interests of Wales on a much broader front. In 1971 he became a member of the National Parks Commission for Wales—again a post that he retains to this present day. In 1970 he designed and edited the publication *Welsh Wildlife in Trust* which, in part, had resulted from his involvement with other conservation bodies throughout the Principality. He was delighted that H.R.H. Prince Charles, Prince of Wales, graciously accepted his request to write the foreword for that publication. He felt that the interest shown by the Prince heightened the importance placed upon conservation in Wales.

All of this effort received due recognition when, in her Jubilee Year, 1977, Queen Elizabeth awarded him the Queen's Silver Jubilee medal for services to Nature Conservation in Wales.

During his time at Bangor Bill Lacey has been an influence and a stimulus to many people. It must also be said that he was happy to learn from others—early bryological excursions or forays into the mountains were heightened by the presence of many professional and amateur field naturalists from in and around Bangor. It is impossible to pick all of them out by name but mention must be made of Evan Roberts, to whom many are most grateful. Bill must be counted amongst these. Also to Dick Roberts with whom Bill did so much botanizing. It was a particularly gratifying moment for Bill when in 1980 the University of Wales honoured Dick Roberts by awarding him an M.Sc. *honoris causa* made even more memorable when he was presented to the congregation by Professor W. S. Lacey. It is this concern for people as well as for plants that Bill Lacey has, that is cherished and will be remembered by many people from many walks of life. He is a natural teacher of botany with an almost inexhaustible knowledge of things around him. Whether in a damp oak woodland surrounded by the luxuriant growth of mosses and bryophytes—or on his beloved Cors Goch Fen on Anglesey—he appears to be completely at home. The same may be said when, with lens in hand, he is holding forth about some Lower Carboniferous fossil.

Generations of students, together with colleagues and friends, would join in thanking him for the inspiration that he has given to them over the years. We would all wish him many more years 'botanizing'—perhaps in Wales?

Professor F. A. Hibbert
Department of Biological Sciences,
Portsmouth Polytechnic.

WILLIAM J. LACEY—WORK PUBLISHED AND IN PRESS

(1) 1941. On *Calamostachys oldhamia* Hick and Lomax and its inclusion in *C. casheana* Williamson. *Annals and Magazine of Natural History, 7:* 536–540.

(2) 1943. The sporangiophore of *Calamostachys. New Phytologist, 42:* 98–102.

(3) 1946. Notes on the Orb-cockle, *Sphaerium corneum* L., and its behaviour in toxic solutions. *North Western Naturalist, 21:* 188–197.

(4) 1947. The chough breeding in North Wales. *North Western Naturalist, 22:* 107.

(5) 1947. *Galinsoga parviflora* Cav. in Leicestershire. *North Western Naturalist, 22:* 114–115.

(6) 1947. *Helleborus foetidus* L. in the Wirral. *North Western Naturalist, 22:* 116.

(7) 1947. *Epilobium pedunculare* A. Cunn. in Caerns. *North Western Naturalist, 22:* 116.

(8) 1948. The genus *Galinsoga* Ruiz and Pavon in Leicestershire. *North Western Naturalist, 23:* 162–166.

(9) 1949. Ecological studies on Puffin Island. *Proceedings of the Llandudno, Colwyn Bay and District Field Club:* 26–34.

(10) 1950. A note on the fruiting of *Mnium undulatum* L. *Transactions of the British Bryological Society, 1:* 370–372.

(11) 1950. The examination of fossil plants. *Proceedings of the Dyserth and District Field Club:* 49–52.

(12) 1950. A rich plant-bed in the Millstone Grit, near Blackburn, Lancashire. *The Naturalist:* 49–50.

(13) 1951. Further notes on the fruiting of *Mnium undulatum* L. *Transactions of the British Bryological Society, 1:* 488–489.

(14) 1952. Correlation of the Lower Brown Limestone of North Wales with part of the Lower Carboniferous Succession in Scotland and Northern England. *Report of the International Geological Congress, 18th Session, Great Britain (1948) 10:* 18–25.

(15) 1952. Additions to the Lower-Carboniferous flora of North Wales. *Compte Rendu du 3ème Congrès pour l'avancement des études de Stratégraphie et de Géologie du Carbonifère, Heerlen (Holland) 1951, 2:* 375–377.

(16) 1952. Additions to the Millstone Grit flora of Lancashire. *Compte Rendu du 3ème Congrès pour l'avancement des études de Stratégraphie et de Géologie du Carbonifere, Heerlen (Holland) 1951, 2:* 379–383.

(17) 1953. A boring in the Millstone Grit, near Darwen, Lancashire. *Liverpool and Manchester Geological Journal, 1:* 194–199.

(18) 1953. Methods of palaeobotany. *North Western Naturalist, (New Series), 1:* 234–249.

(19) 1953. Scottish Lower Carboniferous plants: *Eristophyton waltoni* sp. nov. and *Endoxylon zonatum* (Kidston) Scott in Dunbartonshire. *Annals of Botany, 17:* 579–596.

(20) 1954. Notes on the flora of the Chorley District of South Lancashire. *North Western Naturalist (New Series), 2:* 526–558.

(21) 1955. *Orchis traunsteineri* Saut. in Wales. *Proceedings of the Botanical Society of the British Isles, 1:* 296–300.

(22) 1956. (with K. W. Joy and A. J. Willis). A rapid cellulose peel technique in palaeobotany. *Annals of Botany, 20:* 635–637.

(23) 1957. A comparison of the spread of *Galinsoga parviflora* and *G. ciliata* in Britain. In J. E. Lousley (Ed.), *Progress in the Study of the British Flora* (Botanical Society of the British Isles): 109–115.

(24) 1957. *Campylopus introflexus* (Hedw.) Brid. in Anglesey. *Transactions of the British Bryological Society, 3:* 302.

(25) 1957. The flora of Ynys Seiriol, with some notes on the effects of myxomatosis. *Nature in Wales, 3:* 1–7.

(26) 1957. (with K. W. Joy and A. J. Willis). Observations on the aphlebiae and megasporangia of *Stauropteris burntislandica* P. Bertrand. *Annals of Botany, 21:* 621–625.

(27) 1957. Some effects of myxomatosis on vegetation, with special reference to Ynys Seiriol. *Transactions of the Anglesey Antiquarian Society and Field Club:* 26–33.

(28) 1958. (with R. H. Roberts). Further notes on *Dactylorchis traunsteineri* (Saut.) Vermeulen in Wales. *Proceedings of the Botanical Society of the British Isles, 3:* 22–27.

(29) 1959. Occurrence of presumed glossopteridean fructifications in Rhodesia and Nyasaland. *Nature, London, 184:* 1592–1593.

(30) 1961. Some aspects of palaeo-ecology in the Karroo of Rhodesia and Nyasaland. *Proceedings of the Linnean Society of London (Session 1960), 172:* 7–8.

(31) 1961. Report on fossils from Chalala and Mangulane in the province of Moçambique. *Serviços de Geologia e Minas (Lourenço Marques) Boletim No. 27:* 5–14.

(32) 1961. Studies in the Karroo floras of Rhodesia and Nyasaland. Part 1. A geological account of the plant-bearing deposits. *Proceedings and Transactions of the Rhodesia Scientific Association, 49:* 26–53.

(33) 1962. New records of fossil plants in the Lower Karroo of Southern Rhodesia and Nyasaland. *Compte Rendu du 4ème Congrès Stratégraphique et de Géologie du Carbonifère, Heerlen (Holland) 1958, 2:* 367–368.

(34) 1962. Welsh Lower Carboniferous Plants. 1. The flora of the Lower Brown Limestone in the Vale of Clwyd, North Wales. *Palaeontographica (Stuttgart), B, 111:* 126–160.

(35) 1963. Palaeobotanical techniques. In Carthy & Duddington (Eds), *Viewpoints in Biology*, Vol. 2: 202–243. London: Butterworths.

(36) 1964. (with D. A. Eggert). A flora from the Chester Series (Upper Mississippian) of Southern Illinois. *American Journal of Botany, 51:* 976–985.

(37) 1966. (with D. Huard-Moine). Karroo floras of Rhodesia and Nyasaland. Part 2. The *Glossopteris* flora in the Wankie District of Southern Rhodesia: 13–25. *Symposium on the Floristics and Stratigraphy of Gondwanaland, December 1964.* Birbal Sahni Institute of Palaeobotany, Lucknow, India.

(38) 1966. (with K. Trewren). A cutinized seed in the Indiana Paper Coal. *American Journal of Botany, 53:* 604.

(39) 1966. Conservation and the North Wales Naturalists' Trust. *Proceedings of the Dyserth Field Club for 1965:* 43–48.

(40) 1967. Bryophyta. In *The Fossil Record:* 211–217. Symposium Publication of the Geological Society and the Palaeontological Association.

(41) 1967. Sphenopsida. In *The Fossil Record:* 219–233. Geological Society and Palaeontological Association.

(42) 1967. (Review) Indian Fossil Pteridophytes, by K. R. Surange. *New Phytologist, 66:* 509–510.

(43) 1968. Fossil plants. In R. Wagstaffe & J. H. Fidler (Eds), *The Preservation of Natural History Specimens:* 243–260. H. F. & G. Witherby Ltd.

(44) 1968. (Review) Gymnosperms, Structure and Evolution, by C. J. Chamberlain; Structure and Reproduction of the Gymnosperms, Fossil and Living, by B. S. Trivedi and D. K. Singh; An Introduction to Gymnosperms, by S. C. Datta. *New Phytologist, 67:* 456–458.

(45) 1968. The flora of Anglesey. In *The Natural History of Anglesey:* 32–49. Anglesey Antiquarian Society.

(46) 1968. (Review) The genus *Pinus*, by N. T. Mirov. *Journal of Ecology, 56:* 914–915.

(47) 1968. (Review) Palaeoecology of Africa, 1950–1967, by E. M. van Zinderen-Bakker. *Journal of Ecology, 56:* 916–917.

(48) 1969. Conservation in North Wales. In F. E. Clegg (Ed.), *Introduction to Field Studies in North Wales*, Vol. 4: 1–8. Rivingtons Ltd.

(49) 1969. (with S. Kulkarni). Karroo floras of Rhodesia and Malawi. Part 3. The *Glossopteris* flora in the Tangadzi River Area of Southern Malawi. *J. Sen Memorial Volume:* 259–270. Calcutta.

(50) 1969. Fossil Bryophytes. *Biological Reviews, 44:* 189–204.

(51) 1969. (with F. A. Hibbert). Miospores from the Lower Carboniferous basement beds in the Menai Straits region of Caernarvonshire, North Wales. *Palaeontology, 12:* 420–440.

(52) 1970. (Review) Tribute to a Botanist (J. Sen). *Nature, London, 522.*

(53) 1970. *Welsh Wildlife in Trust* (Editor and Contributor) Bangor: pp. i–xiv, 1–185. Bangor: North Wales Naturalists' Trust Ltd.

(54) 1970. A note on the genus *Gangamopteris* McCoy in Rhodesia. *Arnoldia (Rhodesia), 5(3):* 1–4.

(55) 1970. Some new records of fossil plants in the Molteno Stage of Rhodesia. *Arnoldia (Rhodesia), 5(4):* 1–4.

(56) 1970. (Review) Studies in the Vegetational History of the British Isles, edited by D. Walker and R. G. West. *Scottish Forestry, 24:* 230–231.

(57) 1971. (with C. S. Smith). Karroo floras of Rhodesia and Malawi. Part 4. Karroo floras from the Upper Luangwa Valley, Zambia. *Proceedings of the Second International Gondwana Symposium, South Africa, 1970:* 571–574.

(58) 1972. (with J. M. Pettitt). A Lower Carboniferous seed compression from North Wales. *Review of Palaeobotany and Palynology, 14:* 159–169.

(59) 1972. (with W. G. Chaloner). The distribution of Carboniferous and Permian plants. *Journal of the Geological Society, 128:* 212–213.

(60) 1973. (with W. G. Chaloner). The distribution of Late Palaeozoic floras. In *Organisms and Continents through Time.* Special Papers in Palaeontology, No. 12: 271–289. Palaeontological Association.

(61) 1973. (with L. Lambrecht and C. S. Smith). Observations on the Permian flora of the Law Glacier area, Central Transantarctic Mountains. *Bulletin de la Société belge de Géologie, 81:* 161–167.

(62) In Press. Techniques d'observation des végétaux fossiles, In E. Boureau (Ed.), *Traité de Paléobotanique*, Vol. 1. Paris: Masson et Cie.

(63) 1974. Some new African Gondwana plants. In *Silver Jubilee Volume of the Birbal Sahni Institute of Palaeobotany*, Lucknow, India, 1971.

(64) 1975. Some problems of mixed floras in the Permian of Gondwana. *Proceedings of the Third International Gondwana Symposium.* Canberra, Australia, 1973.

(65) 1975. (with D. E. Van Dijk & K. D. Gordon-Gray). New Permian *Glossopteris* flora from Natal. *South African Journal of Science, 70:* 154–156.

(66) 1975. Fossil plants from the Upper Permian in the Mooi River District of Natal, South Africa. *Annals of Natal Museum, 22:* 349–420.

(67) 1975. (with D. E. Van Dijk and K. D. Gordon-Gray). A new *Glossopteris* flora from Natal. *Abstracts XII International Botanical Congress, Leningrad, 1975, 1:* 116.

(68) 1975. (with D. E. Van Dijk & K. D. Gordon-Gray). Fine structure of fossils from Lidgetton, Natal, South Africa (Lower Beaufort, Upper Permian). *Proceedings of the Electron Microscopy Society of South Africa, 5:* 89–90.

(69) 1975. (with L. C. Matten & D. Edwards). Discovery of one of the oldest gymnosperm floras containing cupulate seeds. *Phytologia, 32:* 299–303.

(70) 1976. Further observations on the Molteno Flora of Rhodesia. *Arnoldia (Rhodsia), 7:* 1–14.

(71) 1976. (with K. D. Gordon-Gray & D. E. Van Dijk). Preliminary report on equisetalean plants from Lidgetton, Natal. *Palaeontographica Africana, 19:* 43–57.

(72) 1976. (with L. C. Matten & D. Edwards). An Upper Devonian/Lower Carboniferous transition flora from South-west Eire. (Abstract) *Courier Forschungs Institut Senckenberg, Frankfurt am Main, 17:* 87.

(73) 1976. (with E. D. Van Dijk & K. D. Gordon-Gray). An Upper Permian *Glossopteris* flora from Natal, South Africa. (Abstract) *Courier Forschungs Institut Senckenberg, Frankfurt am Main, 17:* 84.

(74) 1977. (with W. G. Chaloner & A. J. Hill). First Devonian platyspermic seed and its implications in gymnosperm evolution. *Nature, London, 265:* 233–235.

(75) 1977. Plants of Montgomeryshire. M. Hignett & W. S. Lacey (Eds). *Montgomeryshire Field Society & North Wales Naturalists' Trust.*

(76) 1977. (with R. K. Kar & S. Chandra). Palynological studies in the Lower Karroo of Rhodesia and the Republic of South Africa. *Palaeobotanist, 24:* 71–95.

(77) 1977. *Nonea lutea* (Desr.) DC. in Bangor. *B.S.B.I. News, 17:* 28.

(78) 1978. (with E. D. Van Dijk & K. D. Gordon-Gray). SEM studies of Karroo fossils. *Proceedings of the Electron Microscopy Society of South Africa, 8:* 119–120.

(79) 1978. A review of the Upper Permian *Glossopteris* flora of Western Natal. *Palaeobotanist, 25:* 185–189.

(80) 1978. (with L. C. Matten & R. C. Lucas). Cupulate seeds of *Hydrasperma* from Kerry Head, Ireland. *Botanical Society of America, Miscellaneous Series, 156:* 3.

(81) 1978. (with B. I. May & L. C. Matten). A protostelic stem bearing *Lyginorachis*-like petioles from the Devonian-Carboniferous transition beds of Southern Ireland. *Botanical Society of America, Miscellaneous Series 156:* 30.

(82) 1978. Profile of P. W. Richards. *Nature in Wales, 16:* 129–131.

(83) 1979. (with L. C. Matten, B. I. May & R. C. Lucas). A megafossil flora from the Upper Devonian/Lower Carboniferous transition zone of Southern Ireland. (Abstract) *9th International Congress on Carboniferous Stratigraphy and Geology, Urbana:* 131–132.

(84) 1979. (with W. El-Saadawy). Observations on *Nothia aphylla* Lyon ex Høeg. *Review of Palaeobotany and Palynology, 29:* 119–147.

(85) 1979. (with W. El-Saadawy). The sporangia of *Horneophyton lignieri* (Kidston & Lang) Barghoorn & Darrah. *Review of Palaeobotany and Palynology, 28:* 137–144.

(86) 1979. Bryophyta. In R. W. Fairbridge & D. Jablonski (Eds), *The Encyclopedia of Paleontology:* 141–148.

(87) 1980. (with L. C. Matten, B. I. May & R. C. Lucas). A megafossil flora from the uppermost Devonian near Ballyheigue, Co. Kerry, Ireland. *Review of Palaeobotany and Palynology, 29:* 241–251.

(88) 1980. (Review), by Alfred Runte. *Journal of Applied Ecology, 17:* 520–521.

(89) 1980. (with D. E. Van Dijk, K. D. Gordon-Gray & C. Reid). Contributions to knowledge of the Lower Beaufort (Upper Permian) flora of Natal, South Africa. *IV International Gondwana Symposium, Calcutta, 1977:* 109–121.

(90) 1980. (with L. C. Matten & R. C. Lucas). Studies on the cupulate seed genus *Hydrasperma* Long from Berwickshire and East Lothian in Scotland and County Kerry in Ireland. *Botanical Journal of the Linnean Society, 81:* 249–273.

(91) 1981. (Review) Revision of the Indian species of *Glossopteris*. S. Chandra & K. R. Surange. *International Organisation of Palaeobotanists, Newsletter, 14:* 13–14.

(92) 1981. (Review) Paleobotany: An Introduction to Fossil Plant Biology. T. N. Taylor. *Palaeontological Association Circular, 106:* 15–16.

(93) 1981. (with R. C. Lucas). A permineralized wood flora of probable Early Tertiary age from King George Island, South Shetland Islands. *British Antarctic Survey Bulletin, 53:* 147–151.

(94) 1981. (with R. C. Lucas). A Lower Permian flora from the Theron Mountains, Coats Land. *British Antarctic Survey Bulletin, 53:* 153–156.

(95) 1981. (with R. C. Lucas). The Triassic flora of Livingston Island, South Shetland Islands. *British Antarctic Survey Bulletin, 53:* 157–173.

(96) 1981. (with L. C. Matten). Cupule organization in early seed plants. In R. C. Romans (Ed.), *Geobotany II:* 221–234. London: Plenum Press.

In press 1982

(97)   Scanning Electron Microscopy in Gondwana Palaeobotany. Birbal Sahni Institute of Palaeobotany, Lucknow, India.

(98)   (with R. C. Lucas). Fossil plants from the Luano and Luangwa Valleys of Zambia and their bearing
           on age determination. University of Calcutta, India.
(99)   Lycopsid leaves from the Seaham Formation, New South Wales, Australia. University of Allahabad.

In preparation 1982

(100)   Review of Fossil Bryophytes for British Bryological Society.

Botanical Journal of the Linnean Society (1983), *86*: 1–18. With 22 figures

Proterozoic microfossils from the Mara Dolomite Member, Emmerugga Dolomite, McArthur Group, from the Northern Territory, Australia

MARJORIE D. MUIR*

Bureau of Mineral Resources, P.O. Box 378, Canberra, Australian Capital Territory, Australia 2601

Received January 1982, accepted for publication July 1982

An assemblage of microfossils from the mid-Proterozoic Mara Dolomite Member of the Emmerugga Dolomite, McArthur Group, Northern Territory, Australia, has been studied. The assemblage contains 10 species, of which one is new. The classification of Proterozoic microfossils is reviewed and a morphographic scheme adopted. The new assemblage is compared with other McArthur Group assemblages, and the differences between the assemblages are explained in terms of environmental differences. Comparison with other microfossil assemblages world-wide, suggests that these assemblages have stratigraphic potential.

KEY WORDS:—Australia – microfossils – palaeoecology – palaeontology – Proterozoic – stratigraphy

CONTENTS

INTRODUCTION

Following the extended description of the microfossils from silicified stromatolites in the Lower Proterozoic Gunflint Iron Formation by Barghoorn & Tyler (1965), assemblages are now known from most parts of the Proterozoic (Schopf, 1977). While initially it was sufficient merely to record the occurrence of such an assemblage, more recently hopes have been raised that Proterozoic microfossils may eventually prove to be as stratigraphically useful as are their

*Present address: CRA Exploration Pty/Ltd, P.O. Box 656, Fyshwick, Australian Capital Territory, Australia 2609.

1

0024–4074/83/010001 + 18$03.00/0

Phanerozoic counterparts. Since many microfossil assemblages are associated with stromatolites, and lived fixed to a substrate, they can be very sensitive indicators of facies changes (Golubic, 1976). However, modern cyanobacterial communities are morphologically very conservative, and from the mid-Proterozoic on, the fossil record shows a similar conservatism. Hofman (1976) and later D. Z. Oehler (1978) have both found microfossils isotopically dated at about 1.8 ga and 1.6 ga respectively, which are morphologically identical with *Entophysalis* sp. from cyanobacterial mats of the modern Persian Gulf. Thus any stratigraphic treatment of stromatolitic microfossil communities must be more sophisticated than is much of Phanerozoic microfossil stratigraphy.

A recent development in Proterozoic microfossil studies has been the extension of the search for microfossils to nonstromatolitic assemblages. A dichotomy of morphologies between stromatolite and shale-hosted assemblages is apparent from papers by authors such as Peat *et al.* (1978) and Vidal (1976). Shale-hosted assemblages appear to contain more diverse morphologies in considerably larger size ranges than stromatolitic microbiotas. Evolutionary changes appear to be represented in shale environments in geologically much earlier times than in stromatolitic ones, and the greater diversity of morphologies offers much greater potential for biostratigraphy.

The new assemblage briefly described in this paper will be assessed in the context of the above remarks. The assemblage occurs in flat-laminated, stromatolitic, partly silicified dolomite of the Mara Dolomite Member of the Emmerugga Dolomite (Table 1) of the McArthur Group in the north-east of the Northern Territory of Australia. Samples were collected at the Cooley's lead/zinc prospect east of the McArthur Mine. The Mara Dolomite Member consists of stromatolitic and flat-laminated dolomite, which is variably silicified. At the base of the Mara Dolomite Member, a solution collapse breccia occurs in many localities (Plumb & Brown, 1973), and this is believed to have resulted from solution of evaporites (halite and gypsum) causing fracturing and collapse of overlying strata. Above this, repeated cycles of stromatolitic dolomite capped by laminated dolomite with abundant silicified halite casts make up about 30% of the sequence (Jackson *et al.*, 1978). The remainder of the Mara Dolomite Member consists principally of laminated dolomite with intercalated beds of *Conophyton*. The material sampled at the Cooley's prospect comes from the laminated part of the cyclical sequence. It should be pointed out that there is no connection at Cooley's prospect between the microfossil assemblages and the mineralization which here consists of cross-cutting veins and is clearly epigenetic.

Microfossils have previously been described from the McArthur Group by Croxford *et al.*, (1973), Hamilton & Muir (1974), Muir (1974, 1976, 1978), J. H. Oehler (1977), Oehler & Logan (1977) and D. Z. Oehler (1978). Four assemblages have been delineated: one from the Amelia Dolomite, below the Mara Dolomite Member, and three above the Mara Dolomite Member—two from the Barney Creek Formation (H.Y.C. Pyritic Shale Member, and Cooley Dolomite Member), and one from the Balbirini Dolomite.

CLASSIFICATION OF THE MICROFOSSILS FROM THE MARA DOLOMITE MEMBER

Classification of Proterozoic microfossils has only recently been rationalized (Diver & Peat, 1979), with the institution of the Group Cryptarcha. Previous

Table 1. Stratigraphy of the rocks of the McArthur Basin, showing old and new environmental interpretations (a compilation from many sources in the references)

Stratigraphy				Old environmental interpretations	New environmental interpretations
Group	Subgroup	Formation	Member		
Roper					
"Nathan"		Dungaminnie		Shallow marine	Shallow, lacustrine, fluvial
		Balbirini Dolomite*		Shallow marine	Shallow, lacustrine, fluvial
McArthur	Batten	Amos		Deep water, marine	Vadose, Calcrete
		Looking Glass		Marine	Shallow, lacustrine
		Stretton Sandstone		Fluvial	Fluvial
		Yalco		Deep water, marine	Very shallow lacustrine
		Lynott	Donnegan	Deep water, marine	Lacustrine, Sabkha, fluvial
	Umbolooga	Reward		Deep water, marine	Lacustrine
		Barney Creek*	(Cooley Dolomite*		Fault scarp deposits, vadose
			(HYC Pyritic Shale*	Deep water, marine	Lacustrine, vadose
			Coxco Dolomite	Deep water, marine	Lacustrine, vadose
		Teena Dolomite	(Mitchell Yard Dolomite	Deep water, marine	Vadose
		Emmerugga Dolomite*	(Mara Dolomite*	Marine	? Lacustrine
		Tooganinie	(Myrtle Shale	Shallow marine	Continental, fluvial, lacustrine
			(Leila Sandstone	Fluvial	Fluvial
		Tatoola Sandstone		Fluvial	Fluvial
		Amelia Dolomite*		Shallow marine	Lacustrine or marine, sabkha
		Mallapunyah		Shallow marine	Continental, playa, sabkha
Tawallah					

* Indicates fossiliferous strata listed in Table 2.

workers had used either a so-called biological classification, or a loosely based, informal morphological classification, neither of which can be applied to all organic-walled microfossils.

While it would clearly be desirable to be able to assign Proterozoic microfossils to modern taxonomic categories, Diver & Peat (1979) have cogently argued that this is not possible, because living micro-organisms are classified on their ultrastructure; energy pathways; nature of the life-cycle; biochemistry, and nutritional and ecological requirements.

With well-preserved microfossils, gross morphology is the only criterion usually available for classification (with the rare exception of microfossils that concentrate manganese or iron in and around their sheaths giving some chemical information in addition to morphology (Muir, 1978)). Even with well-preserved microfossils, preservation is often too poor to permit an accurate assessment of the nature of the wall of the organism. Using the light microscope, it is often impossible to determine whether the organic envelope is a cell wall, or a sheath (or perhaps the membrane enclosing an organelle from a eukaryotic cell). Most microfossils are not well-preserved, and this compounds the difficulty of recognition of significant features which could be used to fit them into a biological classification. Furthermore, before becoming incorporated into the sediment, many of the microfossils in any assemblage have undergone various forms of degradation such as oxidation, bacterial attack, in some cases leading to the precipitation of pyrite within cells and in and on cell walls (Muir, 1976), autolysis, or loss of chelated metals. Post-depositional diagenetic and metamorphic degradation may further alter morphological and chemical features of microfossils.

Attempts to use biological classifications are unlikely to be successful and may be misleading. While it appears that in stromatolitic microbiotas, extreme morphological conservatism has been maintained, there is no justification for assuming that the simple morphologies represent the remains of strains of organisms that exist today. Rather is it likely that particular environmental pressures encourage particular morphological features.

Although I do not believe that a biological classification can be applied to Proterozoic microfossil assemblages, some classification is necessary. Until the introduction of the Cryptarcha (Diver & Peat, 1979), there was a superabundance of poorly described and often invalid taxa. Many of the taxa are biostratigraphically useful but cannot be so employed because of initial poor definition followed by validations and emendments, until the original description of the grains becomes obscure. In addition, following normal procedures applied in Phanerozoic stratigraphy, dimensions and stratigraphic ranges of microfossil species become extended with every new occurrence that is reported, making it difficult or impossible to specify stratigraphic position or dimensions of taxa. These problems are well-known in Phanerozoic palynostratigraphy (particularly in the Mesozoic and Tertiary), and several solutions have been put forward. Hughes & Moody-Stuart (1966), and Germeraad, Hopping & Muller (1968) have put forward similar practical solutions to the problem. Their solutions are designed to obtain as much stratigraphic information as possible. They involve treating the microfossils as units which are morphologically strictly circumscribed.

Firstly, if a microfossil is described as having a double wall, then this feature must be visible in all examples from a particular population. Interpretations, however valid, based on poorly-preserved specimens which do not show all the features, cannot be used in classification. Secondly, there must be a statistically valid population. Taxa should not be founded on one or two specimens, but a population of more than 100 individuals is acceptable. Thirdly, dimensions of taxonomic entities must be strictly defined.

The group Cryptarcha, with its subgroups Nematomorphitae, Synaplomorphitae, and Sphaeromorphitae (Diver & Peat, 1979), provides the hierarchical framework for the classification of these morphologically restricted taxa. Definition and typification of taxa at generic and specific levels is according to the procedures of the International Code of Botanical Nomenclature.

Use of rigorous morphological classification does not mean that biological inferences cannot be drawn. Germeraad *et al.* (1968) were able to use their strictly defined morphological entities to produce several levels of zonation for the tropical Tertiary. They found that long-term, widely applicable zones were a function of slow evolutionary changes in assemblages. Shorter term, more localized zones were a product of the effects of climatic variations on plant communities, while very local short-term zonation was a product of local ecological variations, such as changes in water table or edaphic conditions. It was also possible in some cases to relate some of the morphologic taxa to their progenitors.

Proterozoic microfossil assemblages are amenable to this kind of morphological treatment, and I believe will also give rise to biological, stratigraphic, and ecological information when properly handled, and with sufficient numbers of specimens.

The microfossils from the Mara Dolomite Member of the Emmerugga Dolomite can be classified under the group Cryptarcha, and are assigned to the following subgroups:

GROUP:	Cryptarcha
SUBGROUP:	Nematomorphitae Diver & Peat, 1979
GENERA:	*Gunflintia* Barghoorn, 1965
	Eomycetopsis Schopf, 1968
SUBGROUP:	Sphaeromorphitae Diver & Peat, 1979
GENUS:	*Huroniospora* Barghoorn, 1965
SUBGROUP:	Synaplomorphitae Diver & Peat, 1979
GENERA:	*Bisacculoides* J. H. Oehler, 1977
	Clonophycus J. H. Oehler, 1977

DESCRIPTION OF THE MICROFOSSILS

Although microfossils are abundant in the thin sections of the chert from the Mara Dolomite Member, they are not morphologically diverse. There are the two principal morphological forms: filaments and spheroidal forms. The microfossils are described purely as form genera and species. Their possible biological affinities and significance are discussed. This follows the accepted practice for the description of fossil spores and pollen grains (Potonié & Kremp, 1954; Muir & Sarjeant, 1977).

Filamentous microfossils

Gunflintia Barghoorn

Gunflintia minuta Barghoorn, in Barghoorn & Tyler, *Science, 147:* 576 (1965).

SYNONYM: *Gunflintia septata* Schopf, *sensu* Oehler, *Alcheringa, 1:* 345 (1977).

The specimens from the Mara Dolomite Member (see Fig. 1) are, as are those in the Amelia Dolomite, poorly preserved. Although septa can be distinguished, they are irregular in distribution along the length of the filaments, giving the impression that they may not have been preserved in all parts of the filament. In the Mara Dolomite Member specimens, length to width ratios of the cells (as defined by the septa) are extremely variable, from length to width of 0.5:1 to 1:3. Thus, the distinction made by J. H. Oehler (1977) cannot be maintained in this new assemblage. In poorly-preserved material, if all septa are not preserved (which is likely to be the case) measurements of cell proportions will be misleading, and if used to separate morphologies, will only lead to a proliferation of confusing and meaningless names. I propose that *Gunflintia septata* (Schopf, 1968) J. H. Oehler 1977 be recombined with *Gunflintia minuta* Barghoorn 1977, as the differences between the two species cannot be maintained.

DIMENSIONS: Filament width 1.1–1.5 μm (x̄ 1.35 μm, 53 specimens), cells 0.62–3.2 μm long (x̄ 1.45 μm, 82 examples).

Eomycetopsis Schopf

Eomycetopsis filiformis Schopf, *Journal of Paleontology, 42:* 685 (1968).

These non-septate filaments are common in the Mara Dolomite Member (Fig. 2). They tend to occur as short fragments which are aligned indicating that they may have been originally connected. The break-up was undoubtedly post-mortem and may have been caused by diagenetic processes, such as compaction or recrystallization. Identical forms have been recorded by D. Z. Oehler (1978) as *Eomycetopsis* sp., and by J. H. Oehler (1977: Fig. 14F) as "poorly preserved sheath or algal trichome".

DIMENSIONS: Width ranges from 2.4 to 3.8 μm (x̄ 3.3 μm, 25 specimens). Length is variable, apparently the result of accidental breakage.

Eomycetopsis robusta Schopf, *Journal of Paleontology, 42:* 685 (1968).

These filamentous microfossils (Fig. 3) occur in most samples and are identical with those previously described from the Amelia Dolomite (Muir, 1976), and by J. H. Oehler (1977: Fig. 14H) as "hollow tubular structure", and as "unnamed sinuous filament" by D. Z. Oehler (1978: fig. 12H).

DIMENSIONS: Width ranges from 4 to 8 μm, and length up to 400 μm (43 specimens measured).

Poorly-preserved filamentous material

Figures 4–8. In the thin sections, the commonest form is a poorly-preserved mass of twisted, subparallel, almost straight filaments. These tend to occur at right

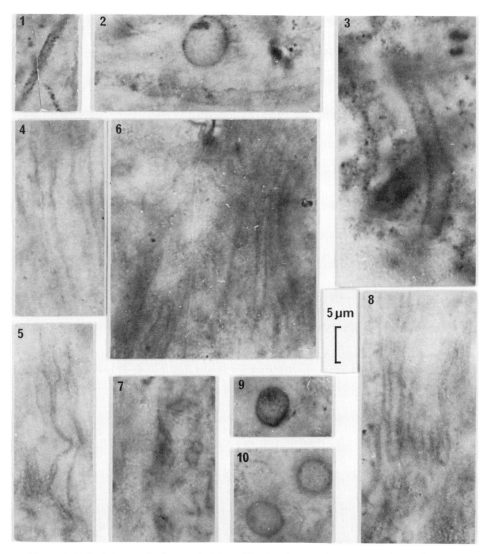

Figures 1–10. Scale bar to all micrographs is 5 μm. Numbers in parentheses are negative numbers, and stage co-ordinates. Slides are deposited in the collections of the Bureau of Mineral Resources in Canberra, and co-ordinates refer to those on the mechanical stage of the Zeiss Photomicroscope III in the care of Dr M. R. Walter; slide numbers 77100026 a and b. Fig. 1. cf. *Gunflintia minuta* (MCA37/23–27 136.3:10.8). Fig. 2. *Eomycetopsis filiformis* (MCA40/22 134.2:4.7). Fig. 3. *Eomycetopsis robusta* (MCA39/12 135.3:12.1). Figs 4–8. Poorly preserved filamentous material. Figs 4, 5, show similarities to *Biocatenoides pertenuis* (MCA39/4 135.5:10.0 and MCA40/28 134.3:11.2). Figs 6, 7, 8 are similar to *Eomycetopsis filiformis* (MCA37/31 136.3:9.1, MCA39/7 135.5:10.1, MCA40/17 134.8:10.8). Fig. 9. *Huroniospora microreticulata* (MCA45/25 129.6:12.8). Fig. 10. *Huroniospora psilata* (MCA45/25 129.6:12.8).

angles to bedding, and, in some cases, they appear to occur in an early lithified crust which had an irregular surface, although this surface may, in fact, be due to pressure solution effects, and the apparent abrupt termination of the ghost filaments may be the result of stylolite formation.

Despite the poor preservation, two morphotypes can be recognized. In the first (Figs 4, 5) the filaments are very thin, and tend to twist together in pairs. They are

2

very similar to *Biocatenoides pertenuis* J. H. Oehler (J. H. Oehler, 1977) although they are much more abundant and have a strikingly different mode of occurrence. J. H. Oehler's specimens appear to be solitary, and to occur more or less parallel to bedding, whereas the Mara Dolomite Member specimens occur perpendicular to bedding and in very large numbers. The Barney Creek specimens also appear to be less straight than those in the present study.

The second morphotype is similar to *Eomycetopsis filiformis* Schopf (Schopf, 1968), but with poorly preserved walls (Figs, 6, 8). Some cross-sections can be observed and indicate that these structures are indeed hollow tubes (Fig. 7). In addition, coccoid cells occur within the filament masses, giving the impression that they have been trapped within them.

Coccoid microfossils

There are three principal kinds of coccoid microfossils which are described below.

Huroniospora Barghoorn

Huroniospora microreticulata Barghoorn, in Barghoorn & Tyler, *Science, 147:* 576 (1965).

These specimens (Fig. 9) resemble in all respects the specimens described from the Amelia Dolomite (Muir, 1974, 1976).

Huroniospora psilata Barghoorn in Barghoorn & Tyler, *Science, 147:* 576 (1965).

These specimens (Fig. 10) resemble in all respects the specimens described from the Amelia Dolomite (Muir, 1974, 1976). It should be pointed out that solitary unicells are not abundant in the Mara Dolomite Member microfossil assemblage.

Bisacculoides J. H. Oehler

Bisacculoides grandis J. H. Oehler, *Alcheringa, 1:* 342 (1977).

This morphotype (Fig. 11) is moderately abundant and closely resembles specimens illustrated by J. H. Oehler (1977: figs 12B, 12C). Most examples contain possible degraded cytoplasmic material within the inner walls.

DIMENSIONS: Mean diameter (22 cells) 7.8 μm: sheaths about 1–2 μm thick.

Clonophycus J. H. Oehler

Clonophycus elegans J. H. Oehler, *Alcheringa, 1:* 346 (1977).

These structures (Figs 12–16) are indistinguishable from the examples in Oehler's type material from the Barney Creek Formation, but whereas in Oehler's examples, these were rare forms, in the Mara Dolomite Member, they are very abundant. Some of the enclosed cells have relatively thick walls (up to 1.5 μm), but others are similar to those described by J. H. Oehler. The exterior sac (?cyst) has variable morphology, from spherical to irregular. Irregular forms are larger than spherical ones and contain more cells (cf. Fig. 13).

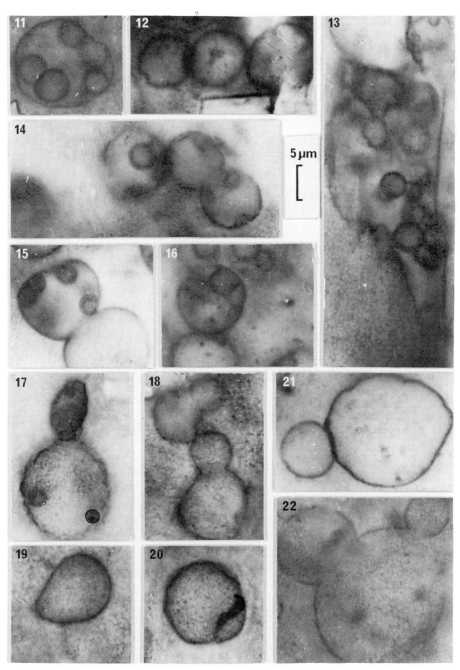

Figures 11–22. Scale bar to all micrographs is 5 μm. Numbers in parentheses are negative numbers, and stage co-ordinates. All slides are deposited in the collections of the Bureau of Mineral Resources in Canberra, and co-ordinates refer to those on the mechanical stage of the Zeiss Photomicroscope III in the care of Dr M. R. Walter; slide numbers 77100026 a and b. Fig. 11. *Bisacculoides grandis* (MCA46/32 128.7:15.9). Figs 12–16. *Clonophycus elegans* (MCA38/16 135.7:9.2, MCA44/19 131.1:14.0, MCA44/17 131,2:14,1, MCA44/19 131.2:14.1, MCA44/26 131.2:14.0). Figs 17–20. *Clonophycus laceyi* (MCA40/34 134.2:15.6, MCA46/5 128.5:22.4, MCA46/3 129.5:14.2, MCA46/9 128.5:17.9). Figs 21, 22. *Clonophycus* sp. (MCA44/22 131.2:14.1, MCA43/3 132.6:15.8).

DIMENSIONS: Inner cells 3.2–7.4 μm diameter (x̄ 4.9 μm, 74 specimens). Outer sacs, 14–35 μm (x̄ 20.3 μm, 20 specimens). Number of cells within each sac variable.

Clonophycus laceyi M. D. Muir **sp. nova**

DERIVATION: Named after Professor W. S. Lacey.

SYNONYMY: *Clonophycus* sp. D. Z. Oehler, *Alcheringa, 2:* 269–309 (1978).

Cells with granular to coarsely granular surface texture. Small inner bodies occasionally present. Outer spheroid irregular in shape, varies from elliptical to spheroidal. Necking occurs in some specimens as if a budding process was possible. Some specimens have a lid-like aperture (similar to the operculum of many dinoflagellate and acritarch species). The cells are solitary and not associated with an organic matrix.

TYPE MATERIAL: Figures 17–20.

TYPE LOCALITY AND STRATIGRAPHY: Balbirini Dolomite, on Balbirini Station (see D. Z. Oehler (1978) fig. 1B for location of sample 106 in which the type specimen occurs).

This is the second most abundant form in the Mara Dolomite Member microfossil assemblage. It is identical with the *Clonophycus* sp. described by D. Z. Oehler (1978) from the Balbirini Dolomite and examples from this latter formation are assigned to the new species.

DIMENSIONS: Outer spheroids 5.0–35.0 μm in diameter (92 specimens measured: x̄ 15.6 μm). Inner bodies up to 7 μm.

? Clonophycus sp.

These large (15–20 μm diameter) sacs (Figs 21, 22) occur in close association with *C. elegans* specimens, although they have no inner cells. It is possible that any inner cells have been discharged previously. They seem to have produced daughter cell(s) by budding.

POSSIBLE BIOLOGICAL AFFINITIES OF THE MICROFOSSILS FROM THE MARA DOLOMITE MEMBER

Subgroup Nematomorphitae: The filamentous microfossils are not well-preserved, but J. H. Oehler (1977), using primarily morphological features, tried to assign species to prokaryotic groupings. Among the abundant but poorly-preserved filamentous material from the Mara Dolomite Member are thin filaments very similar to *Biocatenoides pertenuis* J. H. Oehler. Oehler considered this species to be bacterial in origin because of its small size and simple morphology.

Gunflintia minuta Barghoorn 1965 is regarded as an oscillatorian cyanophyte (J. H. Oehler, 1977). A larger non-septate filament illustrated in J. H. Oehler (1977: fig. 14H) is described as merely a "Hollow tubular structure". Morphologically, this cannot be separated from *Eomycetopsis robusta* Schopf which was originally believed to be possibly fungal in origin. D. Z. Oehler (1978) illustrates a similar filament and describes it as an "unnamed sinuous filament". She also illustrates *Eomycetopsis* sp. (which is morphologically indistinguishable from *Eomycetopsis filiformis* Schopf (Schopf, 1968)), and comments that, because it forms tangled mats in the Bitter Springs Formation, and because in TEM micrographs there is evidence of sheath-like material outside the wall, it is more likely to be algal or bacterial than fungal. Poorly preserved examples of *Eomycetopsis filiformis* Schopf also occur in tangled mat-like masses in the present assemblage.

Thus although there is no compelling evidence for the biological affinities of the filamentous microfossils from the Mara Dolomite Member, they can be regarded as having bacterial and cyanobacterial affinities.

Subgroup Sphaeromorphitae: Two species of *Huroniospora* represent this subgroup and the comments under *Clonophycus* below apply.

Subgroup Synaplomorphitae: J. H. Oehler (1977) regards *Bisacculoides* as a possible chroococcacean alga because he interprets the morphology to indicate that the organisms were sheathed unicellular algae in which "cytoplasmic constituents have condensed to form relatively large intracellular masses", which he believes are not the remains of discrete cellular organelles. On this interpretation, I agree that the Chroococcaceae is a likely taxon for these fossils. It is perhaps also significant that, in the shale assemblage described by Hamilton & Muir (1974) from the shale portion of the H.Y.C. Pyritic Shale member, similar morphologies are abundant and dominate the assemblage. Chroococcacean blooms producing vast volumes of algal matter, occur periodically in freshwater and salt lakes (Muir, 1981) and indicate a possibly non-marine origin for the microfossils. Although similar algae occur in marine settings also, they rarely bloom, and even when they do, little of the organic matter produced is likely to be preserved. These fossils may, in fact, be valuable palaeoenvironmental indicators.

Clonophycus is described by J. H. Oehler (1977) as belonging to either the Cyanophyceae, Chlorophyceae or Rhodophyceae. Based on much more extensive material, D. Z. Oehler (1978) came to the conclusion that this genus represents the remains of spore-like cells and sporangia. However, she believed that there was insufficient evidence available to demonstrate a pro- or eukaryotic origin for the genus. Darby (1974) suggested a possible fungal origin for *Huroniospora*. Such an origin seems not incompatible with the morphologies of *Clonophycus*.

In general, although the suggested biological assignations of these microfossils are based only on morphology, a community composed of filamentous bacteria, cyanobacteria and possible eukaryotic (higher algal or fungal) cells is indicated.

COMPARISON OF THE MARA DOLOMITE MEMBER MICROFOSSIL ASSEMBLAGE WITH PENECONTEMPORANEOUS ASSEMBLAGES

Middle Proterozoic microfossil assemblages have been described from Australia, Canada, India and the Soviet Union (Schopf, 1977).

Several of these assemblages have been described from the McArthur Group, and it is with these that the most useful comparisons can be made. D. Z. Oehler (1977) compared the microfossil assemblages from the Balbirini Dolomite with the assemblages from the Amelia Dolomite (Muir, 1974, 1976) and the H.Y.C. Pyritic Shale Member. Table 2 shows the comparisons, and adds the information on three

Table 2. Distribution of microfossils in units of the McArthur Group. Sources are (1) Muir, 1974, 1976, (2) this paper, (3) J. H. Oehler 1977, (4) Hamilton & Muir, 1974, Muir, 1981, (5) Muir, 1978, (6) D. Z. Oehler 1978, (7) Diver, 1974, and unpublished information

Cryptarch subgroup	Amelia Dolomite (1)	Mara Dolomite Mbr. (2)	H.Y.C. chert (3)	H.Y.C. shale (4)	Cooley Dolomite Mbr (5)	Balbirini Dolomite (6)	Bungle Bungle Dolomite (7)	Preservation state
Nematomorphitae								
Gunflintia minuta	X	X	X		X		X	Poor
Eomycetopsis filiformis	X	X	X			X	X	Poor, may form mats
Eomycetopsis robusta	X	X	X			X		Poor
Biocatenoides pertenuis	X	X	X		X	?	X	Poor
Synaplomorphitae								
Bisacculoides grandis	?	X	X	X		?	X	Fair–good
Clonophycus elegans		X	X					Good
Clonophycus laceyi		X				X		Good
Clonophycus sp.		X						Good
Nannococcus vulgaris			X					Good
Sphaeromorphitae								
Huroniospora microreticulata	X	X	X	X		?	X	Fair–good
Huroniospora psilata	X	X		X		?	X	Fair–good
Incertae sedis								
Eoastrion simplex			X	X	X			Good
Lithology	chert, stromatolite dolomite	chert laminate & stromatolite dolomite	chert laminate dolomite	dolomite shale	chert conglomerate dolomite	chert laminate & stromatolite dolomite	chert stromatolite	

other McArthur Group assemblages (shale assemblage from the H.Y.C. Pyritic Shale Member (Hamilton & Muir, 1974); Mara Dolomite Member; and Cooley Dolomite Member (Muir, 1978)). In addition, species described by Diver (1974) from the age equivalent Bungle Bungle Dolomite of Western Australia are included.

The Mara Dolomite Member assemblage has a number of species in common with all of the assemblages. *Clonophycus* sp. is the only microfossil not found in other assemblages. As might be expected from its stratigraphic position, most of the species are in common with species from the H.Y.C. Pytitic Shale Member chert assemblages (J. H. Oehler, 1977), and three of the four species described from the impoverished assemblage from the shale samples of the H.Y.C. Pyritic Shale Member also occur in the Mara Dolomite Member assemblage. The Cooley-Dolomite Member is a very restricted assemblage consisting of the *Metallogenium*-like organism *Eoastrion simplex* Barghoorn and numerous manganese encrusted filaments of several sizes. Significantly, all three of the Cooley Dolomite Member morphotypes occur in the H.Y.C. Pyritic Shale Member, although here they are iron encrusted. The H.Y.C. Pyritic Shale Member and Cooley Dolomite Member are lateral, interfingering, stratigraphic equivalents.

The Mara Dolomite Member assemblage is poor in species (only 10 have been identified), but rich in numbers of fossils. Considerably more species have been reported from the Amelia Dolomite (19 species), H.Y.C. Pyritic Shale Member chert (21 species), and Balbirini Dolomite (26 species). Although the Amelia and Balbirini Dolomite assemblages are similar, there are some species in the Amelia Dolomite that do not occur in the Balbirini Dolomite and vice versa. J. H. Oehler's H.Y.C. chert assemblage also contains species in addition to those in common with the Mara Dolomite Member assemblage. However, systematic variation of microfossil assemblages cannot be traced through the McArthur Group, as might be anticipated if the changes were the result of evolutionary development of the microflora. Instead microfossil assemblages appear to be more responsive to environmental controls, which are marked by lithological changes.

DEPOSITIONAL ENVIRONMENTS IN THE McARTHUR GROUP, AND BALBIRINI DOLOMITE

From 1977 onwards, the Australian Bureau of Mineral Resources, Geology and Geophysics has been undertaking a major research study of the McArthur Basin. As one of the results of this study, a much better understanding of depositional environments has been developed, which is strongly at variance with previous interpretations (e.g. Brown, Claxton & Plumb, 1978 (originally released in 1969)). In addition, the very detailed work of Logan (1979) and Williams & Logan (1981) has overturned previous concepts of ore depositional conditions. As a result, the earlier environmental conclusions drawn, in particular by J. H. Oehler (1977) and Oehler & Logan (1977) require revision.

It is now clear that the overall stratigraphy of the McArthur Basin is relatively simple(Plumb & Brown, 1973). The depositional environments of all units were assumed to be marine, and the causes of varying lithologies were taken to be the results of fluctuations in water level. Some of the units appear to be continental in origin (red beds of the Mallapunyah Formation, Myrtle Shale Member, Donnegan Member of the Lynnott Formation, and the Balbirini Dolomite: fluvial sandstones of the Tatoola Sandstone, Leila Sandstone Member, Stretton

Sandstone, and Balbirini Dolomite). From the upper part of the Emmerugga Dolomite (the Mitchell Yard Dolomite Member) to the Reward Dolomite (overlying the H.Y.C. deposit) a deep water phase was postulated (300–800 m) by Brown & Plumb (in Plumb & Derrick, 1976). This information was used by J. H. Oehler (1977) as a basis to suggest that many of his presumed bacterial fossils lived on the sediment surface in deeper water beneath the photic zone. Oehler believed that most of the remainder of the assemblage either lived as plankton and sunk post-mortem to the bottom, which he believed to be anoxic, or were swept from marginal regions during periods of turbulent current activity. However, the more recent work has required drastic revision of these interpretations.

The depositional environment of the lowermost units of the McArthur Group the Mallapunyah Formation and the Amelia Dolomite ranges from fluvial, through continental playa, to sabkha (assumed by Muir, 1979, to be marine marginal) followed by marine (?) or lacustrine carbonates. The sequence is abundantly evaporitic with pseudomorphs after halite, gypsum (in various forms) and anhydrite (now replaced by chert). Muir (1979) was able to make a detailed comparison with the Holocene sediments of the Persian Gulf. On the basis of this comparison, a marine marginal sabkha and marine depositional basin for the Amelia Dolomite seemed reasonable. One of the main reasons for assuming a marine environment was that much of the sulphate which was precipitated in the Abu Dhabi sabkhas was believed to be derived by periodic flooding of intertidal and supratidal zones. However, a recent publication by Patterson & Kinsman (1981) has shown that the gypsum and anhydrite in the Persian Gulf are precipitated from continental groundwaters, and a marine influence can only be detected very near the present tidal limits. If the sulphates in the Persian Gulf are derived from continental groundwater, then there is no essential difference between so-called marine and continental sabkhas. The depositional environment of the Amelia Dolomite and Mallapunyah Formation can therefore be readily interpreted as either marine or continental.

The units between the Amelia Dolomite and Mara Dolomite Member are continental (fluvial and playa lake environments) arenites and argillites with abundant halite casts and rarer gypsum, with occasional thin carbonate beds (some stromatolitic) indicating a more subaqueous environment. In these units, too, there is no evidence to indicate whether the carbonates are marine or lacustrine.

The solution collapse breccia at the base of the Emmerugga Dolomite has been mentioned in the introduction. It indicates strongly evaporitic conditions at this level, with precipitation of abundant halite and/or gypsum. Evaporite casts in the Emmerugga Dolomite are more commonly halite, although some gypsum casts do occur (unpublished observations). The Mara Dolomite Member was deposited in an environment that was shallow water, susceptible to evaporation to dryness, precipitation of halite (at least seasonally), and growth of stromatolites. The depositional basin may have had periods of restricted circulation and still water when very large conical stromatolites (up to 3 m tall) became abundant. The Mara Dolomite Member passes up into the unbedded massive Mitchell Yard Dolomite Member. The massive nature of this latter unit was one of the lines of evidence used to indicate a deepening of the basin water (on the assumption that it was unbedded because it had been deposited below wave base). However, the massive nature of the Mitchell Yard Dolomite Member is due to intense

Proterozoic vadose alteration, which recrystallized the dolomite, and formed caves and high porosity zones, many of which were subsequently mineralized.

Between the Emmerugga Dolomite and the overlying Teena Dolomite, there is a break in sequence—probably due to minor uplift—when the Mitchell Yard Dolomite Member underwent vadose alteration. The overlying Teena Dolomite varies from thick to thin bedded dolomite which is flat bedded. It is not stromatolitic. It contains a wide variety of evaporite casts, ranging from acicular gypsum clusters (Walker *et al.*, 1977) to moulds after trona ($Na_2CO_3.NaHCO_3.2H_2O$) and shortite ($Na_2Co_3.CaCO_3$). In addition, the Teena Dolomite has locally suffered intense vadose alteration forming the bizarrely patterned rock type 'paisleyite'. Blocks of 'paisleyite' occur in overlying breccias indicating that the vadose alteration of the Teena Dolomite is not a recent weathering phenomenon, and pre-dates the deposition of the breccias. The sodium carbonate evaporite minerals are of great environmental significance since they cannot precipitate from evaporation of sea water. They are, in fact, indicators of continental alkaline lacustrine systems.

The overlying Barney Creek Formation which contains the H.Y.C. Pyritic Shale Member, and Cooley Dolomite Member was originally believed to be deep water (Brown *et al.*, 1978). The grounds for this were the euxinic nature of the Pyritic Shale, and the abundant graded beds which were compared with oceanic turbidites. The graded conglomeratic breccias of the Cooley Dolomite were believed to have been deposited from slumping from a submarine fault scarp. However, there is abundant evidence for vadose alteration in breccia blocks of the Cooley Dolomite, and in its source rocks, indicating that the fault scarp must have been subaerial. In addition, Williams & Logan (1981) have found evidence of mud flake breccias, of sabkha-type anhydrite nodules, indicating subaerial emergence, and of vadose pisoliths in the Barney Creek Formation. Thus while undoubtedly much of the Barney Creek Formation formed subaqueously, the water had to be shallow enough to permit occasional drying out and subaerial exposure of the surface of the sediments.

Thus the interpretation of J. H. Oehler (1977) is unacceptable. The bacterial assemblage lived in the shallow pools and lagoons but the presumed planktonic algal remains were not rained down through a long water column, but were perhaps washed in from supralittoral and littoral flats during periods of flooding. Small shallow lakes are ideal for the flourishing and preservation of blooms of cyanophytes, and the abundant small cells described by Hamilton & Muir (1974) were probably the result of these algal blooms.

In the sequences overlying the Barney Creek Formation, evidence for continental fluvial and playa deposits, shortite moulds indicating continental alkaline lakes, gypsum and anhydrite moulds indicating sabkhas and brine pools, and desiccation and emergence features such as mud cracks, teepee structures (Muir, Lock & von der Borch, 1979), fossil calcrete (Muir, in press), vadose weathering and karstic activity (Donnelly, *et al.*, in press) is abundant. The depositional environment for the Balbirini Dolomite is now interpreted as continental, fluvial and lacustrine. The Balbirini Dolomite is conspicuously less evaporite than the Amelia Dolomite and the differences in the two microbiotas are probably related to these environmental differences rather than to stratigraphic or evolutionary changes.

It is also significant that the three Barney Creek Formation assemblages, which formed in different environmental conditions from the Amelia and Balbirini Dolomites, have strikingly different microbiotas with an adundance of probable bacterial forms.

Thus within the McArthur Basin itself, different microbiotas occur in varying environmental conditions. Any new assemblage in the McArthur Basin similar to that described by J. H. Oehler (1977) might be used to indicate the presence of conditions similar to those prevailing when the major lead-zinc mineralization was emplaced.

STRATIGRAPHIC POTENTIAL

At this stage, so little is known of the microfossil assemblages of this age that few generalities are possible. D. Z. Oehler (1978) compared the assemblage from the Balbirini Dolomite with one from the slightly older mid-Proterozoic Belcher Subgroup of Canada and with one from the much younger Bitter Springs Formation of central Australia. Schopf (1977) was able to show a major morphological discontinuity at about 1.4 ga and suggested that this indicated an important biological change, such as, perhaps the evolution of the eukaryotes. There is no compelling evidence for the presence of eukaryotes in the McArthur Group or the probable age equivalent, the Bungle Bungle Dolomite, although small tetrahedral tetrads have been described from both (Oehler, Oehler & Muir, 1975; Diver, 1974). Unfortunately isotopic dating for mid-Proterozoic rocks is poor world-wide. A recent date for the Barney Creek Formation gives a very accurate 1.69 ga for contemporaneous zircons deposited in tuff beds within the H.Y.C. Pyritic Shale Member (Page, 1981). Thus the assemblages described by J. H. Oehler (1977), Muir, (1977), and Hamilton & Muir (1974) can be accepted as being of this age. The Mara Dolomite Member and Amelia Dolomite assemblages are older and the Balbirini Dolomite assemblage somewhat younger. Rocks in the overlying Roper Group are approximately 1.5 ga (Kralik, personal communication), thus giving a minimum age for the McArthur Group assemblages. Assemblages from possibly age equivalent stromatolitic assemblages in the Soviet Union (Schopf & Sovietov, 1976; Schopf et al., 1977) and India (Schopf & Prasad, 1978) have somewhat similar assemblages to the Mara Dolomite Member and Balbirini Dolomite. Eventually, when sufficient information is available, overall stratigraphic zonation of the Middle Proterozoic may be possible.

However, assemblages described from shales of roughly the same age (Peat, 1979; Peat et al., 1978) are much more diverse, contain large numbers of specimens and are potentially much more amenable to rigorous stratigraphic treatment (Vidal, 1976).

Thus although stromatolitic microfossil assemblages will contine to be sought because they contain at least potentially useful biological and evolutionary information, from a geological point of view, they will never be as useful as assemblages extracted from a carbonaceous shale. Little progress can be expected to be made in either biological or stratigraphic understanding, until many more assemblages are discovered, and carefully assessed in relation to the geology of the host rocks. At present the number of Proterozoic micropalaeontologists is too small to provide this fuller understanding of the evolution of microfossil assemblages for many years to come.

ACKNOWLEDGEMENTS

I wish to acknowledge the support of M. J. Jackson and K. J. Armstrong of the Australian Bureau of Mineral Resources, for much help in the field and laboratory. I have benefitted greatly from lively discussions with N. Williams, R. G. Logan, R. N. Walker, N. Runnals and many other geologists from Carpentaria Exploration Company Pty. Ltd. N. J. W. Croxford and D. F. Ward have both contributed greatly to my understanding of McArthur geology.

REFERENCES

BARGHOORN, E. S. & TYLER, S. A., 1965. Microorganisms from the Gunflint Chert. *Science, 147:* 563–577.

BROWN, M. C., CLAXTON, C. W. & PLUMB, K. A., 1978. The Proterozoic Barney Creek Formation and some associated carbonate units of the McArthur Group, Northern Territory. *Bureau of Mineral Resources, Geology and Geophysics, Record, 1969/145:* 1–59.

CROXFORD, N. J. W., JANECEK, J., MUIR, M. D. & PLUMB, K. A., 1973. Microorganisms of Carpentarian (Precambrian) age from the Amelia Dolomite, McArthur Group, Northern Territory, Australia. *Nature, London, 245:* 28–30.

DARBY, D. G., 1974. Reproductive modes of *Huroniospora microreticulata* from cherts of the Precambrian Gunflint Iron Formation. *Bulletin of the Geological Society of America, 85:* 1595–1596.

DIVER, W. L., 1974. Precambrian microfossils of Carpentarian age from Bungle Bungle Dolomite of Western Australia. *Nature, London, 247:* 361–362.

DIVER, W. L. & PEAT, C. J., 1979. On the interpretation and classification of Precambrian organic-walled microfossils. *Geology (Boulder), 7:* 401–404.

DONNELLY, T. H., MUIR, M. D., WILKINS, R. W. T. & ARMSTRONG, K. J., (in press). Genesis of the Eastern Creek Pb-Ba prospect and implications for some stratabound base metal deposits of Northern Australia. *Economic Geology.*

GERMERAAD, J. H., HOPPING, C. A. & MULLER, J., 1968. Palynology of Tertiary sediments from tropical areas. *Review of Palaeobotany and Palynology, 6:* 189–384.

GOLUBIC, S., 1976. Organisms that build stromatolites. In: M. R. Walter (Ed.), *Stromatolites,* 113–126. Amsterdam: Elsevier Scientific Publishing Company.

HAMILTON, L. H. & MUIR, M. D., 1974. Precambrian microfossils from the McArthur River lead-zinc-silver deposit, Northern Territory, Australia. *Mineralium Deposita, 9:* 83–86.

HOFMANN, H. J., 1976. Precambrian microflora, Belcher Islands, Canada: Significance and systematics. *Journal of Paleontology, 50:* 1040–1073.

HUGHES, N. F. & MOODY-STUART, J., 1966. Description of schizeaceous spores taken from early Cretaceous macrofossils. *Palaeontology, 9:* 274–289.

JACKSON, M. J., MUIR, M. D., PLUMB, K. A., LARGE, D. E., BROWN, M. C. & ARMSTRONG, K. J., 1978. Field work report, McArthur Basin Project, 1977. *Bureau of Mineral Resources, Geology and Geophysics, Record, 1978/54:* 1–53.

LOGAN, R. G., 1979. *The Geology and Mineralogical Zoning of the HYC Ag-Pb-Zn Deposit, McArthur River, N.T.* M.Sc. Thesis, Australian National University.

MUIR, M. D., 1974. Microfossils from the Middle Precambrian McArthur Group, Northern Territory, Australia. *Origins of Life, 5:* 105–118.

MUIR, M. D., 1976. Proterozoic microfossils from the Amelia Dolomite, McArthur Basin, Northern Territory. *Alcheringa,* 1: 143–158.

MUIR, M. D., 1978. Microenvironments of some modern and fossil iron- and manganese-oxidising bacteria. In W. E. Krumbein (Ed.), *Environmental Biogeochemistry and geomicrobiology. Volume 3: Methods, Metal and Assessment:* 937–944. Ann Arbor: Ann Arbor Science Publishers, Inc.

MUIR, M. D., 1979. A sabkha model for deposition of part of the Proterozoic McArthur Group of the Northern Territory, and implications for mineralisation. *Bureau of Mineral Resources Journal of Australian Geology and Geophysics, 4:* 149–162.

MUIR, M. D., 1981. The microfossils from the Proterozoic Urquhart Shale, Mount Isa, Queensland, and their significance in relation to the depositional environment, diagenesis, and mineralisation. *Mineralium Deposita, 16:* 51–58.

MUIR, M. D., (in press). A proterozoic calcrete in the Amos Formation, McArthur Group, Northern Territory, Australia. In T. M. Peryt (Ed.). *Coated Grains.* Amsterdam: Elsevier Scientific Publishing Company.

MUIR, M. D. & SARJEANT, W. A. S., 1977. *Palynology, Volume 1.* Stroudburg: Bendmark Books, Dowden, Hutchison, & Ross.

MUIR, M. D., LOCK, D. & von der BORCH, C. C., 1979. The Coorong-Model for penecontemporaneous dolomite formation in the Middle Proterozoic McArthur Group, Northern Territory, Australia. In D. H. Zenger (Ed.), *Concepts and Models of Dolomitization—Their Intricacies and Significance. Special Publication 28:* 51–67. Society for Economic Palaeontologists and Mineralogists.

OEHLER, D. Z., 1978. Microflora of the middle Proterozoic Balbirini Dolomite (McArthur Group) of Australia. *Alcheringa, 2:* 269–309.

OEHLER, J. H., 1977. Microflora of the H.Y.C. Pyritic Shale Member of the Barney Creek Formation (McArthur Group), middle Proterozoic of Northern Australia. *Alcheringa, 1:* 315–349.

OEHLER, J. H. & LOGAN, R. G., 1977. Microfossils, cherts and associated mineralisation in the Proterozoic McArthur (HYC) lead-zinc-silver deposit. *Economic Geology, 72:* 1393–1409.

OEHLER, J. H., OEHLER, D. Z. & MUIR, M. D., 1975. On the significance of tetrahedral tetrads of Precambrian algal cells. *Origins of Life, 7:* 259–267.

PAGE, R. W. 1981. Depositional ages of the stratiform base metal deposits of Mount Isa and McArthur River, Australia, based on U-Pb zircon dating of concordant tuff horizons. *Economic Geology, 76:* 648–658.

PATTERSON, R. J. & KINSMAN, D. J. J., 1981. Hydrologic framework of a sabkha along Arabian Gulf. *Bulletin of the American Association Petroleum Geologists, 65:* 1457–1475.

PEAT, C. J., 1979. *A Proterozoic microbiota from the Roper Group, Australia.* Ph.D. Thesis, University of London.

PEAT, C. J., MUIR, M. D., PLUMB, K. A. McKIRDY, D. M. & NORVICK, M. S., 1978. Proterozoic microfossils from the Roper Group, Northern Territory, Australia. *Bureau of Mineral Resources Journal of Australian Geology and Geophysics, 3:* 1–17.

PLUMB, K. A. & BROWN, M. C., 1973. Revised correlation and stratigraphic nomenclature in the Proterozoic carbonate complex of the McArthur Group, Northern Territory. *Bulletin Bureau of Mineral Resources, Geology and Geophysics, 139:* 103–115.

PLUMB, K. A. & DERRICK, G. M., 1976. Geology of the Proterozoic rocks of the Kimberly to Mount Isa region. In C. L. Knight (Ed.), *Economic Geology of Australia and Papua New Guinea, Australasian Institute of Mining and Metallurgy, Monograph Series, 5, 1. Metals:* 217–252.

POTONIÉ, R. & KREMP, G. O. W., 1954. Die Gattungen der paläozoischen Sporae Dispersae und ihre Sratigraphie. *Geologisches Jahrbuch, 69:* 111–194.

SCHOPF, J. W., 1968. Microflora of the Bitter Springs Formation, Late Precambrian, Central Australia. *Journal of Paleontology, 42:* 651–688.

SCHOPF, J. W., 1977. Biostratigraphic usefulness of stromatolitic Precambrian microbiotas: a preliminary analysis. *Precambrian Research, 5:* 143–173.

SCHOPF, J. W. & PRASAD, K. N., 1978. Microfossils in *Collenia*-like stromatolites from the Proterozoic Vempalle Formation of the Cuddapah Basin, India. *Precambrian Research, 6:* 347–366.

SCHOPF, J. W. & SOVIETOV, Yu. K., 1976. Microfossils in *Conophyton* from the Soviet Union and their bearing on Precambrian biostratigraphy. *Science, 193:* 143–146.

SCHOPF, J. W., DOLNIK, T. A., KRYLOV, I. N., MENDELSON, C. V., NAZAROV, B. B., NYBERG, A. V., SOVIETOV, Yu. K. & YAKSHIN, M. S., 1977. Six new stromatolitic microbiotas from the Proterozoic of the Soviet Union. *Precambrian Research, 4:* 269–284.

VIDAL, G., 1976. Late Precambrian microfossils from the Visingsö Beds in southern Sweden. *Fossils and Strata, 9:* 1–57.

WALKER, R. N., MUIR, M. D., DIVER, W. L., WILLIAMS, N. & WILKINS, N., 1977. Evidence of major sulphate evaporite deposits in the Proterozoic McArthur Group, Northern Territory, Australia. *Nature, London, 265:* 526–529.

WILLIAMS, N. & LOGAN, R. G., 1981. Depositional environments of the sediments hosting the McArthur River stratiform Pb-Zn deposits. *Abstracts, Geological Society of Australia, 5th Australian Geological Convention, Perth, Sediments through the ages, 8.*

Botanical Journal of the Linnean Society (1983), *86*: 19–36. With 54 figures

A late Wenlock flora from Co. Tipperary, Ireland

D. EDWARDS, F.L.S.

Department of Plant Science,
University College, P.O. Box 78, Cardiff CF1 1XL

J. FEEHAN*

Department of Geology,
Trinity College, Dublin 2, Ireland

AND

D. G. SMITH

British Petroleum Co. Ltd.,
Exploration & Production Division,
Sunbury on Thames, Middlesex TW16 7LN

Received November 1981, accepted for publication June, 1982

An assemblage of macroplants preserved as highly coalified compressions which lack anatomy is described from a Wenlock locality in County Tipperary, Ireland. Most of the fertile specimens are assigned to *Cooksonia* Lang. The taxonomic status of this genus is discussed. Some poorly preserved palynomorphs, including miospores, acritarchs, chitinozoans and a variety of tubes, have been isolated from associated sediments, but the age of the flora is based on graptolites. Sedimentological and palaeontological studies of the region are summarized. They provide little direct evidence for the habitats of the plants which are considered to have been terrestrial. The relevance of this flora to the current debate on the colonization of the land is evaluated and it is concluded that these plants provide the earliest record of erect fertile land plants of possible pteridophyte affinity.

KEY WORDS:—*Cooksonia* – land plants – palynomorphs – Silurian.

CONTENTS

* Present address: Kamuzu Academy, P.O. Box 1, Mtunthama, Kosungu, Malawi.

0024–4074/83/010019+18$03.00/0

INTRODUCTION

The discovery of a presumed terrestrial flora, comprising mainly plants of *Cooksonia*-type morphology, in marine sediments of late Wenlock age from Co. Tipperary was briefly reported by Edwards & Feehan (1980). Here we propose to present more detailed information on the geological background (J.F.), to illustrate the diversity of sporangial forms in the macroflora (D.E.) and to describe the palynomorphs and tubes recovered on bulk maceration of associated sediments (D.G.S.). The collection of further material has permitted a more intensive search for anatomical evidence for vascular plants, but as yet we have failed to find any tracheids or indeed any well-preserved cellular detail in either fertile or sterile axes.

GEOLOGICAL SETTING

The late Wenlock (mid-Silurian) rocks of the Devilsbit Mountain area constitute part of the largest Lower Palaeozoic inlier in central Ireland. The rocks in the southern part (Slieve Felim) were described by Doran (1974); the north-western part of the inlier (the Keeper Hill area) has more recently been described by Archer (1981). The Devilsbit area itself occupies the eastern end of the inlier which was originally described by Baily, Jukes & Wynne (1860), and subsequently by Cope (1955, 1959). The rocks in all these areas appear to constitute a continuous sedimentary sequence of restricted age span. The strata in the Slieve Felim area were described by Doran as the Hollyford Formation, and the rocks in the Keeper Hill area to the north were referred by Archer to the same formation. The Devilsbit rocks closely resemble those of the Keeper Hill—Slieve Felim area in lithology and inferred environment of deposition, and they are largely of similar (*lundgreni* Zone) age. The Silurian rocks over most of the Devilsbit area of the inlier can accordingly be referred to Doran's Hollyford Formation. A small area of slightly younger rocks in the north-east constitutes a separate formation here named the Cloncannon Formation.

The Hollyford and Cloncannon Formations comprise a monotonous sequence of marine sandstones, siltstones and mudstones which attain a thickness in excess of 2500 m. The siltstones occur as distinctively banded or laminated sequences, sometimes alternating with thin sand bodies. The mudstones are typically 2–9 m featureless blocky or slabby rock units which sometimes contain narrow silt laminae. There are local variations in distribution of these lithologies, resulting in the predominance of sands at certain levels in the succession and laminated silt and mud at other levels. The overall character of this succession indicates deposition in a submarine fan system. Depositional mechanisms, so far as they relate to the plant fossils, are considered in a later section.

The laminated siltstones in the succession are frequently fossiliferous, yielding a nekto-planktonic fauna of graptolites, orthoconic nautiloids, phyllocarids and other arthropods such as *Aptychopsis*, with occasional epifaunal bivalves (Cope,

1959; Palmer, 1970). Over most of the inlier in the Hollyford formation, this fauna is of *Cyrtograptus lundgreni* Zone age (Cope, 1954, 1959; Palmer, 1970; Doran, 1974; Edwards & Feehan, 1980). However, the Cloncannon Formation, a small area of rocks occurring in the core of a syncline near the north-east end of the inlier, is of *Monograptus ludensis* Zone (youngest Wenlock) age. The plant-bearing sequence described here occurs on top of Borrisnoe Mountain (1.7 km south-east of Moneygall: Grid reference 18S 052782) near the south-eastern end of this syncline and towards the base of the Cloncannon Formation, which probably attains a thickness of about 900 m. *Monograptus ludensis* Murchison and *M. auctus* Rickards, the diagnostic graptolites of the *ludensis* Zone, occur both above and below the plant-bearing horizons.

MATERIAL AND METHODS

Small fragments of sterile and fertile axes occur as coalified compressions among irregular patches of coalified material of both plant and animal origin. The latter include pieces of *Dictyocaris* test and graptolite thecae, the former, wefts of wide tubes. Also present are *Pachytheca* and wider, striated, less regular axes probably assignable to *Prototaxites*. The axes were barely distinguishable on untreated rock surfaces, becoming visible on wetting with 50% ethanol. They were photographed under xylene. Specimens were developed (dégagement) by loosening grains of sediment with tungsten wire needles (Leclercq, 1960). Cellulose nitrate film pulls were prepared from coalified axes and less well-preserved sporangia. Larger pieces of organic debris including wefts of tubes were picked out of the residues from bulk maceration before centrifugation and mounted on stubs for SEM. Finally samples of mudstone from within a few metres of the plant-bearing horizon were processed by standard palynological techniques. Strew mounts in glycerine jelly were examined and photographed with a Leitz Dialux microscope.

All specimens and slides will be housed at the Department of Geology, Trinity College, Dublin.

DESCRIPTION OF MACROPLANTS

Sterile axes: These are infrequent and fragmentary. They are smooth, dichotomously branched and range from 0.3 to 1.7 mm in diameter, but are rarely strictly parallel-sided (Figs 25, 35). Some of the longer lengths suggest that branching was frequent, for example an axis 15 mm long has three dichotomies, but both daughter branches are seldom preserved (Fig. 25). Such features also characterize the longer fertile specimens. Film pulls of these coalified compressions yield little cellular detail, except for occasional remnants of longitudinally aligned cell walls.

Fertile axes: At the outset, it should be emphasized that as spores have not been isolated from the terminal expansions of the axes, they can only be assumed to be sporangia.

Edwards (1979) when describing a Downton assemblage from S Wales outlined some of the problems encountered in the measurement, description and identification of small sporangia of simple morphology. Those comments are equally relevant here where the sporangia are even smaller and less well-preserved in a matrix which sometimes proved unsuitable for the uncovering technique.

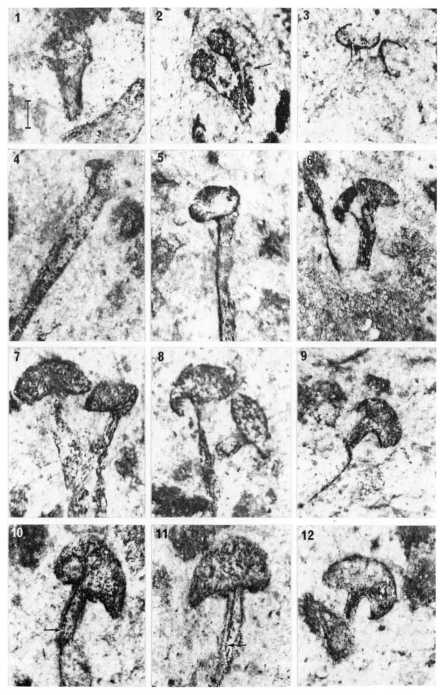

Figures 1–12. All figures to same magnification as Fig. 1; scale bar equivalent to 450 μm. Fig. 1. *TCD 22946b*. Fig. 2. *TCD 22945c;* arrow indicates possible branching point. Fig. 3. *TCD 22949a*. Fig. 4. *TCD 22946c*. Fig. 5. *TCD 22951;* axis attached to one side of base of sporangium. Fig. 6. *TCD 22946g*. Fig. 7. *TCD 22946d;* note changes in axis diameter immediately above branch point. Fig. 8. *TCD 22946f*. Fig. 9. *TCD 22950*. Fig. 10. *TCD 22946a*. Fig. 11. *TCD 22945a;* counterpart of sporangium in Fig. 10. Vertical lines (arrowed) in subtending axis probably represent folds. Fig. 12. *TCD 22947a;* heavily coalified line may represent a vascular strand.

Following photography, the specimens were measured. The main dimensions noted were sporangium width, height, the diameter of the subtending axis at its junction with the sporangium and any changes below. Some of the sporangia are almost circular in shape, but most are considerably wider than high and are elliptical to less regular in outline. None of the sporangia shows a dehiscence line. As few are identical in form, generalizations are impossible. About a third of these fertile specimens are branched, often immediately below the sporangia.

Film pulls were taken from the less well-preserved specimens.

Isodiametric forms: Two examples were found. The first (*TCD 229466;* Fig. 1) comprises a short axis which widens gradually into an almost circular sporangial region (0.65 mm high × 0.7 mm wide), the latter not clearly delimited from the subtending axis. In contrast the axes below the sporangia in the second specimen (*TCD 22945c*; Fig. 2) are parallel-sided so that the sporangia are easier to measure. That terminating the left hand branch, is circular in outline (0.45 × 0.4 mm). The right hand one (which was uncovered) is slightly longer than wide. It is possible that there is another branch point (arrowed in Fig. 2) below each sporangium, although uncovering revealed no further sporangia.

Sporangia wider than high: Figure 3 shows a branching system (*TCD 22949a*) where stout parallel-sided axes terminate in oval sporangia. Although most of the coalified material has disappeared, the outline of each sporangium is clearly delimited and shows a straight contact between sporangium and axis. Specimen *TCD 22951* (Fig. 5) is of similar size (0.85 × 0.5 mm) and shape, but with more coalified material remaining. Here there is no branching close to the sporangium and the subtending axis, most atypically, appears attached to the side of the base of the sporangium. The sporangium illustrated in Fig. 4 is also oval, but less complete. The subtending axis is again unbranched, of varying diameter and bears a longitudinal dark line. A small piece of the axis was removed on a film pull but showed no cellular detail. While it is possible that this line represents a vascular strand, further observations on similar structures on other fertile axes suggest that it results from a compressed longitudinal fold.

One particularly well-preserved specimen (*TCD 22946g*, Fig. 6) has sporangia which are wider and, because of their proximity to a branch point, overlie each other. The right hand sporangium is 0.75 mm wide × 0.4 mm high, subtended by a parallel-sided axis 0.2 mm in diameter. The second sporangium is probably similar but uncovering was stopped to preserve the first intact. The three dimensional shape of such sporangia remains obscure.

Some of the branching specimens have a differently-shaped sporangium terminating each branch of the dichotomy, although axis characteristics are more uniform. The left branch of the uncovered system illustrated in Fig. 7 (*TCD 22946d*) comprises a conspicuous reniform sporangium (1.3 × 0.55 mm) and narrow parallel-sided axis, while the right hand one bears a more irregularly shaped sporangium with asymmetric insertion of the axis. Irregular outlines distinguish some nearby sporangia (Fig. 8) which are otherwise similar in size and axis characteristics. These sporangia are quite flattened when compared with the specimens illustrated in Figs 10–12. They have markedly convex distal margins and either lobed (Fig. 12) or straight bases (Figs 10, 11). In specimen *TCD 22946a* (Figs 10,11) the axis is attached slightly eccentrically and bears a longitudinal striation, but this is in a different position on part and counterpart suggesting that such lines represent compressed longitudinal folds.

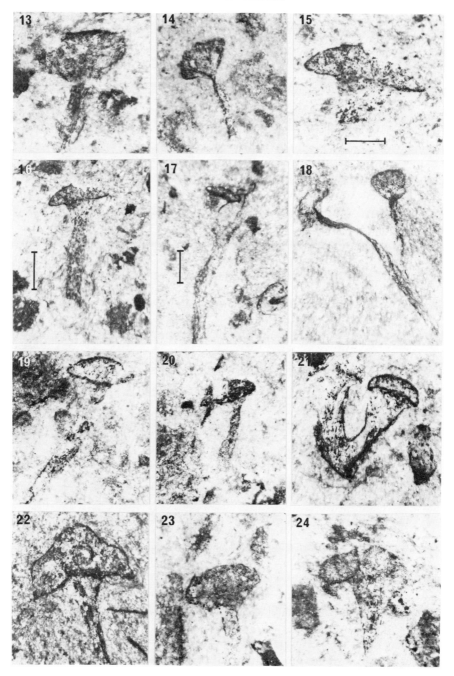

Figures 13–24. All figures except Figs 16, 17 to same magnification as Fig. 15, scale bar equivalent to 650 μm; scale bar in Fig. 16 equivalent to 1.25 mm; scale bar in Fig. 17 equivalent to 1.4 mm. Fig. 13. *TCD 22946f.* Fig. 14. *TCD 22945b.* Fig. 15. *TCD 22949b* before development. Fig. 16. *TCD 22949b* after uncovering. Fig. 17. *TCD 22953.* Fig. 18. *TCD 22952a.* Fig. 19. *TCD 22952b.* Fig. 20. *TCD 22952c.* Fig. 21. *TCD 22947b.* Fig. 22. *TCD 22954b.* Fig. 23. *TCD 22950a.* Fig. 24. *TCD 22945d.*

Specimen *TCD 22952a* (Fig. 18) was also uncovered. The sporangium terminating the longer left hand branch has a very irregular shape and is probably incomplete, but that on the shorter right hand branch appears intact and is subtriangular in outline. Here the subtending axis remains parallel-sided while the sporangium increases in diameter to 0.6 mm and has a flattened distal margin. Similarly shaped isolated sporangia are illustrated in Figs 13 and 14.

All the previous examples lack any pronounced change in axis diameter below the sporangium. In the following specimens the axis gradually widens in this region with variation resulting from degree of expansion and extent and shape of contact as well as sporangium shape. Specimen *TCD 229496* (Figs 15, 16) comprises an unbranched axis 3.2 mm long which gradually increases in diameter beneath one of the largest sporangia found (2.2 × 0.5 mm). The latter is strongly flattened: its contact with the axis is long and straight. Smaller, less completely preserved sporangia of slightly greater height when compared with width are shown in Figs 19 & 20. These were destroyed in the preparation of film pulls, which were unsuccessful in that no spores were recovered. Further examples of similar shape (e.g. Figs 23, 30) lack a clear distinction between sporangium and stalk.

In a few cases, the tip of the subtending axis is almost as wide as the sporangium itself. Figure 21 shows a branching specimen *TCD 229476b* in which the right hand branch was uncovered, revealing a squat sporangium (0.8 × 0.25 mm). A wide subtending axis characterizes the more completely preserved left hand branch in Fig. 17 (*TCD 22953*) where the sporangium itself shows a pronounced lobed or pouched relief (1.0 × 0.25 mm).

In contrast the sporangium (*TCD 22954*) illustrated in Fig. 22 is almost semicircular in outline (1.6 × 0.7 mm wide). Contact between sporangium and axis is extensive and appears as a straight line. The axis itself tapers abruptly. The change in axis diameter is less pronounced in specimen *TCD 22948* (Fig. 27), where the sporangium terminates a 13.5 mm long axis with two branching points. The sporangium was uncovered and during the process a considerable amount of sediment was removed from between upper and lower surfaces. The latter, illustrated here, is more heavily coalified than the subtending axis. It is fan-shaped with an irregular distal margin and longitudinally dissected into three lobes (Fig. 28). Lying alongside the ultimate dichotomy is a second, possibly detached, sporangium (Fig. 29). This has a clearly defined oval outline and pronounced relief. It was film-pulled but yielded no spores. Film pulls made from another isolated dissected fan-shaped sporangium were also unproductive.

In the above account, a major distinguishing feature has been the presence or absence of any widening of the subtending axis. The branch system (*TCD 22944*) illustrated in Fig. 26 appears to contradict the validity of this distinction. The more or less parallel-sided left hand axis is attached asymmetrically to a subtriangular sporangium, but the right hand sporangium appears quite different with a wide tapering base in which it is impossible to distinguish the junction between sporangium and axis. The latter is very poorly-preserved and incomplete. Another noteworthy feature of the specimen is the marked decrease in width of the branches immediately above the branching point (see also Fig. 7). It was unfortunately impossible to develop the specimen further, by removing more matrix from around the base of the left hand sporangium, because it was beginning to disintegrate.

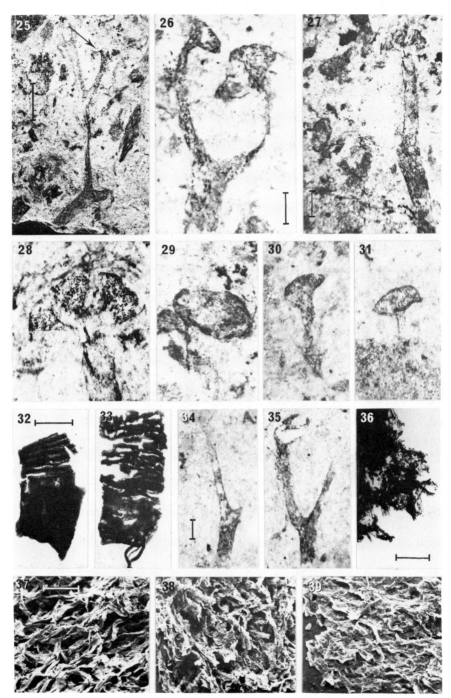

Figures 25–39. Fig. 25. *TCD 22955;* possible sterile tip is indicated by arrow; scale bar equivalent to 2.2 mm. Fig. 26. *TCD 22944;* scale bar equivalent to 500 μm. Fig. 27. *TCD 22948;* arrow indicates possible detached sporangium; scale bar equivalent to 1 mm. Figs 28–31 to same magnification as Fig. 26. Fig. 28. *TCD 22948a;* terminal sporangium from Fig. 27. Fig. 29. Detached sporangium from Fig. 27. Fig. 30. *TCD 22952b.* Fig. 31. *TCD 22945e.* Figs 32 & 33; *TCD 22960 & TCD 22961;* fragments of ornamented tubes; scale bar equivalent to 22 μm. Figs 34 & 35; scale bar equivalent to 900 μm. Fig. 34. *TCD 22957;* sterile axis. Fig. 35. *TCD 22956;* counterpart of Fig. 34. Fig. 36. *TCD 22962;* edge of dense mass of tubes; scale bar equivalent to 23.6 μm. Figs 37–39; to same magnification scale bar in Fig. 37 equivalent to 40 μm; scanning electron micrographs of wefts of tubes. Fig. 37. *TCD 22963a.* Fig. 38. *TCD 22963b.* Fig. 39. *TCD 22964.*

AFFINITIES OF THE MACROPLANTS

Short and wide sporangia terminating dichotomously branched smooth axes characterize the genus *Cooksonia* as erected by Lang (1937). Thus, on the evidence of comparative morphology, the branching fertile specimens will be assigned to *Cooksonia* as will many of the isolated sporangia terminating short lengths of unbranched axis. The taxonomic status of the genus itself will be discussed below.

Recognition of species in an assemblage which is notable for its considerable diversity in sporangial shape presents greater problems. Some forms, although smaller, fall within the limits of variability already described for certain species in younger geological formations such as *C. pertoni* Lang and *C. cambrensis* D. Edwards, while others are unique to the locality. In addition there are a few examples where differently shaped sporangia occur on the same branching system (e.g. Figs 7, 18, 25). In some cases this may have been produced by distortion or damage during transport or in subsequent burial and diagenesis or indeed may be an artefact of investigation. It is also possible that it may result from the preservation of sporangia in different stages of development. Edwards (1979) speculated that some elongation of the subtending axis may have occurred as the sporangium matured. The uncovered specimen (Fig. 18) provides additional support for this hypothesis in that the ultimate branches are of unequal length, the shorter wider one bearing an intact sporangium of well-defined shape while the left hand one is more spindly and ends in a poorly-preserved sporangium, possibly in the dehisced state. The fan-shaped dissected sporangium (Fig. 28) may also represent a dehisced sporangium.

Forms resembling *Cooksonia pertoni* Lang

TCD 22945d (Fig. 24), *TCD 22947b* (Fig. 21), *TCD 22950a* (Fig. 23), *TCD 22952b & c* (Figs 19, 20), *TCD 22953* (Fig. 17), *TCD 22954* (Fig. 22). ? *TCD 22948* (Fig. 27), ? *TCD 22949b* (Fig. 16).

Lang based the specific diagnosis on specimens collected from just above the base of the Downton Series at Perton Lane Quarry, Herefordshire (Lang, 1937; Edwards, Bassett & Rogerson, 1979). The main diagnostic characters of the species are smooth, dichotomously branching axes terminating in sporangia which are broader than high, the axis widening just below the sporangium. Lang's specimens are 1.5 × 1.0 mm–3.5 × 3 mm and, from his illustrations, appear quite variable in shape. New collections from Perton Lane are currently being investigated in Cardiff, but the following observations are based on comparison with Lang's illustrations and diagnosis. The specimen illustrated in Lang, 1937: pl. 8, fig. 12 resemble those in Figs 17 and 21 here, while Lang, 1937: pl. 8, fig. 6 is similar to Fig. 24. The specimens illustrated in Figs 19, 20, 22 & 23 conform to Lang's diagnosis, but are smaller. More problematic is specimen *TCD 22949b* (Figs 15, 16) but this is closer to *C. pertoni* than to any other *Cooksonia* species as is the specimen shown in Figs 27 & 28.

Forms resembling *C. hemisphaerica* Lang

? *TCD 22945b* (Fig. 1).

In the original specific diagnosis (Lang, 1937), sporangia are described as hemispherical, being almost as wide as high. The type locality, a small quarry called Targrove alongside the drive to Downton Hall, near Ludlow, occurs in strata some distance above the main *Psammosteus* limestone and is hence of Ditton not Downton age. Edwards (1979) extended the diagnosis of the species to encompass oval sporangia when describing a Downton assemblage from Dyfed.

Only one of the Irish specimens (Fig. 1) is similar to *C. hemisphaerica sensu* Lang and this was very fragmentary.

Forms resembling *C. cambrensis* D. Edwards

Forma α: *TCD 22945c* (Fig. 2).
Forma β: *TCD 22946c & g* (Figs 4, 6), *22949a* (Fig. 3), *22951* (Fig. 5). ? *TCD 22946f* (Fig. 13); ? *TCD 22952a* (Fig. 18).

Cooksonia pertoni and *C. hemisphaerica* are both characterized by a gradual increase in diameter of the subtending axis and thus differ from *C. cambrensis* where the axis is parallel-sided or increases slightly in diameter immediately below the sporangium. Sporangial outline in *C. cambrensis* ranges between circular (forma α) and elliptical (forma β). The type locality is in Downton strata at Freshwater East in Dyfed, S Wales (Edwards, 1979). Parallel-sided subtending axes are present in some of the Irish specimens, for example, that shown in Fig. 2 subtends a sporangium with circular outline while in Figs 3 & 6, sporangia are oval. They differ from *C. cambrensis* in that branching occurs close to the sporangia and axis diameter is quite wide when compared with sporangial width. This latter feature characterizes the specimens illustrated in Figs 4 and 5, but for the moment all these specimens will be included in *C. cambrensis*.

The right hand sporangium shown in Fig. 18 closely resembles the left hand one in Edwards, 1979: pl. 2, Fig. 10, and so this, together with the specimen illustrated in Fig. 13, is also assigned to the *C. cambrensis* complex.

Forms of uncertain affinity.

A few specimens cannot be assigned to the above or remaining species of *Cooksonia* (such as *C. crassiparietilis* Yurina and *C. caledonica* D. Edwards). Figures 7, 8 & 31 illustrate sporangia of variable shape ranging from regular to irregular in outline. Some resemble *C. pertoni* but are subtended by parallel-sided axes. It is possible that they are distorted or mature examples of *C. cambrensis*, but more probably they represent a new *Cooksonia* species. In the light of this uncertainty, plus a reluctance to erect a new species on a small number of specimens, it is proposed that they be left as *Cooksonia* sp.

Specimen *TCD 22944* (Fig. 26) also presents problems. The right hand sporangium if found isolated would be assigned to *C. pertoni* and the left hand one included in the small group just mentioned. Finally there is a small number of sporangia (Figs 9–12, 14) each terminating an unbranched axis which is parallel-sided and passes without apparent interruption into the sporangial region. None of the sporangial shapes is identical. It is proposed to leave them as *incertae sedis* until further more complete specimens are found.

DESCRIPTION OF PLANT MICROFOSSILS

The organic debris resulting from bulk maceration included triradiate miospores together with much rarer acritarchs (Figs 40–42, 45) chitinozoans (Fig. 48), scolecodonts, banded tubes (Figs 32, 33) and wefts of narrow tubes (Figs

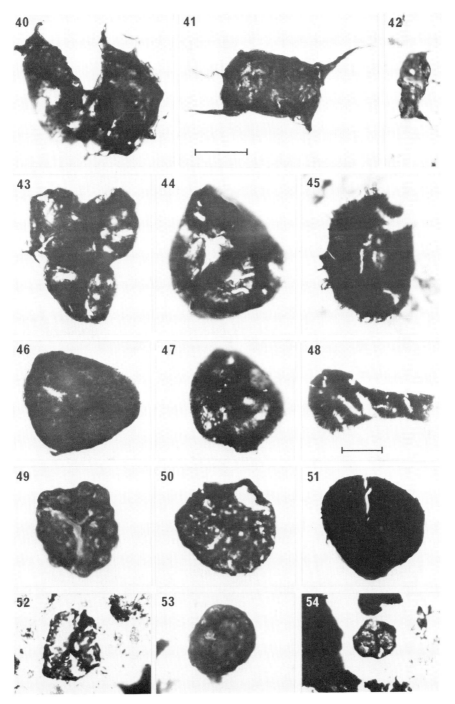

Figures 40–54. All figures except Fig. 48 to same magnification; scale bar in Fig. 41 equivalent to 20 μm. Fig. 40. *TCD 22965;* unidentified acritarch. Fig. 41. *TCD 22966;* acritarch of diacrodian type, possibly reworked. Fig. 42. *TCD 22967; Domasia* sp. indet. Fig. 43. *TCD 22968;* disintegrating tetrad of trilete spores of *Ambitisporites* type. Fig. 44. *TCD 22969;* tetrad of spores of *Ambitisporites* type with equatorial thickenings. Fig. 45. *TCD 22970;* unidentified acritarch. Fig. 46. *TCD 22971;* triradiate spore possibly of *Ambitisporites* type. Fig. 47. *TCD 22972;* triradiate spore of *Ambitisporites* with subequatorial thickening. Fig. 48. *TCD 22973;* chitinozoan, possibly *Sphaerochitina,* scale bar equivalent to 38.8 μm. Fig. 49. *TCD 22974;* trilete spore in which pyrite crystallization has distorted its appearance and opened the laesurae. Fig. 50. *TCD 22975;* spore with some pyrite damage. Fig. 51. *TCD 22975;* triradiate spore with tiny coni apparent around outline. Fig. 52. *TCD 22977;* trilete spore with laesurae extending to equator. Fig. 53. *TCD 22978;* spore-like body possibly resembling *Retusotriletes* ?sp. nov. in Holland & Smith, 1979 (pl. 2, figs 11, 13, 14). Fig. 54. *TCD 22979;* trilete spore possibly referable to *Retusotriletes.*

36–39). All the palynomorphs are rather poorly preserved and dark brown in colour, with much damage from pyrite crystal growth (Fig. 49).

Miospores: Most of the spores observed occur singly, but some tetrads (Fig. 44) and partial tetrads (Fig. 43) were also found. The spores vary in outline from almost circular to triangular with well-rounded apices. Nearly all are smooth, though a few appear to have minimal ornament of sparse low coni (a possible artefact of pyrite crystal growth: Fig. 51). The dissociated spores range from 20 to 59 µm in overall diameter ($\bar{x} = 44$ µm, 77 measurements); those in tetrads are 22–38 µm ($\bar{x} = 33$ µm, 14 measurements). Many of the spores are too dark to allow any further observations, but some, both separated (Fig. 44) and in tetrads (Fig. 43) have a clear equatorial crassitude, 2.5–6 µm wide. Some others equally clearly lack such a thickening (Fig. 50). Where seen, the simple triradiate laesurae appear to extend to the equator (Fig. 52). While positive identification of such indifferently preserved specimens is avoided here, it seems probable that some at least belong to the genus *Ambitisporites* (Figs 46, 47). It is quite possible that more than one taxon, (perhaps including the genus *Retusotriletes:* Figs 53, 54) is represented in the assemblage.

Tubes; ornamented forms: The fragments with annular or spiral internal thickenings illustrated in Figs 32 & 33 are typical. Longer specimens were not found. Such tubes were originally described by Lang (1937) as part of the *Nematothallus* complex. More recently published records are summarized in Edwards (1982) together with a somewhat inconclusive discussion on the structure and affinity of the tubes.

Tubes; smooth forms: There are two sizes. The larger ones (15–40 µm in diameter) are visible under a dissecting microscope as straight threads which form a very loose meshwork. They proved very difficult to detach from the rock on film pulls and when isolated in this way, comprised a mosaic of amber to black, featureless, irregular fragments. It was not surprising therefore that transfer preparations, made by coating the surface with a thick layer of cellulose nitrate and then dissolving the rock in hydro-fluoric acid, yielded no intact tubes. Branching was never observed and the walls appear to be of uniform thickness. Featureless tubes of similar dimensions are recorded in palynomorph assemblages throughout the Silurian (Pratt, Phillips & Dennison, 1978, Strother & Traverse, 1979; Edwards, 1982), however the organisms, be they plant or animal, from which they derive remain obscure.

Wefts of more closely packed and much smaller tubes (<9 µm diameter) were picked out of maceration residues. Some were treated with Schulze's solution and mounted for light microscopy while others were prepared for SEM. Figure 36 shows an example of the former; the tubes are smooth-walled and aseptate, but the specimens are so fragmentary that little confidence is placed on these observations. SEMs show some examples where the tubes appear fused or tightly adpressed to amorphous coalified material, the whole presenting a highly disorganized appearance (Figs 38, 39). In others the specimen consists almost entirely of entangled tubes (Fig. 37). They are of uniform diameter and appear unbranched.

Wefts of smooth tubes have been recovered from Silurian strata elsewhere (Lang, 1937; Pratt *et al.*, 1978; Edwards, 1982). The septa which led Pratt *et al.* (1978) to suggest their hyphal nature have not been observed in Irish and Welsh specimens, whose affinities remain unknown.

STRATIGRAPHIC AND BIOLOGICAL SIGNIFICANCE OF THE PALYNOMORPHS

These microfossils do not add significantly to the stratigraphical information provided by the graptolites, although the acritarchs include several specimens of *Domasia* (Fig. 42), a characteristically early to mid-Silurian genus. However, the presence of acritarchs, chitinozoa and scolecodonts confirms the fully marine origin of the sediments, while the miospores are of palaeobotanical interest because of their association with the macroplant occurrences.

Table 1 summarizes records of trilete spores from independently dated formations in Ireland, showing the graptolite zone to which each has been assigned. Further details of localities and stratigraphy have been given elsewhere (see Smith 1980, 1981 and references therein). The generally poor quality of palynomorph preservation in these occurrences would make them rather uninteresting except for the presence of graptolites in the same formations allowing the most refined stratigraphic dating which could be hoped for. The Irish records are thus of some value in providing support for the schemes of evolutionary development of Silurian spore floras outlined by Richardson (1974) and Allen (1981). According to these authors, and confirmed by the data in Table 1, the earliest isolated trilete spores found anywhere are of *Ambitisporites*-type and their earliest occurrence is within the Llandovery Series. Gray, Massa & Boucot (1982) have noted a corresponding change from exclusive occurrence of tetrahedral tetrads to a dominance of isolated trilete spores within the Fronian stage of the

Table 1. A summary of records of trilete spores

Series	Stage	Graptolite zone	Ambitisporites avitus Hoff.	A. dilutus (Hoff.)	Ambitisporites sp. indet.	cf. Archaeozonotriletes chulus-nanus	Retusotrilets cf. warringtonii	R. sp.	R. sp. indet.	Indeterminate trilete spores	Spore tetrads	Locality
Base of Ludlow												
Wenlock	Homerian	ludensis				●				●	●	Devilsbit locality
		lundgreni	●		●		●				●	Hollyford Fm, Slieve Felim
	Sheinwoodian	ellesae linnarssoni rigidus riccartonensis murchisoni	●	●		●	●	●	●			{ Capard Fm, Slieve Bloom { Knockshigowna Fm
		centrifugus	●								●	Lettergesh Fm, Kilbride
Llandovery	Telychian	crenulata griestoniensis crispus			●	●						Killanena Fm, Slieve Aughty
	Fronian Idwian Rhuddanian		Earliest *Ambitisporites* in Wales									

Llandovery based on material from Libya. In Wenlock strata a slightly greater diversity of still basically simple spore forms can be found, the Slieve Bloom locality providing a good example (Holland & Smith, 1979). The development of further sculptural features in the Ludlow epoch is not represented in securely dated rocks in Ireland.

There is unfortunately nothing to prove any direct connection between the *Cooksonia*-type plant fossils of the Devilsbit locality and the dispersed spores extracted from the associated sediments. *Ambitisporites* has been reported *in situ* in younger *C. pertoni* sporangia from the Welsh Borderland (Lang, 1937; Allen, 1980), but this is not enough to prove an exclusive relationship between these two taxa. The downward extension of the range of *Ambitisporites* into the Llandovery Series either disproves such a relationship, or suggests the possibility of finding even older *Cooksonia*. The diversity, albeit minimal, of trilete spores from Wenlock strata also indicates that there must be further taxa of Silurian plants yet to discover; whether or not they were vascular is an unanswered question.

PALAEOENVIRONMENTAL RECONSTRUCTION

The thick, deep-water Wenlock sequences of central Irish inliers form a typical flysch succession in the classic sense; a thick succession of poorly fossiliferous muds, silts and sands laid down in the late stages of development of a geosyncline. Continuing subsidence, overlapping in time with uplift of ridges, signalled the early phases of the major tectonic event that was to follow. Sedimentological considerations suggest that the Devilsbit area was virtually devoid of relief in late Wenlock times with deposition taking place principally on the distal regions of broad deep sea fans. The succession is dominated by sedimentary associations which are particularly characteristic of the lower fan and fan fringe environments of the submarine canyon-fan system, with occasional incursions by sequences of a more proximal nature, and possibly by abyssal plain sequences. Outer fan environments have also been postulated for the Hollyford Formation in the Keeper Hill area by Archer (1981).

Abundant palaeocurrent evidence from Devilsbit and neighbouring areas demonstrates easterly or north-easterly derivation of the sediments. A study of the turbidite conglomerates found in the central Irish inliers also suggests a source area to the north: they become much coarser northwards, and the proportion of volcanic clasts decreases southwards (Jackson, 1972).

However, the above conclusions regarding the environment of deposition of the Devilsbit rocks are of little relevance to the palaeoecology of the rhyniophyte or rhyniophytoid plants contained in those rocks. The fauna associated with the plant remains is predominantly pelagic, and it was earlier pointed out by one of us (Smith, 1980) before the discovery of any macroplant remains, that there was nothing to rule out a pelagic habitat for the then unknown plants producing the abundant trilete spores encountered in Irish Silurian rocks. However, the plant remains in Devilsbit are also associated with the fossil crustacean *Dictyocaris* (Størmer, 1935), which is generally believed to be characteristic only of brackish water to marginal marine environmental conditions. *Dictyocaris* has evidently been transported with the sediment, probably over a considerable distance, and it is a perfectly acceptable conclusion that the cooksonias, sterile axes, the spores, tubes and other dispersed microfossils have likewise been transported from a marginal marine or even continental setting.

It has been proposed that the source of sediment for the central Irish area in the Silurian period could have been the Derwen Ridge at the north-east end of the vestigial Irish Sea landmass (Holland, 1969). The coastal flats of this and other low-lying ridges in the narrow remnant of the Iapetus Ocean might well have provided a typical site for growth and evolution of the immediate protovascular ancestors of *Cooksonia* and indeed for the earliest representatives of that genus. The evident ease of transport of the small spores presumably produced by such plants might have been a factor in the remarkable uniformity and wide spread of the early terrestrial floras from North America and Europe (Banks, 1980; Edwards, 1980).

A number of lines of evidence suggest that these rhyniophyte or rhyniophytoid plants were erect land plants. *Cooksonia*, for example, is found in assemblages which contain undoubted vascular, and hence land, plants from late Silurian times onwards. In these late Silurian representatives, the terminal sporangia contain simple spores with sporopollenin-impregnated walls suggestive of aerial transport and the axes themselves are cutinized. The stiff non-flexuous nature of the Wenlock axes suggest that they were self-supporting. However, there does, of course, remain the possibility that the basal regions of the plant were permanently covered by water with only the fertile sporing regions emergent, or that they grew in very shallow perhaps ephemeral pools (Raven, 1977). We have no direct evidence relating to the habits of plants such as those in the *Cooksonia* flora we have described. The majority of records, as this one, comes from marine sediments. The preservation of a Downton flora from Dyfed in sediments "accumulated on coastal sandflats influenced by the sea" (Edwards, 1979: 49) raised the possibility that the plants grew on nearby saltmarshes, as the tiny fossils, although fragmentary, were not badly damaged, thus suggesting limited transport. However, the Ditton record of *C. hemisphaerica* (Lang, 1937, but see discussion below) is from typical fluviatile sediments of the Old Red Sandstone (Allen & Tarlo, 1963) so that by this time, if not earlier, cooksonias had migrated inland perhaps colonizing river banks or mudflats around lakes.

GENERAL OBSERVATIONS

Whether or not *Cooksonia* was a vascular plant remains conjectural as convincing tracheids have never been demonstrated in the fertile axes, which are usually preserved as highly coalified compressions. The genus has been reported in association with vascularized axes at Whitcliffe (Capel Horeb, S Wales; Edwards & Davies, 1976) and Ditton (Targrove, Salop; Lang, 1937) localities, but as there are further fertile plants at both, this cannot be accepted as evidence for the vascular status of the genus. Thus there is no incontrovertible evidence that *Cooksonia* is a member of the Rhyniophytina (Banks, 1975a) and it is perhaps better described as rhyniophytoid, a term used by Pratt *et al.* (1978) when describing plants of rhyniophyte morphology, but lacking vascular tissue, which were probably early colonizers of the land, intermediates between vascular plants and their aquatic algal ancestors (Raven, 1977). The hypothesis is an attractive one, but as already indicated (Edwards *et al.*, 1979) anatomical detail is rarely preserved in highly coalified compression fossils, so that lack of tracheids may simply result from preservation failure. We are forced to conclude that the Irish Wenlock macroplant assemblage includes the remains of erect, fertile, land plants, most of which can be assigned to the genus *Cooksonia*, but we have no direct evidence that these plants were vascular.

This is all the more disappoinging as the flora predates other macrofloras suggestive of land vegetation except for one from Llandovery strata in the United States, which contains sterile axes named *Eohostimella* (Schopf *et al.*, 1966). In general aspect the Irish flora resembles those from later Ludlow and Downton localities in Europe and N America (for summary see Edwards, 1979; Banks, 1980) except that greater diversity is recorded from the younger floras (e.g. Ishchenko, 1969, 1975). Absent from the Irish Wenlock flora are cuticles of *Nematothallus* recovered from most localities in the late Silurian (Lang, 1937; Edwards, 1982) and considered further evidence for a land vegetation, but of non-vascular, possibly thallophytic forms, although here again preservation failure is a strong possibility. The dispersed spore record is meagre, but does not conflict with that based on better-preserved microfloras from rocks of similar age elsewhere in Ireland and in Britain (Smith, 1981; Richardson, 1974).

CONCLUSIONS

The major importance of this Irish Wenlock flora lies in its age. To date it is the earliest record of erect fertile land plants. Unfortunately in the absence of anatomy we can make no contribution to the continuing debate on the time of origin of vascular plants (Banks, 1975a, b; Gray & Boucot, 1977, 1980; Schopf, 1978; Garratt, 1978; Edwards *et al.*, 1979). From plant assemblages reported from N America and Europe has emerged a general picture of land surfaces colonized since mid Silurian times by plants of rhyniophyte morphology associated with possibly thallophytic non-vascular plants such as those of the *Nematothallus* complex and further enigmatic forms such as *Prototaxites*. There is a gradual increase in numbers of taxa and further morphological innovation towards the end of the Silurian. Late Silurian assemblages from Libya (Klitzsch, Lejal-Nicol & Massa, 1973) and Australia (Garratt, 1978, 1981; Douglas & Lejal-Nicol, 1981) appear at variance with this scenario as they contain plants of more advanced organization including lycopods, indicating perhaps a far longer previous history of vascular plants in these areas. The Libyan (late Silurian – early Devonian) record is particularly puzzling as there are some lines of evidence which indicate similarities with trends in N America and Europe, and indeed its dating has been queried elsewhere (Chaloner & Sheerin, 1979; Edwards *et al.* 1979; Allen, 1981). For example, a *Cooksonia* flora (Daber, 1971) has been reported from a borehole in sediments not older than Pridoli (approximately equivalent to the Downton) in age (Jaeger, personal communication) and while the very well-preserved Silurian-Lower Devonian spore succession (Richardson, Rasul & Al-Ameri, 1981) comprises a far larger number of taxa than that in Europe but with a sufficient number in common to allow correlation between the two areas, there is none of the morphological innovation (but see comment on spore size in Gray *et al.*, 1982) which would be anticipated from the levels of organization in the macrofloras. In our opinion the probability that the Lower Plant Assemblage at Yea, Victoria (containing *Baragwanathia*) occurs in late Silurian strata, the age being based to a large extent on the identification of graptolites, presents a far greater challenge to current ideas on the evolution of vascular plants.

ACKNOWLEDGEMENTS

We are grateful to the Irish Forest and Wildlife Service for permitting and financing the opening up of a new section and, in particular, thank Mr P. Bane, Forester-in-charge in the area for his personal help and interest in the project. Edwards' research on the Silurian of the British Isles was supported by a NERC Research Grant.

Mrs V. Rose and Mr A. Harry were of great assistance in the preparation of this paper.

REFERENCES

ALLEN, K. C., 1980. A review of *in situ* late Silurian and Devonian spores. *Review of Palaeobotany and Palynology, 29:* 253–270.

ALLEN, K. C., 1981. A comparison of the structure and sculpture of *in situ* and dispersed Silurian and early Devonian spores. *Review of Palaeobotany and Palynology, 34:* 1–9.

ALLEN, J. R. L. & TARLO, L. B., 1963. The Downtonian and Dittonian facies of the Welsh Borderland. *Geological Mazazine, 100:* 129–155.

ARCHER, J. B., 1981. The Lower Palaeozoic rocks of the northwestern part of the Devilsbit-Keeper Hill Inlier and their implications for the postulated course of the Iapetus Suture Zone in Ireland. *Journal of Earth Sciences (Dublin) 4:* 21–38

BAILY, W. H., JUKES, J. B. & WYNNE, A. B., 1860. Explanations to accompany Sheet 135 of the memoirs of the Geological Survey of Ireland, comprising parts of County Tipperary, and the King's and Queen's Counties. Dublin: H.M.S.O.

BANKS, H. P., 1975a. The oldest vascular land plants: a note of caution. *Review of Palaeobotany and Palynology, 20:* 13–25.

BANKS, H. P., 1975b. Early vascular land plants: proof and conjecture. *Bioscience, 25:* 730–737.

BANKS, H. P., 1980. Floral assemblages in the Siluro-Devonian. In D. L. Dilcher & T. N. Taylor (Eds), *Biostratigraphy of Fossil Plants:* 1–24. Pennsylvania: Dowden, Hutchinson & Ross, Inc.

CHALONER, W. G. & SHEERIN, A., 1979. Devonian macrofloras. In M. R. House, C. T. Scrutton & M. G. Bassett (Eds), *The Devonian System, Special Papers in Palaeontology, 23:* 145–161. London: Palaeontological Association.

COPE, R. N., 1954. Cyrtograptids and retiolitids from County Tipperary. *Geological Magazine, 96:* 209–220.

COPE, R. N., 1955. *The Silurian rocks of the Devilsbit Mountains, County Tipperary, Eire.* Ph.D. Thesis. University of Liverpool, U.K.

COPE, R. N., 1959. The Silurian rocks of the Devilsbit Mountain District, County Tipperary. *Proceedings of the Royal Irish Academy, 60B:* 217–242.

DABER, R., 1971. *Cooksonia*—one of the most ancient psilophytes—widely distributed, but rare. *Botanique, 2:* 35–39.

DORAN, R. J. P., 1974. The Silurian rocks of the southern part of the Slieve Phelim inlier, County Tipperary. *Proceedings of the Royal Irish Academy, 74B:* 193–202.

DOUGLAS, J. G. & LEJAL-NICOL, A., 1981. Sur les premières flores vasculaires terrestres datées du Silurien: Une comparaison entre la "Flore à *Baragwanathia*" d'Australie et la "Flore à Psilophytes et Lycophytes" d'Afrique du Nord. *Comptes rendus hebdomadaires des Séances de l'Académie des sciences, Paris, Série D, 292:* 685–688.

EDWARDS, D., 1979. A late Silurian flora from the Lower Old Red Sandstone of south-west Dyfed. *Palaeontology, 22:* 23–52.

EDWARDS, D., 1980. Early land floras. In A.L. Panchen (Ed.), *The Terrestrial Environment and the Origin of Land Vertebrates:* 55–85. *The Systematics Association Special Volume No. 15,* London: Academic Press.

EDWARDS, D., 1982. Fragmentary non-vascular plant microfossils from the late Silurian of Wales. *Botanical Journal of the Linnean Society, 84:* 223–256.

EDWARDS, D. & DAVIES, E. C. W., 1976. Oldest recorded *in situ* tracheids. *Nature, London, 263:* 494–495.

EDWARDS, D. & FEEHAN, J., 1980. Records of *Cooksonia*-type sporangia from late Wenlock strata in Ireland. *Nature, London, 287:* 41–42.

EDWARDS, D., BASSETT, M. G. & ROGERSON, E. C. W., 1979. The earliest vascular plants: continuing the search for proof. *Lethaia, 12:* 313–324.

GARRATT, M. J., 1978. New evidence for a Silurian (Ludlow) age for the earliest *Baragwanathia* flora. *Alcheringa, 2:* 217–224.

GARRATT, M. J., 1981. The earliest vascular land plants: comment on the age of the oldest *Baragwanathia* flora. *Lethaia, 14:* 8.

GRAY, J. & BOUCOT, A. J., 1977. Early vascular land plants: proof and conjecture. *Lethaia, 10:* 145–174.

GRAY, J. & BOUCOT, A. J., 1980. Microfossils and evidence of land plant evolution. *Lethaia, 13:* 174.

GRAY, J., MASSA, D. & BOUCOT, A. J., 1982. Caradocian land plant microfossils from Libya. *Geology, 10:* 197–201.

HOLLAND, C. H., 1969. Irish counterpart of Silurian of Newfoundland. *Memoirs of the American Association of Petroleum Geologists, 12:* 298–308.

HOLLAND, C. H. & SMITH, D. G., 1979. Silurian rocks of the Capard inlier, County Laois. *Proceedings of the Royal Irish Academy, 79B:* 99–110.

ISHCHENKO, T. A., 1969. The *Cooksonia* palaeoflora in the Skalski horizon of Podolia and its stratigraphical significance. *Geological Journal, Kiev, 29:* 101–109 [Translation from Russian by Geological Survey of Canada].

ISHCHENKO, T. A., 1975. *The Late Silurian Flora of Podolia.* Kiev. Naukova Dumka.

JACKSON, A. A., 1972. *The Geology of the Silurian and basal Old Red Sandstone of the Galty Mountain area.* Ph.D. Thesis. University of Dublin, Ireland.

KLITZSCH, E., LEJAL-NICOL, A. & MASSA, D., 1973. Le Siluro-Dévonien a psilophytes et lycophytes du bassin de Mourzouk (Libye). *Comptes rendus hebdomadaires des Séances de l'Académie des sciences, Paris, Série D, 277:* 2465–2467.

LANG, W. H., 1937. On the plant remains from the Downtonian of England and Wales. *Philosophical Transactions of the Royal Society of London, B, 227:* 245–291.

LECLERCQ, S., 1960. Refendage d'une roche fossilifère et dégagement de ses fossiles sous binoculaire. *Senckenbergiana Lethaea, 41:* 483–487.

PALMER, D., 1970. *Monograptus ludensis* Zone graptolites from the Devilsbit Mountain District, Tipperary. *Scientific Proceedings of the Royal Dublin Society, 3A:* 335–342.

PRATT, L. M., PHILLIPS, T. L. & DENNISON, J. M., 1978. Evidence of non-vascular land plants from the early Silurian (Llandoverian) of Virginia, U.S.A. *Review of Palaeobotany and Palynology, 25:* 121–149.

RAVEN, J. A., 1977. The evolution of vascular land plants in relation to supracellular transport processes. In H. W. Woolhouse (Ed.), *Advances in Botanical Research*, Vol. 5, 154–219. London & New York: Academic Press.

RICHARDSON, J. B., 1974. The stratigraphical utilization of some Silurian and Devonian miospore species in the northern hemisphere; an attempt at a synthesis. *International Symposium of Belgian Micropalaeontological Limits (Namur 1974). Publication 9:* 1–13. Geological Survey of Belgium.

RICHARDSON, J. B., RASUL, S. M. & AL-AMERI, T., 1981. Acritarchs, miospores and correlation of the Ludlovian-Downtonian and Silurian-Devonian boundaries. *Review of Palaeobotany and Palynology, 34:* 209–224.

SCHOPF, J. M., 1978. *Foerstia* and recent interpretation of early, vascular land plants. *Lethaia, 11:* 139–143.

SCHOPF, J. M., MENCHER, E., BOUCOT, A. J. & ANDREWS, H. N., 1966. Erect plants in the early Silurian of Maine. *U.S. Geological Survey Professional Paper, 550-D:* D69–D75.

SMITH, D. G., 1980. The distribution of trilete spores in Irish Silurian rocks. In A. L. Harris, C. H. Holland & B. E. Leake (Eds), *The Caledonides of the British Isles Reviewed:* 423–431. London: Geological Society of London.

SMITH, D. G., 1981. Progress in Irish Palaeozoic Palynology. *Review of Palaeobotany and Palynology, 34:* 137–148.

STØRMER, L., 1935. *Dictyocaris* Salter, a large crustacean from the upper Silurian and Downtonian, *Norsk Geologisk Tidsskrift, 15:* 267–298.

STROTHER, P. K. & TRAVERSE, A., 1979. Plant microfossils from Llandoverian and Wenlockian rocks of Pennsylvania. *Palynology, 3:* 1–21.

Botanical Journal of the Linnean Society (1983) *86*: 37–55. With 36 figures.

Algae from the Rhynie Chert

DAVID S. EDWARDS

Department of Botany, University of Cape Coast, Cape Coast, Ghana, West Africa

AND

A. G. LYON

17, The Square, Rhynie, by Huntly, Aberdeenshire AB5 4HD

Received January 1982, accepted for publication July 1982

New material collected from the Rhynie Chert Bed (Siegenian age) has revealed new filamentous and unicellular algae. Two of the filamentous forms and one palmelloid form are described. Filamentous and unicellular algae are particularly common in a white kind of chert not previously reported and which is interpreted as having been formed by the *in situ* silicification of a silicate-rich pond.

Some rhizoid-nodes of *Palaeonitella cranii* (Kidston & Lang) Pia and a possible rhizoid-borne pro-embryo are also described and compared with similar structures in living Charophyta.

KEY-WORDS:—Algae – Charophyta – Chroococcaceae – Devonian – fossil – Ulotrichaceae.

CONTENTS

INTRODUCTION

Knowledge of Devonian algae is almost entirely confined to calcified forms of the Cyanophyta, Chlorophyta and Rhodophyta. Non-calcified filamentous and unicellular algal microfossils are far less well-known, the most widespread being the organic-walled cysts and resting spores of certain Prasinophyta (families Cymatiosphaeraceae, Tasmanitaceae, Leiophaeridiaceae—following the classification by Tappan, 1980) which were previously included in the heterogeneous group Acritarcha (see Downie, 1973; Tappan, 1980). The few other

0024–4074/83/010037 + 19$03.00/0

forms known have been assigned to the Ulotrichales (Grüss, 1928; Baschnagel, 1942; Fairchild, Schopf & Folk, 1973; Wicander & Schopf, 1974), Zygnematales (Baschnagel, 1966), Oedogoniales (Baschnagel, 1966), Stigonematales (Croft & George, 1959; Tappan, 1980).

Previous studies of the algal flora of the Rhynie Chert, which is regarded as being of early Devonian, possibly Siegenian, age (see Richardson, 1967; Westoll, 1977) have recognized five taxa of cyanobacteria and a possible charophyte. Kidston & Lang (1921) described two species similar to modern *Oscillatoria*— *Archaeothrix oscillatoriformis* and *A. contexta*. The narrow filaments (3–4 μm diameter) of discoid cells of *A. oscillatoriformis* were found within a stem of *Rhynia gwynne-vaughanii* while the narrower (*c.* 2 μm diameter) filaments of *A. contexta* occurred in large masses lying loose in the matrix. Croft & George (1959) described three species of cyanobacteria—*Langiella scourfieldi*, *Kidstoniella fritschii* and *Rhyniella vermiformis* from a single chip of chert which also contains a vascular axis. *Langiella* and *Kidstoniella* had a heterotrichous growth habit with well-developed prostrate and erect systems, heterocysts and akinetes; Croft & George (1959) assigned both genera to the Stigonemataceae. The unbranched, non-heterotrichous mucilaginous filaments of *Rhyniella* were regarded by Croft & George (1959) as being of uncertain systematic position within the cyanobacteria; Tappan (1980) includes *Rhyniella* in the Scytonemataceae.

Possible algal unicells were described by Kidston & Lang (1921); additional specimens of similar unicells, together with a filamentous alga were found, but not described, by Lyon (1962).

Kidston & Lang (1921) also described some remains which they regarded as those of a probable charophyte. *Palaeonitella cranii* (Kidston & Lang) Pia includes branch whorls, possible bulbils and rhizoid-nodes; but no reproductive structures were found and Tappan (1980) considers its status questionable.

We describe a number of previously unknown filamentous and unicellular algae from the Rhynie Chert and provide additional information on *Palaeonitella cranii* (Kidston & Lang) Pia to support the interpretation of that plant as a charophyte.

MATERIAL AND METHODS

The essential features of the Rhynie Chert have been described by Kidston & Lang (1917, 1921). In general terms, the deposit is up to 2.5 m in thickness and consists of alternating beds of chert and sandstone. The chert is dark grey or black in hand specimens and contains well-preserved plant and animal fossils. Kidston & Lang (1921 : 891), in a detailed study of sections through the deposit, noted that algae were limited to certain horizons (designated A″ and F); Tasch (1957) has also pointed out that animal (principally crustacean) remains are also limited in occurrence, being most common in horizon A″, the most south-easterly of the chert horizons found by Tait (in Horne *et al.*, 1916) and regarded by him as the basal bed of the section.

Some of the material used in the present study was collected from the surface of the field which overlies the bedded chert and hence its relationship with the *in situ* chert is uncertain. Other material was obtained from a trench (designated No. 2a) which was excavated in the approximate position of Tait's trench No. 2 (in Horne *et al.*, 1916). While no algae were encountered in samples of bedded chert collected from the exposure, remains of such organisms were found in some loose blocks, a lens of which was discovered embedded in an ochreous clay 31 m from the top (north-west end) of the bedded rock and 20.5 m from its lower (south-eastern) end. Hand specimens of this chert are white in colour and their identity with the deposit is confirmed by the occurrence within the blocks of *Rhynia major* axes bearing sporangia. For the most part these axes are aligned parallel to one another with the arms of their dichotomies all pointing in the same direction. This tends to suggest *situ* preservation. The matrix between the axes includes light grey and black areas which are usually localized in bands perpendicular to the *Rhynia* axes. Unlike the typical black chert of the deposit, the white chert, as a whole, is very porous, the grey areas having the highest porosity, the pure white ones, the lowest. Large (up to 5 mm diameter) cavities lined with crystalline quartz are also present. The algae are localized within the low porosity chert, most commonly in close proximity to *Rhynia* axes. (Fig. 7). Animal remains, largely disarticulated specimens of *Lepidocaris*, are more frequent in the areas of grey, high porosity, chert. Fungal hyphae and resting spores, isolated nematophytalean tubes and 'branch-knots' (Lyon, 1962) and *Rhynia major* spores also occur scattered throughout the matrix.

Material from Trench 2a was studied by means of ground sections; chip preparations were made from blocks gathered on the surface of the fields. All slides prepared from the white chert will be deposited in the Palaeobotanical Collection, Hunterian Museum, Glasgow, U.K., on completion of this study and are identified by the Hunterian Museum numbers allocated to them. Chip preparations made from the black chert are in the A. G. Lyon Collection. Stage co-ordinates for the individual fossils are not given in this paper since it is felt that they would be of limited usefulness for future workers. The general positions of the figured specimens has been indicated with a diamond scribe and outlined in ink on the bottom of the glass slides to which the sections are attached. Specimens in the Kidston Collection (Hunterian Museum) and the Lang Collection (Manchester Museum) have also been examined.

DISCUSSION AND SYSTEMATIC PALAEONTOLOGY

The affinities of many fossil algae are uncertain since modern classifications (for example that of Bold & Wynne, 1978) emphasize biochemical and ultrastructural features which are rarely, and in most cases could never be, preserved (for example the number of flagella of motile stages of otherwise non-motile forms). The parallel trends in morphological complexity in many algal classes and the variations in form both within and between different stages of the life-cycle of a single species (for example *Stichococcus bacillaris* Naegeli (Hayward, 1974) make comparisons of fossil and living taxa extremely speculative. In addition, post-mortem diagenetic changes may obscure, or so alter, cell morphology, that the diagnostic features of the original organism are lost (Golubic & Barghoorn, 1977; Francis, Barghoorn & Margulis, 1978). Thus, although indications of cell contents can be seen in some of the Rhynie Chert algae, it is difficult to assess the relationship of these features to the original organelles and their relevance in indicating the affinities of the organisms. In accordance with current palaeobotanical practice (Schopf, 1968; Horodyski, 1980; Tappen, 1980) diagnosed taxa are referred to higher categories, but the doubtful usefulness of such a procedure must be emphasized.

Filamentous algae

DIVISION: Charophyta
SPECIES INCERTAE SEDIS: *Palaeonitella cranii* (Kidston & Lang) Pia

In their account of the Thallophyta of the Rhynie Chert, Kidston & Lang (1921) described some fragmentary remains which they regarded as probably representing the vegetative organs of a charophyte. These consisted primarily of branched septate filaments with dark contents which, in some cases, showed evidence of a nodal structure, as in modern charophytes. In the apparent absence of corticating cells, a particular comparison was suggested and implied in the provisional generic name *Palaeonitella*. This name has been accepted by most authors (Walton, 1953; Emberger, 1968; Tappan, 1980) rather than *Algites* Seward as originally proposed by Kidston & Lang (1921). Pia (1927) first used the generic designation *Palaeonitella* in referring to *Algites* (*Palaeonitella*) *cranii* Kidston & Lang; the currently accepted name is thus *Palaeonitella cranii* (Kidston & Lang) Pia.

Associated with these axial structures were tubular elements lacking obvious contents, but with occasional oblique septa. In connection with these, groups of cells were developed and, in some cases, large oval or spherical vesicles. The former were interpreted as rhizoid-nodes, the latter as bulbils such as are formed by certain contemporary charophytes.

Some rhizoid-nodes from the new material were so orientated and preserved that a very close comparison with similar structures in living material has been possible.

In the development of a charophyte rhizoid-node, a curved oblique septum is laid down across the rhizoid in such a way as to give the ends of the resultant cells the appearance of two feet placed sole-to-sole (Groves & Bullock Webster, 1920). The 'toe' of one of these subsequently enlarges and divides to form a group of cells, some of which form secondary rhizoids.

Figure 3 shows a fossil node (Specimen No. 1) in side view. Although the cells of the nodal cluster are obscured by black granular material, the undivided 'toe' is prominently displayed and conforms very closely in relative size and position to comparable structures similarly orientated in living material. An interpretation of all visible structures is given in Fig. 1A and for purposes of comparison, a photograph of a rhizoid-node of *Nitella* sp. appears as Fig. 2.

Like specimen No. 1., specimen No. 2 (Figs 4, 8) is also orientated with its long axis parallel to the surface of the chip. In spite of its excellent preservation, it has proved difficult to obtain a satisfactory overall photograph, as slight differences in plane, coupled with the relatively high magnification required, fail to show clearly all the details of structure which can be resolved by differential focusing. An interpretation of all visible structure is given in Fig. 1B. In this specimen, three branch rhizoids can be seen arising from the cells of the nodal cluster and the longest of these seems, itself, to have developed a septum (seen in face view) with associated slight swelling of the ends of the two cells thus formed. While the undivided 'toe' of the larger node occupies a rather more prominent position than might be expected from the orientation of the speciment as a whole, this could be accounted for by slight twisting or displacement prior to petrifaction. Although the primary rhizoid can be traced, passing obliquely through the matrix for some

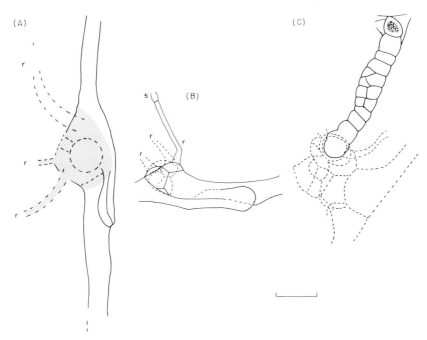

Figure 1. *Palaeonitella cranii*. A. Rhizoid-node No. 1. r,r,r, = branch rhizoids. cf. Fig. 3. B. Rhizoid-node No. 2. r,r,r = branch rhizoids. s = septum. cf. Fig. 4 & Fig. 8. C. ? rhizoid-borne pro-embryo. cf. Fig. 6. Scale bar equivalent to 100 μm.

distance on one side of the node, no connection with any axial structure can be detected.

In specimen No. 3 (Fig. 5), one of the 'toes' has divided tranversely and formed a branch rhizoid although only the proximal part of this can be seen.

One of the chips obtained from the *Palaeonitella*-containing blocks was found to include a well-preserved filament of somewhat barrel-shaped cells (Fig. 6). Examination from both sides has indicated that this distinctive structure is borne at the junction between two large tubular elements in association with an irregular cluster of cells. Arising from one of these cells is a small rhizoid-like outgrowth. Some of the cells of the filament appear to have undergone vertical divisions and one of the (?) two conical cells at its apex contains a brownish mass which may reflect former cell contents. The appearance of the specimen as a whole (Fig. 1C) is somewhat suggestive of a vegetative pro-embryo developed at a rhizoid-node, although the filament shows no clear indication of incipient nodal structure. However, the small size of the object and the rather granular texture of the matrix, make it difficult to distinguish between true cells walls and what might be merely superficial creases. It is also possible that development might have been abnormal. Although the septum between the two large tubular elements appears transverse, the junction cannot be interpreted as an axial node owing to the absence of an internodal cell. Occasional rhizoid-nodes with apparently transverse septa have been encountered and an example of such a node was figured by Kidston & Lang (1921: fig. 10).

In the absence of any direct evidence of continuity between the organs which have been grouped as *Palaeonitella cranii*, the question as to whether they all

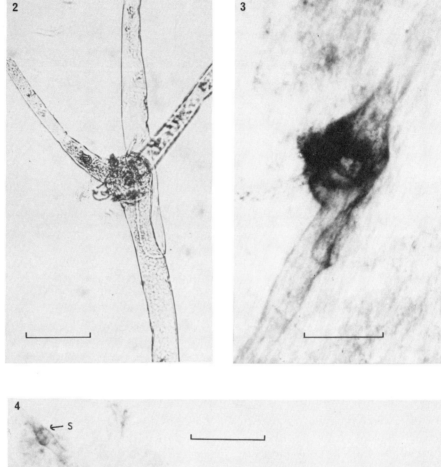

Figures 2–4. Fig. 2. Rhizoid-node of *Nitella* sp. Scale bar equivalent to 100 µm. Fig. 3. *Palaeonitella cranii.* Rhizoid-node No. 1. cf. Fig. 1A & Fig. 2. A. G. Lyon Collection Slide R.144. Scale bar equivalent to 200 µm. Fig. 4. *Palaeonitella cranii.* Rhizoid-node No. 2. r,r,r = branch rhizoids; s = septum on branch rhizoid. Scale bar equivalent to 100 µm.

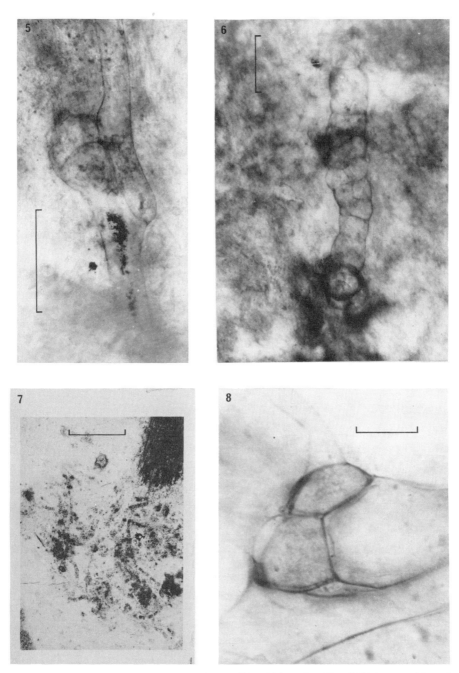

Figures 5–8. Fig. 5. *Palaeonitella cranii*, Rhizoid-node No. 3. Enlarged 'toe' has divided once and formed a branch rhizoid. Scale bar equivalent to 200 μm. Fig. 6. (?) *Palaeonitella cranii*, ? rhizoid-borne pro-embryo. cf. Fig. 1C. A. G. Lyon Collection; Slide 137. Scale bar equivalent to 100 μm. Fig. 1. General view of filaments in white chert close to *Rhynia major* axes (bottom left of figure); some of the filaments are *Mackiella rotundatus*, others have not been described in this paper; Slide FSC 2202. Scale bar equivalent to 500 μm. Fig. 8. *Palaeonitella cranii*, rhizoid-node No. 2 enlarged to show details of divided 'toe'. Scale bar equivalent to 50 μm.

represent parts of a single species cannot yet be settled, although it remains likely. Furthermore, as no trace of reproductive organs of the characteristic charophyte type have ever been found in the Chert, conclusive evidence as to the nature of the remains is still lacking. Nevertheless, the highly distinctive structure of the rhizoid-nodes shown so clearly in these specimens, may serve to maintain and give added support to the belief that *Palaeonitella cranii*, either in whole, or in part, was probably a charophyte.

DIVISION:	Chlorophyta
CLASS:	Chlorophyceae
ORDER:	Ulotrichales
FAMILY:	Ulotrichaceae

Mackiella D. S. Edwards & A. G. Lyon **genus novum**

DERIVATION: Named after Dr W. Mackie who discovered the Rhynie Chert.

Unbranched, unattached, uniseriate filaments with straight or rarely curved cross walls. Individual cells cylindrical, length approximately equal to diameter, rarely half that of diameter. Terminal cells rounded, longer than median cells of filament. Cell walls thin, non-stratified, mucilaginous sheath absent. Where preserved the cell contents are largely uniform and include fine granular material and a single darker body. Reproduction unknown.

TYPE SPECIES: *Mackiella rotundata* D. S. Edwards & A. G. Lyon (Figs 9–16).

Mackiella rotundata D. S. Edwards & A. G. Lyon **sp. nova**

DERIVATION: The specific epithet refers to the proportions of the individual cells.

Cell width 28–40 µm ($\bar{x}$ 34 µm); cell length; normal intercalary cells 23–41 µm ($\bar{x}$ 31.5 µm); terminal cells, 29–49 µm; short cells, 16–18 µm ($\bar{x}$ 17 µm) filament length, 3–25 cells.

TYPE: Holotype   from   slide   F.S.C.   2204,   Hunterian   Museum Palaeobotanical Collection, Glasgow. Illustrated in Fig. 10.

LOCALITY: Rhynie, Aberdeenshire.

HORIZON: Rhynie Chert Bed, Scottish Lower Old Red Sandstone (Siegenian).

Filaments of this alga are fairly common in the white chert (to which it is confined), particularly in close proximity to *Rhynia major* axes, although it has not been found attached to that plant (Fig. 9). While cell width is fairly constant between different filaments, cell length varies both between and within filaments. Most filaments include one or two short cells amongst longer cells, possibly indicating a generalized capacity for cell division; in addition to this variation, some filaments consist of cells which are generally rather shorter. A few filaments have been found which are shorter than most—the shortest being of three cells (Fig. 13) and this may indicate a tendency for the filaments to fragment.

In most cases cell contents are a uniform light-brown with included granular areas; a regular feature is a larger spherical body (about 3.5 µm diameter), usually located towards one end of the cell. The best-preserved examples (Figs 10, 11, 14) show the brown background material (here interpreted as cytoplasmic remnants) to have contracted from the wall in some areas. The nature of the single dark body

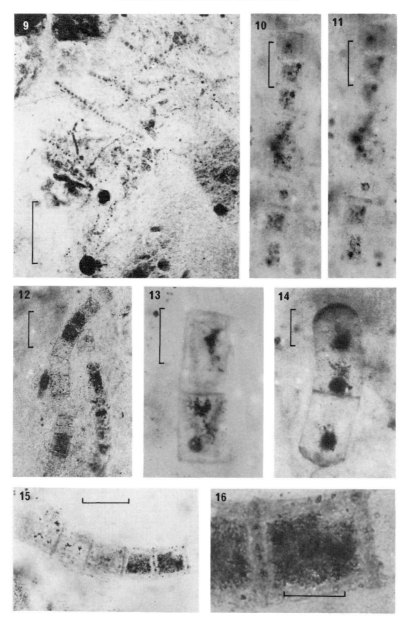

Figures 9–16. Fig. 9. General view of filaments close to *Rhynia major* axis (off picture to bottom right);
mostly *Mackiella rotundata*, the large dark circular bodies are spores of *R. major*; Slide FSC 2204. Scale bar
equivalent to 300 μm. Figs 10–16. *Mackiella rotundata*. Figs 10 & 11. Views at two different focal planes of
the type specimen, showing the single spherical body and dispersed cytoplasm in each cell; Slide FSC
2204. Scale bars equivalent to 50 μm. Fig. 12. 12-celled broken filament at left with all but three cells
empty; these contain dense granular material; two short cells in the lower part of filament; Slide FSC
2203. Scale bar equivalent to 50 μm. Fig. 13. Two (of three) cells of a narrow incomplete filament,
showing contraction of the cytoplasm which remains attached to the cross wall between the cells; Slide
FSC 2229. Scale bar equivalent to 20 μm. Fig. 14. Three-celled (?) young filament, showing
prominent dark spots; the cytoplasm is less conspicuous than in most filaments and is dispersed
throughout the cell. Slide FSC 2204. Scale bar equivalent to 20 μm. Fig. 15. Short, incomplete
filament with two granular cells; unlike the filament shown in Fig. 12, the remaining cells contain
poorly preserved contents; Slide FSC 2204. Scale bar equivalent to 50 μm. Fig. 16. Granular cells of
Fig. 15; the granules are not resolvable; Slide FSC 2204. Scale bar equivalent to 20 μm.

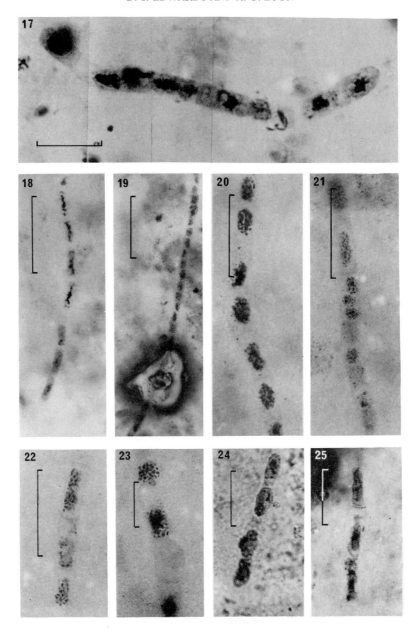

Figures 17–25. Figs 17–24. *Rhynchertia punctata*. Fig. 17. Composite picture of a vegetative filament, each cell with dispersed cytoplasm filling the lumen and containing a single large body (? chloroplast); in one cell this body appears to have uncoiled and broken through the cell wall; Slide FSC 2204. Scale bar equivalent to 40 µm. Fig. 18. Vegetative filament with contracted cell contents and granular contents (lower portion) and carbonized cells (upper portion); Slide FSC 2229. Scale bar equivalent to 100 µm. Fig. 19. Long (37-celled) reproductive filament, each cell of which contains large numbers of ovoid bodies (? spores); Slide FSC 2234. Scale bar equivalent to 100 µm. Fig. 20. Type specimen; a 13-celled reproductive filament; Slide FSC 2235. Scale bar equivalent to 40 µm. Fig. 21. Filament with cells of varying lengths and contents, some containing ovoid bodies, others with a more uniform appearance; Slide FSC 2235. Scale bar equivalent to 40 µm. Fig. 22. Three-celled (incomplete) reproductive filament with well-defined cell walls; Slide FSC 2235. Scale bar equivalent to 40 µm. Fig. 23. Filament with protuberant cells, provisionally assigned to *R. punctatus*. Cells with ovoid bodies alternate with empty cells in this filament; Slide FSC 2235. Scale bar equivalent to 20 µm. Fig. 24.

is unclear although three possibilities exist—it may be a collapsed chloroplast, a pyrenoid or a nucleus. Comparable bodies within fossil unicellular algae from the Bitter Springs Formation (Precambrian) have been interpreted as nuclei (Schopf, 1968, 1974) or as pyrenoids and associated chloroplasts (Oehler, 1977). While there is little direct evidence in support of either of these interpretations in *Mackiella*, the existence of discrete black areas within the spots could suggest a pyrenoidal rather than a nuclear nature. The lack of a distinct chloroplast may be a result of collapse and integration into the darker body. Some filaments contain dense granular material which is not resolvable into discrete units (Figs 11, 15). While it is possible that the granular nature of these cells reflects the presence of zoospores within them, it is equally likely that their appearance merely reflects a difference in preservation, particularly since filaments with this kind of cell have not been found in close proximity to those with indications of organelles.

Cell size, the presence of possible organelles and the distinctive contraction of cell contents suggest that *Mackiella* was a eukaryotic alga. No comparable fossil material has been found, but among extant algae, members of the orders Ulotrichales, Oedogoniales, Zygnematales (Chlorophyta) and Tribonematales (Xanthophyta) may be considered similar. Assignment of *Mackiella* to any of these orders is not possible since in each case *Mackiella* lacks at least one of the diagnostic vegetative features of these extant orders: Ulotrichales—fossil lacks the distinctive chloroplast; Oedogoniales—fossil lacks cap cells (although certain living members produce these only rarely) (Fritsch, 1935); Zygnematales—fossil lacks the distinctive chloroplasts; Tribonematales—fossil lacks 'H-pieces' in the cell wall (although they are frequent in some living taxa (*Bumilleria*) and absent in others (*Heterothrix*) (Fritsch, 1935: Smith, 1950)). Of these orders, the most similar forms to *Mackiella* are amongst the Ulotrichales—*Klebsormidium* has unattached filaments, cells with a single pyrenoid, and may fragment into short filaments (Bold & Wynne, 1978); however, cell width is rather narrow—up to 14 μm in *Hormidium* (*Klebsormidium*) *flaccidum* (Kütz.) A. Braun (from Ramanathan, 1964). Filaments of *Ulothrix zonata* (Weber & Mohr.) Kütz. have a similar cell width/length ratio with cells 11–45 μm wide, but typically the cell wall is thickened in older filaments (young filaments do not show this feature) (Ramanathan, 1964: 30). On the basis of these similarities, *Mackiella* is tentatively assigned to the Ulotrichales.

Rhynchertia D. S. Edwards & A. G. Lyon **genus novum**

DERIVATION: Named after the Rhynie Chert beds.

Unbranched, unattached, uniseriate multicellular filaments with cells of narrow diameter (8–17 μm) and variable length, most commonly twice the width of the filament, rarely much shorter. Terminal cells of filament slightly longer than median cells, bluntly rounded. Cells thin-walled, lacking a mucilage sheath. Cell contents uniform or with a single dark body or containing many small ovoid bodies. Reproduction (?)by the production of many spores per cell.

TYPE SPECIES: *Rhynchertia punctata* D. S. Edwards & A. G. Lyon

---

Short vegetative filament provisionally assigned to *R. punctatus* with cell walls not preserved but showing the protuberant nature of the cells; found in the gelatinous zone of *Nematoplexus;* Peel 38B/B. 17 (A. G. Lyon Collection). Scale bar equivalent to 150 μm. Fig. 25. Large-celled filament showing contraction of the cell contents; Slide FSC 2233. Scale bar equivalent to 150 μm.

Rhynchertia punctata D. S. Edwards & A. G. Lyon **sp. nova**

DERIVATION: The epithet refers to the spots in some of the cells.

Cell width 8–16.5 μm ($\bar{x}$ 11 μm), cell length about twice width 16–40 μm ($\bar{x}$ 21.5 μm). Filaments vegetative, 'reproductive' or mixed.

TYPE: Holotype from slide FSC 2235, Hunterian Museum Palaeobotanical Collection, Glasgow. Illustrated in Fig. 20.

ILLUSTRATIONS: Figures 17–24.

LOCALITY: Rhynie, Aberdeenshire.

HORIZON: Rhynie Chert Bed, Scottish Lower Old Red Sandstone (Siegenian).

The narrow filaments of *R. punctata* are fairly common in the white chert and they have also been found in association with *Nematoplexus rhyiensis* Lyon where they are confined to a possible gelatinous (post-mortem degradation ?) zone of the plant (Lyon, 1962).

Most vegetative cells have poorly-defined walls and amorphous contents, but a few are better preserved. In these specimens there are indications of internal differentiation. A number of darker areas may be visible (Fig. 18) or a single rather irregular body may occupy the centre of the cell (Fig. 17). This body may be the remains of a single large organelle, possibly a chloroplast.

Some complete filaments (or individual cells of otherwise vegetative filaments), consist of cells which contain a large number of small (about 0.4 μm maximum length) opaque ovoid bodies (Figs 19, 20). These bodies are interpreted as being reproductive (? zoospores or gametes) since possible differences in preservation seem insufficient to explain the regularity of their occurrence and the mixed nature of some filaments. It has not been possible to count the exact number of bodies within an individual cell but there are about 30–50.

Extant zoospore-producing unbranched filamentous uninucleate algae are found in the Ulotrichales and while details of the chloroplast structure are lacking for *Rhynchertia*, it is reasonable to assign the plant to that order on the basis of gross morphology: in particular, *Ulothrix cylindricum* Prescott has cells of similar shape and size (10–12.5 μm diameter, 2–3 times as long as wide) (Ramanathan, 1964: 36).

Commonly associated with filaments of *R. punctata* are filaments of similar size and general appearance (? spores in some cells, other cells containing one or two discrete bodies), but which include cells with lateral protuberances (Figs 23, 24). Filaments may contain several such cells or just a single cell of this kind. The other cells are often of variable length, frequently slightly shorter than typical *R. punctata* cells, and may even appear spherical rather than cylindrical. The bulges on the sides of the cells are somewhat suggestive of early stages in branching as found in extant Cladophorales and Chaetophorales, although in these taxa the protuberances are confined to one end of the cell. *Hormidium* (*Klebsormidium*) *rivulare* (Kütz.) A. Braun shows a similar variation in cell shape when the filaments undergo false branching or become geniculate (Ramanathan, 1964); these latter features have not been observed in the fossil material.

In the absence of any sharp distinction between cell length or width or cell contents, it seems likely that these filaments represent a different growth form of *R. punctatus* and they are tentatively referred to that taxon.

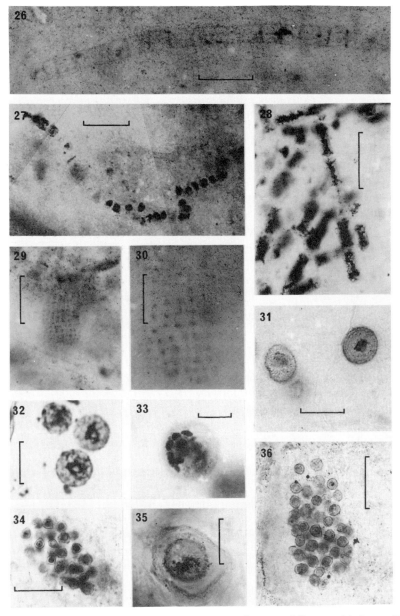

Figures 26–36. Fig. 26. Composite photograph of narrow, well-preserved filament with short and long cells, most of which contain a single darker spot; Slide FSC 2244. Scale bar equivalent to 30 μm. Fig. 27. Composite photograph of filament with spherical cells (to right) each containing a large dark body and elongate empty cells (centre left). The elongate structure (centre top) is a fungal hypha; Slide FSC 2204. Scale bar equivalent to 50 μm. Fig. 28. Five-celled filament of relatively large cells, each containing a carbonised reticulum; Slide FSC 2202. Scale bar equivalent to 100 μm. Figs 29 & 30. *Rhyniococcus uniformis*, type specimen; The spherical cells containing a central dark spot have poorly defined cell walls (visible in Fig. 30 centre); Slide FSC 2235. Scale bars equivalent to 50 μm—Fig. 29, to 20 μm.—Fig. 30. Figs 31 & 32. Unicells with irregular contents A. G. Lyon Collection. Fig. 31— slide FSC 2242. Scale bar equivalent to 20 μm. Fig. 32—Peel 38B/B.8, Scale bar equivalent to 20 μm. Fig. 33. Unicell containing ovoid bodies; a larger spherical body can be seen in the upper left of the cell; Slide FSC 2240. Scale bar equivalent to 10 μm. Figs 34–36. Unicells with two outer coverings; the middle layer remains attached to the thin outer one in one area in many examples (at right in Fig. 35); Figs 34 & 35, chip preparation 83. A. G. Lyon Collection. Fig. 36 Slide R1187, Lang Collection. Scale bars equivalent to 100 μm—Figs 34 & 36; to 10 μm—Fig. 35.

Filamentous algae incertae sedis: The affinities of the large-celled filamentous algae shown in Fig. 25 and Fig. 30 are not clear. The smaller form (cell width 22–32 μm, mean 27 μm; cell length 53–73 μm, mean 60 μm) is confined to two sections from a single block and a single filament from another block, while the larger form (cell width 24–33 μm, mean 30 μm) is found in several blocks of the white chert. The short-celled form is generally poorly-preserved with little indication of cell contents apart from a carbonized reticulum, and is often found as individual cells dispersed through the matrix. The longest filament found has five cells with the terminal cells having bluntly rounded ends. The larger-celled filaments have better-preserved cell contents, with indications of contents but, possibly due to the thickness of the sections relative to cell length, filaments longer than three cells have not been observed.

The large cell size of these filaments invites comparison with certain Chlorophyta, particularly members of the Cladophorales and Zygnematales (for example *Zygnema ornatum* (Li) Transeau, in Randhawa, 1959: 248): but lack of cell contents or reproductive bodies in the fossil material precludes a more definite assignment.

Figure 26 shows the only specimen of a well-preserved filament which has been found. The cells are about 12 μm wide and fall into two length groupings—the longer cells being about 14 μm long and the shorter, (two groups of four cells), about 7.5 μm long. One of the terminal cells is much longer (about 29 μm) and is bluntly rounded, the other appears almost spherical. Cell contents are indistinct but in some cells there is a single faint darker spot. The filament is similar in general appearance to macrandrous species of *Oedogonium* with the shorter cells representing antheridia, but the lack of cap cells limits this comparison.

One specimen of a poorly-preserved filament (Fig. 27) also contains two sizes of cells—the shorter 16 cells are rectangular in section, between 11 and 13 μm wide and include a single dark spherical body with indications of internal differentiation, while the larger five cells are between 26 and 40 μm long and appear empty. Although comparisons based on a single specimen are highly speculative, it is worth noting that the cell diameter is similar to that of *Langiella fritschii* (vegetative cells 9–12 μm, akinetes up to 18 μm) (Croft & George, 1959) but this specimen lacks the mucilage sheath and has longer cells than *Langiella*. In addition, the occurrence of discrete spherical bodies within the cells is more compatible with a eukaryotic rather than a prokaryotic organization, particularly since in *Langiella* the cell contents are extremely uniform and have contracted uniformly (see Oehler, 1977) away from the walls.

Palmelloid and coccoid algae

DIVISION:	Cyanophyta
CLASS:	Schizophyceae
ORDER:	Chroococcales
FAMILY:	Chroococcaceae

Rhyniococcus D. S. Edwards & A. G. Lyon **genus novum**

DERIVATION: A coccus from the Rhynie Chert bed.

Colonial alga consisting of a one cell thick flat, or nearly flat, sheet of cells. Cells regularly arranged in rows, not grouped within the sheet. Reproduction unknown.

TYPE: *Rhyniococcus uniformis* D. S. Edwards & A. G. Lyon

Rhyniococcus uniformis D. S. Edwards & A. G. Lyon **sp. nova**

DERIVATION: Referring to the uniform spacing of the cells in the sheet.

Cells 4 μm diameter, separated by approximately 1 μm. Cell contents indistinct, consisting of a central dark spot.

TYPE: From slide FSC 2235, Hunterian Museum Palaeobotanical Collection, Glasgow. Illustrated in Figs 29 & 30.

LOCALITY: Rhynie, Aberdeenshire.

HORIZON: Rhynie Chert Bed. Scottish Lower Old Red Sandstone (Siegenian).

Both the Chlorophyta and Cyanophyta include forms with a similar organization—*Prasiola* (Ulvales) and *Merismopedia* (Chroococcales) include species with cells arranged in rows and not (as is usual in both genera) arranged in groups of two or more (e.g. *P. furfuraceae* (Nert.) Menegh.—Pascher (1914); *M. trolleri* Bachman—Pascher (1925) and some forms of *M. tenuissima* Lemm.—Desikachary (1959)). The small size of the cells of *R. uniformis* is more typical of a prokaryotic rather than a eukaryotic alga (Schopf & Oehler, 1976) and is comparable with several species of *Merismopedia*; *Rhyniococcus* is therefore referred to the Chroococcales.

Unicellular algae Incertae Sedis: The three kinds of unicellular algae which have been distinguished will not receive formal taxonomic treatment at the present time. The most common kind has been found in both the white and the black chert. Two size classes can be recognized, the larger having a diameter of 11–21 μm ($\bar{x}$ 15 μm) and the smaller with a diameter of 8–11 μm ($\bar{x}$ 10 μm). The larger (Fig. 31) has been found in the white chert and in the zone surrounding *Nematoplexus* (Lyon, 1962); the smaller is confined to the black chert. The '? algal unicells' figured in Kidston & Lang, (1921: Pl. VII, figs 77, 78) fall into the smaller-celled category. Apart from differences in size, unicells of both categories are similar in general appearance. The cell contents have contracted slightly from the wall and in some cases consist of a single large dark body which does not appear uniform. In other cases, smaller irregularly shaped units are dispersed throughout the cell.

The existence of possible evidence of organelles and cytoplasmic contraction is suggestive of a eukaryotic rather than a prokaryotic alga. Possible eukaryotic unicellular algae have been reported from several Precambrian deposits (e.g. *Maculosphaera* from the Beck Spring Dolomite (Licari, 1978); *Caryosphaeroides*, *Glenobotrydion* and *Globophycus* from the Bitter Springs Formation (Schopf, 1968; Oehler, 1977) but most of these consist of individuals with a single prominent central or eccentric spot; rarely is there any indication of discrete contents and the eukaryotic nature of such cells has been questioned (Knoll & Barghoorn, 1975; Colubic & Barghoorn, 1977; Knoll, Barghoorn & Awrmik, 1978; Francis, Barghoorn & Margulis, 1978). The Rhynie Chert unicells differ from typical spot cells in that their contents are made up of several discrete bodies, partially or completely fused into one larger entity and are thus not strictly comparable with Precambrian material (e.g. *Glenobotrydion* Schopf and *Caryosphaeroides* Schopf) which is tentatively accepted as eukaryotic by some authors (Oehler, 1977; Tappan, 1980).

A second kind (Fig. 32) of thin-walled unicell (diameter 16–20 μm) has been found in several sections of the white chert and differs from the first kind in that the cells contain a number of opaque bodies, each about 3.75 μm long and up to 2.0 μm wide. A larger black spherical body is also present in some cells. Although these structures may be merely crystals, they could be gametes or aplanospores, the larger body being two or three of the smaller ones clumped together, or they may represent pyrenoids, the chloroplast having degenerated. The organization of this alga (which may be the larger of the first type in a reproductive state) is similar in many respects to extant Chlorococcaceae which reproduce by zoospores or aplanospores and which have one to several pyrenoids in each cell (Fritsch, 1935; Bold & Wynne, 1978), and its affinities may be in that direction.

While the affinities of the unicells described above may be with Chlorophyta, the relationships of the organisms shown in Figs. 34–36 are less obvious. The central sphere (c. 14 μm diameter) often contains a spot, although sometimes the cell contents appear more diffuse. It is surrounded by an envelope which appears to have contracted from a thin-walled, outer covering 24–34 μm ($\bar{x}$ 27 μm) in diameter but remains attached to it at one point, giving the impression of an exit pore. Specimens of these unicells are found in groups in the black chert, and in two examples (Chip preparation 83 and slide R1187, Lang Collection, Manchester Museum) there is an indication of a boundary layer to the cell group, suggesting that they were embedded within mucilage or developed within a larger structure. Slide R1187 also contains remains of *Lepidocaris* and fragments of *Palaeonitella* but the unicells appear unconnected with either of these organisms. Unicells of similar morphology have been described as "encysted cells of *Caryosphaeroides pristina* ?" by Schopf (1968: 682) cells of *Gleodinopsis lamellosa* described by Schopf & Blacic (1971) and *Cumulasphaera lamellosa* described by Edhorn (1973) also show similarities to the Rhynie unicells. Tappan (1980) regards these as of uncertain systematic position, possibly of chroococcacean affinity.

GENERAL DISCUSSION

It is apparent from the above descriptions that the Rhynie Chert microflora includes a wide range of algae, many of which are of an eukaryotic aspect. However, the algae are localized within the deposit—the prokaryotic forms being largely confined to the black chert and the eukaryotic to the white. Algae are rare in the black chert; *Kidstoniella*, *Langiella* and *Rhyniella* all occur in a single chip (Croft & George, 1959) and we know of no other specimens, *Archaeothrix oscillatoriformis* was also reported from a single group of samples (Kidston & Lang, 1921) although a few additional specimens have since been found. *Archaeothrix contexta* is more common and several specimens of this have been encountered. *Palaeonitella cranii* is much more widespread and occasionally occurs in quite compact masses. Algae are far more common in the white chert with five being confined to this rock. However, filaments of *Rhynchertia* and unicells have also been found in the black chert, particularly in association with nematophytalean remains. It is likely that the differences in the algal flora reflect a difference in the environmental or preservation conditions. In this connection it is interesting to note that cyanobacteria are absent from acid waters with a pH of less than 4 (Brock, 1973), possibly indicating a more acid environment for the algae in the white chert.

Kidston & Lang (1921 : 895) thought that the tracheophytes in the deposit had grown on a substrate saturated with hot siliceous acid water and that the chert as a whole had silicified from the base upwards. Tasch (1957) favoured their original interpretation (Kidston & Lang, 1917 : 764) that the bed silicified after its full thickness had accumulated—but also suggested an influx of hot siliceous water at an early stage in the history of the deposit (Bed A″) to explain the restricted distribution of algal and animal fossils which were washed in amongst the vascular plants. The latter interpretation is more in keeping with the rarity of algae in the black chert—in comparison with present day hot spring waters the deposit is very deficient in cyanobacterial remains.

The algae in the white chert seem more likely to have been growing amongst axes of *Rhynia major*—the stratification of the blocks, the alignment of the axes and their localized occurrence, indicates gradual flooding of a stand of *R. major* (which then started to decay), the growth of the algae and subsequent silicification. This suggests a sequence of chertification similar in some ways to that proposed for the Pleistocene deposits of some East African natron lakes (e.g. Lake Nagadi) in which a sudden lowering of pH in a silica-saturated environment resulted in a deposition of silicates (Nagadiite); subsequent freshwater leaching of the silicates formed bedded chert (Surdam & Eugster, 1976; Collinson, 1978). However, such a sequence commences in an alkaline environment and this in turn suggests either that the pH controlled algal distribution indicated above is incorrect or that the algae in the white chert were very limited temporally within the sequence. Further examination of this possibility and an explanation for the white colour of the algal containing chert requires a detailed petrological examination which we have not attempted.

Fossil floras comparable with that of the Rhynie Chert are rare—most Precambrian and Cambrian microfossil assemblages are stromatolitic with a high proportion of cyanobacteria and were formed under saline or intermittently saline, shallow (rarely deep) water conditions (see Monty, 1977), although some shale-facies microbiotas are also known (Horodyski, 1980; Zhongying, 1982). The flora of the Middle Devonian Onandaga Chert consists of possible freshwater algae which were suggested to have been washed into an off-shore shallow water environment before preservation (Baschnagel, 1942, 1966) and includes forms similar to modern Ulotrichales (cf. *Geminella*, Oedogoniales (*Palaeoedogonium*) and Zygnematales (*Palaeoclosterium*). The algal flora of the Kalkberg Limestone (Lower Devonian) is very poorly-preserved (mainly as pyrite casts) but includes possible Ulotrichaceae, Nostocaceae, Chroococcaceae and Scytonemataceaen forms (Wincander & Schopf, 1974); the depositional environment was similar to that of the Onondaga Chert—a shallow marine shelf. Although the Rhynie Chert contains similar elements to both these floras, the difference in depositional environments limits any meaningful comparison.

Palaeonitella described from the Late Devonian Caballos novaculite (Fairchild, Schopf & Folk, 1973) may be a pseudofossil (Schopf, 1974) as may also be the other filamentous forms from that deposit, particularly since the formation has recently been reinterpreted as a deep-water deposit (Folk & McBride, 1976; Jenkyns, 1978). The algal flora described by Grüss (1928) from the Devonian of Germany includes ulotrichalean forms but, as has been pointed out by Fairchild *et al.* (1973), these filaments may be modern contaminants since they were found in macerates rather than in thin sections. The siphonaceous *Courvoisiella* described by

Niklas (1976) was apparently epiphytic on *Archaeopteris* but no siphonacous or epiphytic forms have been found in the chert.

The algal flora of the Rhynie Chert thus occupies a rather isolated position in relation to other floras but serves to emphasize the diversity of forms in a freshwater (siliceous) habitat during early Devonian time. It also provides one of the earliest records of non-calcareous filamentous chlorophytes and suggests that the vegetative structure of the Charophyta was well-developed by that time.

ACKNOWLEDGEMENTS

We would like to thank Dr J. Franks (Herbarium, Manchester Museum) and Dr W. D. I. Rolfe (Hunterian Museum, Glasgow) for the loan of specimens in their care. Part of this work was carried out at University College, Cardiff, during the tenure of a Science Research Council studentship (awarded to D.S.E.).

REFERENCES

BASCHNAGEL, R. A., 1942. Some microfossils from the Onondaga Chert of Central New York. *Bulletin of the Buffalo Society of Natural Sciences, 17:* 1–8.

BASCHNAGEL, R. A., 1966. New fossil algae from the Middle Devonian of New York. *Transactions of the American Microscopical Society, 85:* 297–302.

BOLD, H. C. & WYNNE, M. T., 1978. *Introduction to the Algae.* New Jersey: Prentice-Hall Inc.

BROCK, T. D., 1973. Lower pH limit for the existence of blue-green algae: evolutionary and ecological implications. *Science, 179:* 480–482.

COLLINSON, J. D., 1978. Lakes. In H. G. Reading (Ed.), *Sedimentary Environments and Facies:* 61–79. Oxford: Blackwell.

CROFT, W. N. & GEORGE, E. A., 1959. Blue-green algae from the Middle Devonian of Rhynie, Aberdeenshire. *Bulletin of the British Museum (Natural History), Geology, 3:* 341–353.

DESIKACHARY, T. V., 1959. *Cyanophyta.* New Delhi: Indian Council of Agricultural Research.

DOWNIE, C., 1973. Observations on the nature of acritarchs. *Palaeontology, 16:* 239–259.

EDHORN, A. S., 1973. Further investigations of fossils from the Animikie, Thunder Bay, Ontario. *Proceedings of the Geological Association of Canada, 25:* 37–65.

EMBERGER, L., 1968. *Les Plantes Fossiles.* Paris: Masson et cie.

FAIRCHILD, T. R., SCHOPF, J. W. & FOLK, R. L., 1973. Filamentous algal microfossils from the Caballos Novaculite, Devonian of Texas. *Journal of Palaeontology, 47:* 946–952.

FOLK, R. L. & McBRIDE, E. F., 1976. The Caballos Novaculite revisited, Part 1: origin of novaculite members. *Journal of Sedimentary Petrology, 46:* 659–669.

FRANCIS, S., BARGHOORN, E. S. & MARGULLIS, L., 1978. On the experimental silicification of microorganisms. III. Implications of the preservation of the green prokaryotic alga *Prochloron* and other coccoids for interpretation of the microbial fossil record. *Precambrian Research, 7:* 377–383.

FRITSCH, F. E., 1935. *The Structure and Reproduction of the Algae.* Cambridge: Cambridge University Press.

GOLUBIC, S. & BARGHOORN, E. S., 1977. Interpretation of microbial fossils with special reference to the Precambrian. In E. Flugel (Ed.), *Fossil Algae, Recent Results and Developments:* 1–14. Berlin: Springer-Verlag.

GROVES, J. & BULLOCK-WEBSTER, G. R., 1920. *The British Charophyta,* Vol. 1. London: Ray Society.

GRÜSS, J., 1928. Zur Biologie Devonischer Thallophyten. *Paläeobiologica, 1:* 487–518.

HAYWARD, J., 1974. Studies on the growth of *Stichococcus bacillaris* Naeg. in culture. *Journal of the Marine Biological Association, 54:* 261–268.

HORODYSKI, R. J., 1980. Middle Proterozoic Shale-facies microbiota from the Lower Belt Supergroup, Little Belt Mountains, Montana. *Journal of Palaeontology, 54:* 649–663.

HORNE, J., MACKIE, W., FLETT, J. S., GORDON, W. T., HICKLING, G., KIDSTON, R., PEACH, B. N. & WATSON, D. M. S., 1916. The plant-bearing cherts at Rhynie. *Report of the British Association for the Advancement of Science, 1916:* 206–216.

JENKYNS, H. C., 1978. Pelagic environments. In H. G. Reading (Ed.), *Sedimentary Environments and Facies:* 314–371. Oxford: Blackwell.

KIDSTON, R. & LANG, W. H., 1917. On Old Red Sandstone Plants showing structure, from the Rhynie Chert Bed, Aberdeenshire. Part 1. *Rhynia Gwynne-vaughanii* Kidston & Lang. *Transactions of the Royal Society of Edinburgh, 51:* 761–784.

KIDSTON, R. & LANG, W. H., 1921. On Old Red Sandstone plants showing structure, from the Rhynie Chert Bed, Aberdeenshire. Part V. The Thallophyta occurring in the peat-bed; the succession of plants throughout a vertical section of the bed, and the conditions of accumulation and preservation of the deposit. *Transactions of the Royal Society of Edinburgh, 52:* 855–902.

KNOLL, A. H. & BARGHOORN, E. S., 1975. Precambrian eukaryotic organisms: A reassessment of the evidence. *Science, 190:* 52–54.

KNOLL, A. H., BARGHOORN, E. S. & AWRAMIK, S. M., 1978. New microorganisms from the Aphebian Gunflint Iron Formation, Ontario. *Journal of Palaeontology, 52:* 956–992.

LICARI, G., 1978. Biogeology of the Late Pre-phanerozoic Beck Spring Dolomite of Eastern California. *Journal of Palaeontology, 52:* 767–792.

LYON, A. G., 1962. On the fragmentary remains of an organism referable to the Nematophytales, from the Rhynie Chert, *"Nematoplexus rhyniensis"* gen. et sp. nov. *Transactions of the Royal Society of Edinburgh, 65:* 79–87.

MONTY, C., 1977. Evolving Concepts on the Nature and the Ecological Significance of Stromatolites. In E. Flugel (Ed.), *Fossil Algae, Recent Results and Developments:* 15–35. Berlin. Springer-Verlag.

NIKLAS, K. J., 1976 Morphological and chemical examination of *Courvoisiella ctenomorpha* gen.et sp.nov., a siphonous alga from the Upper Devonian, West Virginia, U.S.A. *Review of Palaebotany and Palynology, 21:* 187–203.

OEHLER, D. Z., 1977. Pyrenoid-like structures in Late Precambrian Algae from the Bitter Springs Formation of Australia. *Journal of Palaeontology, 81:* 885–890.

PASCHER, A. (Ed), 1914. *Die Süsswasser-Flora Deutschlands, Österreichs und der Schweiz. Heft 6, Chlorophyceae III.* Jena: Fischer.

PASCHER, A. (Ed.), 1925. *Die Süsswasser-Flora Deutschlands, Österreichs und der Schweiz. Heft 12, Cyanophyceae.* Jena: Fischer.

PIA, J., 1927. Thallophyta. In M. Hirmer (Ed.), *Handbuch der Paläobotanik, 1:* 31–136. Munich: Oldenbourg.

RAMANATHAN, K. R., 1964. *Ulotrichales.* New Delhi: Indian Council of Agricultural Research.

RANDHAWA, M. S., 1959. *Zygnemaceae.* New Delhi: Indian Council of Agricultural Research.

RICHARDSON, J. B., 1967. Some British Lower Devonian spore assemblages and their stratigraphic significance. *Review of Palaeobotany and Palynology, 1:* 111–129.

SCHOPF, J. W., 1968. Microflora of the Bitter Springs Formation, late Precambrian, central Australia. *Journal of Palaeontology, 42:* 651–688.

SCHOPF, J. W., 1974. Palaeobiology of the Precambrian: the age of blue-green algae. *Evolutionary Biology, 7:* 1–43.

SCHOPF, J. W. & BLACIC, J. M., 1971. New microorganisms from the Bitter Springs Formation (late Precambrian) of the north-central Amadeus Basin, Australia. *Journal of Palaeontology, 45:* 925 960.

SCHOPF, J. W. & OEHLER, D. Z., 1976. How old are the eukaryotes? *Science, 193:* 47–50.

SMITH, G. M., 1950. *The Fresh-water Algae of the United States,* 2nd ed. New York: McGraw-Hill Book Co.

SURDAM, R. C. & EUGSTER, H. P., 1976. Mineral reactions in the sedimentary deposits of the Lake Magadi region, Kenya. *Bulletin of the Geology Society of America, 87:* 1739–1752.

TAPPAN, H., 1980. *The Palaeobiology of Plant Protista.* San Francisco: W. H. Freeman & Co.

TASCH, P., 1957. Flora and fauna of the Rhynie Chert; A paleoecological reevaluation of published evidence. *University of Wichita, Bulletin, 32:* 3–24.

WALTON, J., 1953. *An Introduction to the Study of Fossil Plants.* London: A. & C. Black.

WESTOLL, T. S., 1977. In M. R. House *et al.* (Eds), A correlation of Devonian rocks of the British Isles. *Special Report of the Geological Society of London. 7: No. 8:* 1–110.

WICANDER, E. R. & SCHOPF, J. W., 1974. Microorganisms from the Kalkberg Limestone (Lower Devonian) of New York State. *Journal of Palaeontology, 48:* 74–77.

ZHONGYING, Z., 1982. Upper Proterozoic microfossils from the Summer Isles, N.W. Scotland. *Palaeontology, 25:* 443–460.

Botanical Journal of the Linnean Society (1983), *86:* 57–79. With 49 figures

A new species of *Baragwanathia* from the Sextant Formation (Emsian) Northern Ontario, Canada

FRANCIS M. HUEBER

Department of Paleobiology, Smithsonian Institution, Washington, D.C. 20560, U.S.A.

Received November 1981, accepted for publication May 1982

Baragwanathia abitibiensis is described from the Sextant Formation of northern Ontario, Canada (middle to upper Emsian age). The plant remains are primarily compressions in which flattened xylem strands, epidermis with stomata, and cuticle are well preserved. Pyrite permineralization of the leaves was found in one specimen. Comparison with *B. longifolia* Lang & Cookson indicates remarkable morphological similarities and probable anatomical similarities although details of the structure of the epidermis and cuticle are lacking in *B. longifolia*. No information is available on the outline of the cauline xylem strand in *B. abitibiensis* nor on the form and position of the sporangia it bore. Comparisons are also made to species of *Drepanophycus*. The age of *B. longifolia* and the *Baragwanathia* Flora, is discussed with particular reference to the putative dating of some specimens as Silurian. The morphological and anatomical complexity of *Baragwanathia* is at a level of advancement typical of Early to Middle Devonian lycopods. There is no evidence of precursors to the genus associated with the fragments of land vascular plants in the well-dated Silurian sediments of Wales, Czechoslovakia, New York State, or Podolia.

KEY WORDS:— *Baragwanathia* – lycopods – palaeobotany.

CONTENTS

INTRODUCTION

On the basis of the original description and illustrations of *Baragwanathia* by Lang & Cookson (1935), I have believed, for many years that the remains described herein belonged with *Baragwanathia*. I have now collected specimens of the type species at the type locality and at new localities in Australia. I have examined the type specimens and excellent collections at the National Museum of Victoria and the Mines Department of the Geological Survey of Victoria. My

0024–4066/83/010057 + 23$03.00/0

confidence in identifying the Canadian remains with the genus is unreserved. This decision has, as its basis, the remarkable identity of morphological characteristics between the Canadian and Australian specimens. The species described below takes into account anatomical differences as well as differences of geological age and geographic occurrence.

MATERIAL AND METHODS

The specimens from Canada were obtained by D. C. McGregor and me during field studies in September, 1961. The plant remains occur abundantly in moderately to well-consolidated arkosic and/or argillaceous sandstones in which mica flakes are well-distributed through the bedding and along the bedding surfaces. The sediments are of continental origin and belong to the Sextant Formation which passes laterally into, and is overlain by, the marine carbonate Stooping River Formation. Miospore analyses of both Formations reported by McGregor & Camfield (1976) and McGregor (1979) indicate a middle to upper Emsian age for the sediments. The collection sites were given Geological Survey of Canada locality numbers and may be described as follows:

GSC Loc. 6437; west side of Abitibi River, about 300 yds north of the north end of large exposure of the Sextant Formation on the east bank of the river, below rapids. Other taxa represented at this site are *Sawdonia ornata* (Dawson) Hueber, *Taeniocrada* sp., and axes referable to the Trimerophytina.

GSC Loc. 6438. No other taxa are represented from this site.

GSC Loc. 6439; west side of Abitibi River, about 100 yds north of large, obvious outcrop at a point near the head of Sextant Rapids; this outcrop 60 ft high, bordered on the south by a very small creek. Other taxa represented at this site are *Taeniocrada* sp. and *Psilophyton dawsonii* Banks, Leclercq & Hueber. The matrix has been metamorphosed as a result of its proximity to an igneous intrusion.

GSC Loc. 6440; west side of Abitibi River, perhaps 500 yds north of north end outcrop on east side of river; 20 ft above the river; not in place, but Sextant outcrop above and to the south. Other taxa represented at this site are *Drepanophycus gaspianus* (Dawson) Kräusel & Weyland and axes referable to the Trimerophytina.

GSC Loc. 6441; west bank of Abitibi River, first outcrop north of spit opposite north end of large east bank exposure; about 20 ft above the river. This is the principal site for the collection of the new species. Other taxa represented from this site, on the basis of fragments macerated from the matrix, are *Psilophyton dawsonii* and *Spongiophyton* sp.

The holotype, paratypes, and remaining specimens in the individual collections are housed in the Geological Survey of Canada, Ottawa, Ontario. They bear Geological Survey of Canada type collection numbers *GSC 69288–69318*.

Specimens of *Baragwanathia longifolia* Lang & Cookson illustrated for purpose of morphological comparison, were collected in 1967 during field studies with Thomas Darragh of the National Museum of Victoria; John Talent, formerly with the Geological Survey of Victoria; and my field assistant Douglas Pasley. The specimens are from the 'Nineteen Mile Quarry' located on the north side of Warburton-Woods Point Road, 1.8 miles west of the junction with Noojee Road, Victoria, Australia. These specimens are deposited in the collections of the National

Museum of Victoria, Melbourne, Australia and bear numbers *NMV, P160511–P160514.*

The plant remains from Canada are preserved as compressions with cuticle and epidermal layers intact. The axial and foliar xylem strands are present but are quite well flattened. Bulk maceration of the matrix, modified after Walton (1925) and Harris (1926) was the principal technique used for obtaining specimens for study. Concentrated hydrofluoric acid (48%) was used as the reagent for maceration. Fragments of plant material up to 60 mm long were freed from the matrix and were subsequently washed 10 times with water. An hour was allowed between each complete change of the wash waters. Individual fragments were chosen for study, and they were dehydrated through an aqueous ethanol series. The specimens were subsequently air-dried and most were mounted on microscope slides using Canada balsam in xylene as the mounting medium. Particularly well-preserved and substantial fragments of stems, leaves and xylem strands were individually oriented and attached in vertical position on circular glass coverslips using white glue (Elmer's Glue-All, Borden Inc., Ohio, U.S.A). The coverslips were attached to aluminium stubs and the specimens were sputter coated with gold-palladium. They were examined and photographed in Cambridge Mark-IIA. Cambridge S4-10, and Coates and Welter 106B SEM. No additional preparations were made from the large quantity of material obtained from the macerations, but it has been stored in distilled water, with thymol to retard bacterial and fungal attack, and may serve in further study of the species.

One specimen, with leaves permineralized by pyrite, was prepared by the dégagement technique described by Leclercq (1960). A thin coating of 'Alvar' was applied to the leaves after the matrix had been removed from around them to serve as reinforcement and hopefully to retard oxidation of the pyrite. Fragments of two of the leaves were removed, embedded in methyl methacrylate resin, and sectioned into 0.375 mm slices using a Leco, Vari/Cut diamond saw. The sections were polished to metallic brightness using a series of abrasive grits with water on glass plates. The polishing process began with 600 mesh Carborundum and progressed through 1000 mesh and 1200 mesh aluminium oxide, ending with chromium oxide polishing compound. The sections were cleaned between each change of grit by immersing them in distilled water in an ultra-sonic cleaner. The finished sections were each then mounted on circular glass coverslips using epoxy 220 resin (Hughes Associates, Minnesota, U.S.A.). A drop of warmed (80°C), concentrated nitric acid was placed on the surface of each of the sections and allowed to remain there until effervescence ceased. The sections were then washed by immersion in distilled water and neutralized in a 1% solution of sodium carbonate. After two rinses in distilled water, the sections were allowed to air-dry. The coverslips were attached to aluminium stubs and the sections were sputter coated with gold-palladium. They were observed and photographed in a Cambridge S4-10 SEM.

DESCRIPTION OF FOSSIL MATERIAL

The remains of the stems of this species lie nearly parallel within the matrix (Fig. 2). None of the specimens of matrix available exceeds 120 mm in width, in the direction of the alignment of the stems; therefore, no valid estimates of the height or length of the whole plant can be given. The stems branch dichotomously (Fig. 1) and anisotomously as evidenced by small lateral buds that are found on

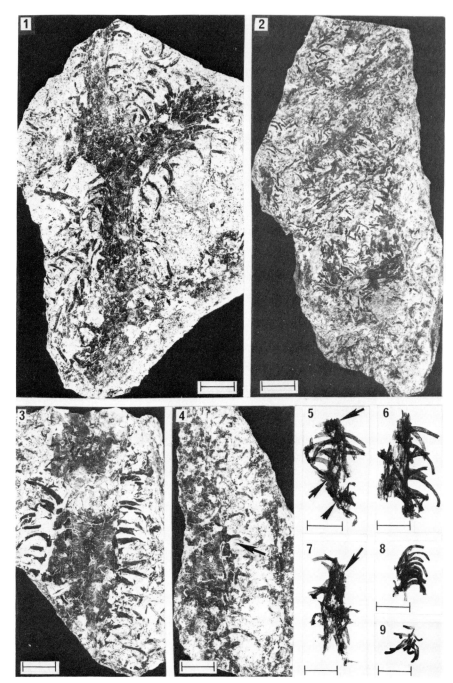

Figures 1–9. *Baragwanathia abitibiensis*. All specimens are from GSC Loc. 6441. Fig. 1. Holotype. Dichotomously branched axis showing closely spaced, downwardly curving leaves; *GSC 69288*. Fig. 2. Paratype. Specimen showing the density in numbers of stems in the matrix; *GSC 69289*. Fig. 3. Paratype. Specimen representing broadest specimen at hand from Loc. 6441; *GSC 69290*. Fig. 4. Paratype. Stem showing pattern of leaf arrangement as a result of removal of the leaves when the matrix was split, broadened base of leaf seen at arrow; *GSC 69291*. Fig. 5. Paratype. Stem with dark areas (arrows) at first thought to represent large sporangia; *GSC 69292*. Fig. 6. Paratype. Fragment of stem with long, truncated leaves; *GSC 69293*. Fig. 7. Paratype. Stem with compressed xylem strand (at arrow) that served as source of Figs 47–49; *GSC 69294*. Fig. 8. Paratype. Fragment of stem with closely spaced leaves; leaf apices rounded, inwardly shrunken, and truncate (Figs 20, 23, 25); *GSC 69295*. Fig. 9. Paratype. Fragment of stem with leaves with apices as illustrated in Fig. 21 *GSC 69296*. Scale bars equivalent to 10 mm, except in Fig. 2 where it is equivalent to 5 mm.

some of the stems. The diameter of the smallest, apparently mature stem or branch isolated from the matrix is 6 mm (Fig. 5). The largest stem from GSC Loc. 6441 is 23 mm in diameter (Fig. 3), and the largest of all of the specimens assigned to the species, collected at GSC Loc. 6439, is 32 mm in diameter (Fig. 12). All of the stems bear spirally arranged leaves that are acutely conical in the juvenile state. At maturity the leaves are elongate and downwardly curved (Figs 1–8, 11, 15). The leaves arise from broad bases (Fig. 4) that may be crowded, as in young stems and well-preserved mature stems (Figs 8, 10), or there can be 4–9 mm between gyres (Figs 20, 21). The length of the leaves varies from 7 mm on lateral buds and small branches to 25 mm on mature stems. Their width ranges from 0.8 to 2 mm and they taper only slightly from base to apex (Figs 6, 8, 9). The apices of the leaves are rounded (Fig. 20); but a truncate form is commonly observed in the matrix as well as in the macerated specimens (Fig. 24). Just over 3000 individual leaves were examined in this study and results from SEM and light microscopy suggest that the apices, because of their dense structure, were perhaps necrotic when they were mature (Fig. 21). Upon death of the plant, the disintegration of the soft inner tissues of the leaves and stems caused an inward shrinkage (Fig. 22) and inward collapse of the leaf apex (Fig. 23). The leaf appears truncated when shrinkage had been completed. Hundreds of truncated specimens were observed in which the apex was obviously broken away (Fig. 25) at the stage of inward shrinkage (Fig. 23). A few specimens exhibit bulbous tips (Fig. 26) that may be the result of extrusion of degraded cells from the centre of the leaf into the apex.

One specimen was found in which the stem was compressed but the leaves were permineralized by pyrite (Fig. 18). Dégagement made possible the removal of one of the leaves at the friable, upper margin of the specimen. Sections of that leaf demonstrate that the leaves were terete in cross-section (Fig. 19). A single layer of cells with heavily thickened periclinal walls and only slightly less thickened anticlinal walls forms the epidermis (Figs 19, 32, 33). The major volume of the tissue in the leaf is parenchyma that consists of what appears to be a random arrangement of large and small cells just beneath the epidermis and a greater proportion of larger cells toward the centre. The centre, which is interpreted as the site of the vein, because that is its location in compressed specimens (Figs 44, 45), is occluded with organic residues that mask the cell details. Such residues also are present in the lumens in the parenchyma and they appear as irregular and angular shapes (Figs 19, 47) that confuse attempts to trace the outlines of some of the cells.

A thin cuticle, varying from 2 to 5 μm, covers the stems and leaves (Figs 42–45). Most of the stem fragments and leaves, obtained by maceration of the matrix, required no oxidative treatments to clear the cuticle of carbonized residues. The impressions of epidermal cells and stomata are easily observed over large areas of the stems and leaves. Fragments of the cuticle adhering to well-preserved but coalified epidermis are common and serve as the source of material to illustrate the epidermal cells and stomata.

The cells of the epidermis are polygonal (Figs 28, 36). They are generally isodiametric, ranging from 25 to 42 μm ($\bar{x}=29μm$) in diameter, or they may be elongate parallel to the long axes of the stems and leaves and range from 40 to 62 μm ($\bar{x}=59$ μm) long and 25 to 34μm ($\bar{x}=32$ μm) wide. The outer, periclinal walls are heavily thickened while the inner ones are thin (Fig. 42). The anticlinal walls are thickened but not as heavily as the periclinals (Figs 33, 42). At or near the bases of the leaves the epidermal cells are radially elongate (Fig. 32), perhaps

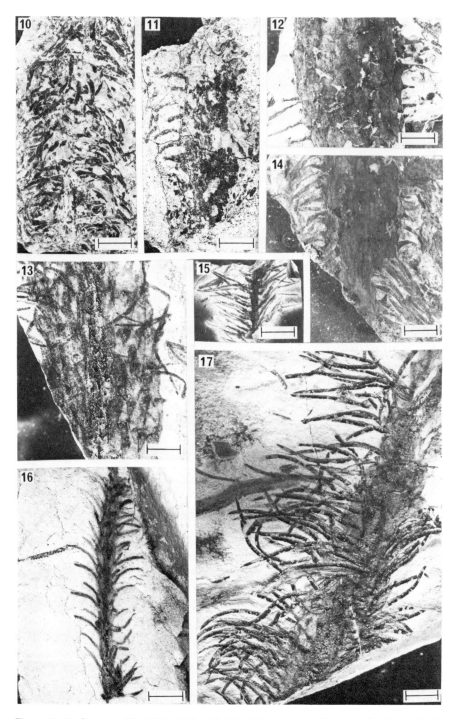

Figures 10–17. *Baragwanathia abitibiensis*. Figs 10–12 & 14 from three additional collecting sites in the Sextant Formation and B. longifolia Lang & Cookson, Figs 13, 14, 16, 17, from the "Nineteen Mile Quarry", Victoria Australia. Fig. 10. Paratype. Stem with dense covering of leaves that are preserved in nearly uncompressed radial arrangement; stem obscured by the matrix, leaves supported by infilling of fine sediment between individuals; GSC Loc. 6437, *GSC 69297*. Fig. 11. Paratype. Stem in coarser sediment but still retaining typical morphological characteristics of the leaves; GSC Loc. 6438, *GSC*

twice the radial diameter of epidermal cells at other levels in the leaves (Fig. 33) or in the stem (Figs 41, 42).

Stomata are widely and randomly scattered over the surface of the leaves (Fig. 28) and densely and randomly scattered over the surfaces of the stems, up to 28 mm^{-2} (Figs 28, 34). The long axes of the stomata are oriented parallel to the long axes of the stems (Fig. 36) and leaves (Figs 29, 30). The stomatal apparatus is anomocytic; that is, it consists of two large guard cells and lacks subsidary cells (Figs 39, 40). The dimensions of the stomatal apparatus vary from 46 to 65 μm ($\bar{x}=55$ μm) long, 62 to 90 μm ($\bar{x}=76$ μm) wide, and the pore is 12 to 21 μm ($\bar{x}=18$ μm) long. The pore is surrounded by a heavy thickening (Fig. 37) that thins to form a ledge at the margin of the pore (Figs 31, 37, 41). The thickening is circular in outline when the pore is open and oval, as in Fig. 38, when the pore is closed. The stomata are not sunken below the level of the neighbouring epidermal cells; but the epidermal cells that abut on the guard cells do overlap the guard cells in an imbricate fashion. In surface view (Fig. 37), the epidermal cells surrounding the stomatal apparatus seem to encroach almost to the raised thickening around the pore. The positions of the walls of the guard cells are marked by bulges crossing at about the middle of the periclinal walls of the abutting epidermal cells. Such bulging is clearly shown on the cell to the lower right of the pore in Fig. 37. Without microtime sections this detail cannot be shown clearly, but the section of the stomatal apparatus in Fig. 41 lends some support to the description. In the illustration, the cavity seen beneath the pore is the result of loss of all but the anticlinal walls of the guard cells and abutting epidermal cells. The imperfect trans-section of the epidermal cell on the right side of the cavity shows the lumen of that cell arched to the left over the anticlinal wall and the lumen of the adjoining guard cell.

No permineralized specimens of the xylem of either the stems or leaves were found, therefore the outlines of the cauline and foliar strands were not determined. The compressed xylem strand of one stem (Fig. 7) served as the source for illustrating, somewhat imperfectly, the kind of secondary thickenings in the tracheids comprising the cauline strand (Figs 47–49). Annular and helical secondary thickenings were the only forms observed. Irregular pits in the primary walls between the secondary thickenings are characteristic of all of the tracheids. The pits were thought at first to be the result of degradation of the primary wall; but the character seems to reflect the fibrillar structure of the wall and the pits are normal elements of the structure. The greatly variable size of the pits, probably in part owing to the state of preservation, would have made measurement of them a meaningless exercise.

Commonly a single vein is found preserved in the leaves (Figs 44, 45). It appears to consist of eight to 12 tracheids, as observed by light microscopy, and extends through the entire length of the leaf. Generally the vein follows a meandering course through the remains of the leaf as a result of displacement after degradation

---

69298. Fig. 12. Paratype. Large stem with most leaves eroded away before burial; GSC Loc. 6439, *GSC 69299* Fig. 13. Large stem with most of its leaves missing; Fig. 12 is quite comparable to this specimen in appearance; *NMV P160511*. Fig. 14. Very small stem to which Figs 5 & 8 are comparable in appearance; *NMV P160512*. Fig. 15. Paratype. Large stem with well preserved long leaves; GSC Loc. 6439, *GSC 69300*. Fig. 16. Small stem to which the specimen in Figs 1–9 is comparable; *NMV 160513*. Fig. 17. Most common size and appearance of *B. longifolia* to which Figs 1, 3, 6, 10, 14 are comparable; *NMV P160514*. Scale bars equivalent to 10 mm.

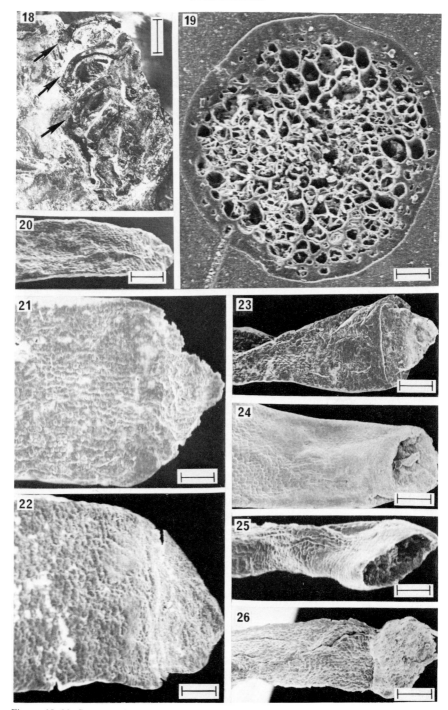

Figures 18–26. *Baragwanathia abitibiensis*. All specimens are from GSC Loc. 6441. Fig. 18. Paratype. Compressed stem with leaves (at arrows) permineralized by pyrite; Scale bar equivalent to 10 mm; *GSC 69301* Fig. 19. Transverse section of a leaf removed from the specimen in Fig. 18, note heavily thickened epidermis and the parenchymatous central portion; organic residues mask details of the vein and create peculiar angular forms within the cell lumens; scale bar equivalent to 100 µm. Fig. 20. Paratype. Portion of leaf showing rounded apex; scale bar equivalent to 333 µm; Stub SS-3, *GSC 69302* Fig. 21. Paratype. Portion of leaf showing pointed, possibly necrotic, apex; scale bar equivalent to

of the parenchymatous groundmass. In coalified specimens that had been compressed so thinly as to be translucent, the vein can be observed along the centre of the leaf.

Sporangia were not found on any of the specimens although large dark areas observed on one axis (Fig. 14) were first thought to be sporangia closely appressed to the stem. These spots appear to be sites of fungal attack or scar tissue resulting from loss of lateral branches or buds, but other techniques would be necessary to determine this.

NOMENCLATURE

DIVISION:	Lycophyta
CLASS:	Lycopsida
ORDER:	Drepanophycales
FAMILY:	Drepanophycaceae Kräusel & Weyland, 1949.

Baragwanathia Lang & Cookson, 1935

"Stems of considerable size clothed with spirally arranged, long, simple leaves. Leaves maintaining a uniform width, not contracted from a markedly expanded base and not spine-like. Sporangia, among the leaf insertions in certain zones of the ordinary shoots; possibly adaxially, near the base of the leaves. Large cylindrical central cylinder with primary xylem, which is stellate in cross-section and composed of uniform tracheides with narrow annular thickenings."

Baragwanathia abitibiensis Hueber **sp. nova**

DERIVATION: The name refers to the geographic origin of the plant, the Abitibi River, James Bay Lowland, Ontario, Canada.

Stems branching dichotomously and anisotomously 6–32 mm in diameter, bearing spirally arranged, broadly based, downwardly curved leaves, 7–25 mm long and 0.8–2 mm wide. Juvenile leaves acutely conical. Mature leaves terete in cross-section, only slightly tapered, and with simple rounded apices; commonly appearing truncated as result of post-mortem changes in morphology. Tissues of leaf comprising an epidermis of a single layer of cells, parenchyma with cells progressively larger toward the centre of the leaf, and xylem as a vein extending to the tip of the leaf. Cuticle on the leaves and stems thin, 2–5 μm. Epidermal cells polygonal, isodiametric, 25–42 μm in diameter ($\bar{x} = 29$ μm) or elongate, 40–62 μm long ($\bar{x} = 59$ μm) and 25–34 μm wide ($\bar{x} = 32$ μm) parallel to the long axes of the stems and leaves, periclinal and anticlinal walls heavily thickened. Stomata anomocytic, widely and randomly scattered on the leaves, densely and randomly scattered on the stems, up to 28 mm^{-2}, long axis of the guard cells orientated parallel to the long axes of the stems and leaves. Width of the stomatal apparatus

---

200 μm; Stub TT-4A, *GSC 69303*. Fig. 22. Paratype. Portion of leaf showing partially, inwardly shrunken apex; scale bar equivalent to 200 μm; stub TT-4C, *GSC 69304*. Fig. 23. Paratype. Portion of leaf with apex more inwardly shrunken than in 3E; scale bar equivalent to 250 μm; Stub TT-3E, *GSC 69305*. Fig. 24. Paratype. Portion of leaf with completely withdrawn apex; scale bar equivalent to 250 μm; Stub TT-1C, *GSC 69306*. Fig. 25. Paratype. Portion of leaf with apex missing, typical of the truncate leaves often observed in compressed specimens; scale bar equivalent to 200 μm; Stub TT-6B, *GSC 69307*. Fig. 26. Paratype. Portion of leaf with bulbous apex, possibly an artifact of preservation or the result of extrusion of the degraded inner contents of the leaf during compression; scale bar equivalent to 250 μm; Stub TT-3B, *GSC 69308*.

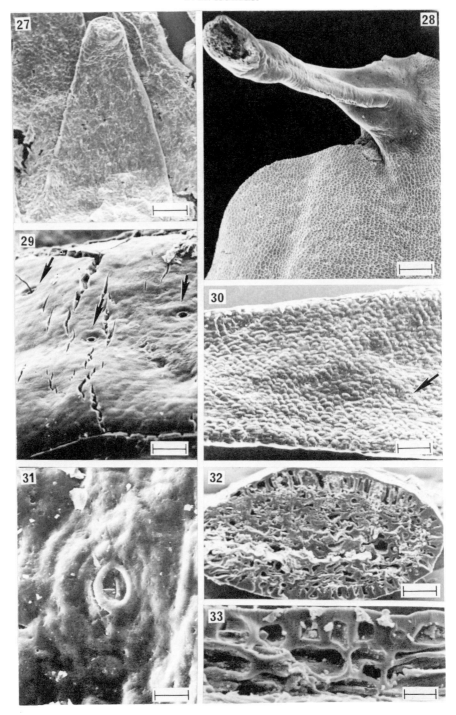

Figures 27–33. *Baragwanathia abitibiensis.* 11 Specimens are from *GSC Loc. 6441.* Fig. 27. Paratype. Juvenile leaf on small lateral bud; scale bar equivalent to 0.2 mm; Stub SS-3; *GSC 69309.* Fig. 28. Paratype. Leaf in attachment to stem surface, note distribution of stomata on leaf and stem; scale bar equivalent to 0.5 mm; Stub LL-4B, *GSC 69310.* Fig. 29. Paratype. Fragment of coalified leaf showing stomata with pores oriented parallel to the long axis of the leaf (at arrows); scale bar equivalent to 100 μm; Stub LL-13, *GSC 69311* Fig. 30. Paratype. Surface of compressed leaf, note stomate at arrow;

46–65μm ($\bar{x}=55$ μm), length 62–90 μm ($\bar{x}=76$ μm), pore 12–21 μm long ($\bar{x}=18$ μm). Xylem comprising tracheids 12.5–18 μm ($\bar{x}=16$ μm) in diameter with annular secondary thickening and 22–42 μm ($\bar{x}=32$ μm) in diameter with helical secondary thickenings; primary walls between thickenings marked by irregular pits. Outline of the xylem strand not known. Sporangia not known.

TYPE: *GSC 69288;* Fig. 1 (Holotype).

TYPE LOCALITY: GSC Loc. 6441, Abitibi River, James Bay Lowland, Ontario, Canada.

HORIZON: Sextant Formation comprising continental clastics that pass laterally into and are overlain by the Stooping River Formation, Onesquethaw Stage, Ulsterian Series, Lower Devonian, equivalent to the middle to upper Emsian of Europe (McGregor, 1979).

SYSTEMATIC POSITION OF *B. ABITIBIENSIS*

The occurrence of herbaceous lycopods can be traced over a period of hundreds of millions of years from the bogs and forests of today to the marshes and river deltas of the Devonian. Their survival as a recognizable group stands as a unique record among vascular land plants. Characteristics that typify many species of *Lycopodium* today, such as creeping to ascendant habit; dichotomous and/or anisotomous branching; numerous spirally arranged microphylls; protostelic xylem strand; and reproduction by spores produced in large sporangia borne adaxially to the leaves, are remarkably unchanged from *Baragwanathia longifolia* Lang & Cookson (Lang & Cookson, 1935) which putatively is the oldest member of the group. Only positive evidence is lacking for the precise position of the sporangia in *Baragwanathia*, although the appearance of fertile specimens strongly suggests that the sporangia are adaxial and very close to the axils of the leaves.

Baragwanathia longifolia and *B. oelheyi* Hundt (Hundt, 1952) are the only published species of the genus. The latter is looked upon by Chaloner (1967) as probably representing some form of marine organism such as a dendroid graptolite. It is placed with other genera of doubtful affinities such as *Saxonia* (Roselt, 1962) and *Boiophyton* (Obrhel, 1962).

Lang & Cookson (1935) described *B. longifolia* as dichotomously branching, leafy shoots in which the stems are 40–65 mm, but mostly 10–20 mm broad, the leaves appear lax and are up to 40 mm long and 1 mm wide, and the sporangia are reniform, 2 mm wide and 2 mm high. Bud-like groupings of leaves are present on the stems in addition to downwardly directed, presumably rhizomatous, slender, lateral branches. The cauline xylem strand is stellate in transverse section and it, as well as the leaf trace and foliar bundle, are composed of tracheids with annular secondary thickenings.

---

scale bar equivalent to 180 μm; Stub LL-6B, *GSC 69312.* Fig. 31. Stomate at arrow in Fig. 30, rotated 90 degrees counterclockwise, pore closed; scale bar equivalent to 20 μm. Fig. 32. Paratype. Transverse section near base of coalified leaf showing radially elongate epidermal cells differing from those seen nearer the apex of the leaf as in Fig. 33; scale bar equivalent to 120 μm; Stub SS-11, *GSC 69313.* Fig. 33. Transverse section taken near the apex of the same coalified leaf as in Fig. 32, epidermal cells similar in morphology to those in the stem surfaces; scale bar equivalent to 40 μm; Stub SS-9.

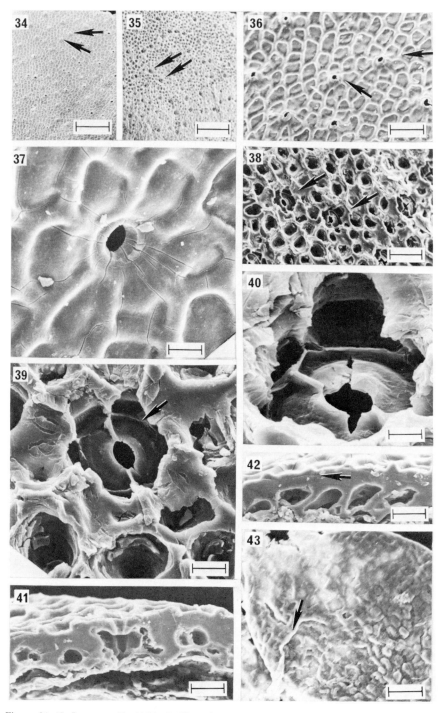

Figures 34–43. *Baragwanathia abitibiensis* All specimens are from GSC Loc. *6441*. Figs 34–40 Stub LL-5A, *GSC 69314;* Figs 41 & 42, Stub LL-4A, *GSC 69315;* Fig. 43, Stub TT-5C, *GSC 69316.* Fig. 34. Outer surface of stem showing epidermal pattern and distribution of the stomata; scale bar equivalent to 0.5 mm. Fig. 35. Inner surface of epidermal layer; scale bar equivalent to 0.5 mm. Fig. 36. Outer surface of stem enlarged for detail, arrows denote stomata marked in Fig. 34; scale bar equivalent to 100 μm. Fig. 37. Stomate at arrow to right in Fig. 36 guard cells not evident; scale bar equivalent to

Comparison of the morphological characteristics of *B. abitibiensis* with those of *B. longifolia* indicates a high degree of identity between the two species. Both are dichotomously branching, leafy stems that may also exhibit anisotomous branching on the basis of the occurrence of lateral buds. The diameter of the stems in both species is mostly 10–20 mm. However, the maximum size given for *B. longifolia* is more than double that of *B. abitibiensis* 65 mm as opposed to 32 mm). The specimen designated by Lang & Cookson (1935) as representing the largest stem available to them for study (Lang & Cookson 1935: plate 30, fig. 12) is, in my interpretation, a fragment of a stem, at the point of its dichotomy, overlapping another stem. The dichotomy is to the right in the illustration and the margin of the overlap passes diagonally on the left. If this interpretation is correct, the stems in the illustration would approximate to 40 mm and 20 mm respectively, thus nearly providing the measurement of 65 mm given in the original description. This would reduce the significant size difference between the two species. The leaves are spirally arranged in both species appearing lax (at least, not rigid and spine-like); nearly uniform in width throughout their length, and have rounded tips. The range in the length of the leaves differs only negligibly between the species. The variation in the form of the leaf tips in *B. abitibiensis* appears to be a post-mortem phenomenon and not a characteristic of growth. Variation in the morphology of the tips of the leaves in *B. longifolia* is difficult or impossible to demonstrate because of the poor preservation of the species. The leaves of *B. abitibiensis* are terete in cross-section; their form in *B. longifolia* is unknown. Lateral, rhizomatous branches have not been found in *B. abitibiensis* but are relatively common on the larger stems of *B. longifolia*. I believe that when longer specimens of *B. abitibiensis* are collected such structures will be present.

Cauline and foliar xylem strands are present in both species. The outline of the cauline xylem could not be demonstrated for *B. abitibiensis* because in all specimens the strand was completely flattened. Tracheids with annular secondary thickenings were occasionally observed on the outer margins of the flattened strands suggesting an exarch development. Tracheids with helical secondary thickenings are the predominant form observed in *B. abitibiensis*. In *B. longifolia* only tracheids with annular secondary thickenings were described. I doubt that that such is the case after analysing the forms of tracheids illustrated by Lang & Cookson (1935: plate 31, fig. 34) where the presence of helical secondary thickenings is clearly suggested. In *B. abitibiensis* the width of the tracheids with annular secondary thickenings varies from 12.5 to 18 μm ($\bar{x} = 16$ μm) and helical, secondary thickenings from 22 to 42 μm. In *B. longifolia* the tracheids, described as all of one form, vary from 12.5 to 42.5 μm, mostly 30–33 μm, in diameter. It can be seen that the overall variation of

25 μm. Fig. 38. Inner surface of epidermal layer, stomata indicated by arrows as in Fig. 35; scale bar equivalent to 100 μm. Fig. 39. Inner view of stomate in D, arrow marks remnant of wall between a guard cell and epidermal cell; the wall between the guard cells and the margins of the pore are thickened; scale bar equivalent to 25 μm. Fig. 40. Inner view of stomate at lower arrows in Figs. 34–36 & 38, remnants of walls of guard cells are clearly evident; scale bar equivalent to 13 μm. Fig. 41. Transection of epidermis of stem with stomate, periclinal walls at same level as those of the epidermal cells; poor preservation precludes interpretation of the cavity beneath the stomate; scale bar equivalent to 45 μm Fig. 42. Transection of the epidermis of the stem showing heavily thickened periclinal walls and the less thickened anticlinal walls of the epidermal cells; arrow marks slight separation near the cell surface that may represent the boundary of the cuticle; scale bar equivalent to 45 μm. Fig. 43. Tip of a leaf showing the separation of the cuticle from the surfaces of the epidermal cells, at arrow; scale bar equivalent to 45 μm.

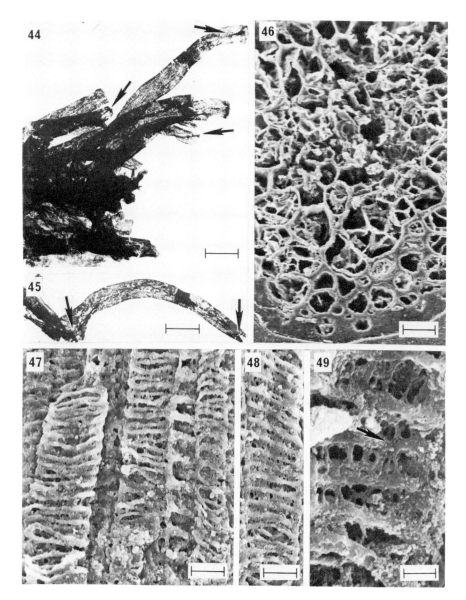

Figures 44–49. *Baragwanathia abitibiensis*. All specimens are from GSC Loc. 6441. Fig. 44. Paratype. Fragment of stem with leaves in which veins are visible at arrows; stem is oriented horizontally, base to right; scale bar equivalent to 2.5 mm; *GSC 69317*. Fig. 45, Paratype. Isolated leaf showing vein along the entire length of the leaf, between arrows, the apex is to the right; scale bar equivalent to 3.5 mm. *GSC 69318*. Fig. 46. Portion of the transverse section of the permineralized leaf shown in Fig. 19. Large parenchyma cells containing organic residues are clearly visible, area of vein is masked by organic residue precluding the demonstration of tracheids by means of SEM; scale bar equivalent to 57 μm; Stub UU-1, *GSC 69301*. Fig. 47. View of surface of fragment of xylem removed from the stem shown in Fig. 7. Tracheids with annular and helical secondary thickenings are clearly evident; scale bar equivalent to 20 μm; Stub TT-7, part of *GSC 69294*. Fig. 48. Large tracheid with helical secondary thickenings; primary wall exhibits irregular pits similar to those seen in *Drepanophycus gaspianus;* scale bar equivalent to 20 μm; Stub TT-7, part of *GSC 69294*. Fig. 49. Enlarged view of the irregular pits in the primary wall of a tracheid; scale bar equivalent to 5 μm; Stub TT-7, part of *GSC 69294*.

tracheid size is nearly identical between the two species. Foliar traces and bundles that extend to the tips of the leaves are present in both species.

Sporangia have not been found on specimens of *B. abitibiensis*. The quality of the preservation and abundance of material at GSC Loc. 6441 suggests that the eventual discovery of fertile specimens can be expected. Efforts should be made to obtain and prepare larger collections. I will note here that the occurrence of fertile specimens of *B. longifolia* amounts to perhaps 2% of the total specimens collected.

The presence of cuticle and stomata on the leaves and stems of *B. abitibiensis* and their absence on *B. longifolia* represent the most significant difference between the two species. Unfortunately, the investigations by Lang & Cookson (1935) as well as recent studies of additional specimens from several new localities in Victoria, Australia, have not produced evidence of the presence of a cuticle and stomata on the stems or leaves of *B. longifolia*. This lack of evidence is most probably a result of the manner in which the remains were preserved. Hundreds of specimens of *B. longifolia* are in collections of the National Museum of Victoria and the Mines Department of the Geological Survey of Victoria and 95% of them represent mineral replacement of the plant remains. The mineral appears to be limonite. Only the morphology of the plant is well-preserved. On very rare occasions tracheids can be seen in the cauline and foliar xylem strands and even more rarely is there evidence of epidermal patterns on stem impressions. In the remaining 5% of the specimens, some coaly matter is present but it is fractured into minute angular fragments from which very limited cell detail has been obtained. My suggestion that the type species had a cuticle and stomata is purely subjective and awaits evidence from better-preserved material.

Cuticle, cuticle and stomata, and structure of tracheary elements are known for only two species of *Drepanophycus* (Göppert, 1852). These are *D. spinaeformis* Göppert, and *D. gaspianus* (Dawson) Kräusel & Weyland, (Kräusel & Weyland, 1948). The stomatal apparatus of *D. spinaeformis* was well-described by Stubblefield & Banks (1978) as being paracytic, that is, the guard cells are accompanied by two subsidiary cells. The SEM studies of the stomata of *B. abitibiensis* did not produce evidence of subsidiary cells even though, with light microscopy, they appeared to be present. When I observed a stomate in surface view, the pore appeared to be encircled by three, concentric, dark bands. The first was the raised thickening immediate to the pore. The second was formed by the margins of the thickened periclinal walls of the epidermal cells that overlap the guard cells. The third was formed by the walls of the large guard cells beneath the imbrication of the neighbouring epidermal cells. I conclude that the stomatal apparatus of *B. abitibiensis* is anomocytic as opposed to the paracytic condition in *D. spinaeformis*. Details of the structure of the epidermis are lacking in all other species of *Drepanophycus*. A thin cuticle was found on stems of *D. gaspianus*; but no impressions of the epidermal cells were preserved (Grierson & Hueber, 1967). The secondary thickenings in the tracheids of *B. abitibiensis* are annular and helical, in *D. gaspianus* annular (Grierson & Hueber, 1967), and in *D. spinaeformis* annular, helical, and scalariform (Fairon-Demaret, 1971). These differences are probably directly related to the mode of preservation of the remains of each of the species. In their respective order, they represent carbonized compression, carbonate-sulphate permineralization, and hydrous iron oxide permineralization. The carbonized compression usually permits viewing of the longitudinal walls of the tracheids, particularly with the SEM, and requires only three-dimensional reconstruction of the otherwise flattened cells. The areas where such detail can be observed are usually limited because of the infilling of the cell lumens with coaly matter. As a result, there is perhaps a lesser range of cells observed. In the case of carbonate-sulphate permineralization, the peel technique was used, but because of the presence of acid-insoluble sulphates, such as barite, crystallized along with calcium carbonate in the plant tissues, only fragments of the xylem strand could be obtained. Also, in most instances, only one wall of a tracheid was included in the peel. A three-dimensional form could not be established that would permit definition of tracheary forms other than annular. Preservation of cell structures by hydrous iron oxide permineralization is amazingly good but requires skilled preparation of the specimens.

One can be assured of obtaining excellent detail from such specimens, as witness that for *D. spinaeformis* (Fairon-Demaret, 1971). Of particular interest in the comparison of *B. abitibiensis*, *D. spinaeformis* and *D. gaspianus* is the occurrence of perforations in the walls between the secondary thickenings in the tracheids of all three species. Such a structural characteristic helps unite the members of the Drepanophycaceae even more firmly. I do not agree with the interpretation of Hartman (1981) that the perforations in the walls of tracheids in *D. spinaeformis* are artifacts resulting from the crystallization of pyrite. The smoothly rounded margins of the perforations preclude such an interpretation.

The leaves of *B. abitibiensis* differ morphologically from all of the forms described for the species of *Drepanophycus*. In *D. spinaeformis*, (see Banks & Grierson, 1968; Fairon-Demaret, 1978) *D. spinosus* (Krejci) Kräusel & Weyland, (Kräusel & Weyland, 1933), *D. colophyllus* Grierson & Banks, (Grierson & Banks, 1963, 1983) and *D. gaspianus* the leaves are falcate and spine- to thorn-like as opposed to the downwardly curved, lax form with rounded tip seen in *B. abitibiensis*. The regular, close-spiral arrangement of the leaves in *B. abitibiensis* differs from the irregular arrangement in *D. spinaeformis* but is similar to some specimens of *D. spinosus* and particularly to *D. colophyllus* and *D. gaspianus*. The presence of a midvein in the leaves is known only for *D. spinaeformis* and *D. gaspianus*, and in those species the midvein extends to the tip of the leaf as it does in *B. abitibiensis*.

AGE OF BARAGWANATHIA

The age of *B. abitibiensis* is Emsian on the basis of the associated spore assemblage. The age of *B. longifolia*, when first described, was given as Silurian, not younger than Lower Ludlow (Lang & Cookson, 1935). The age was determined on the basis of association of the plant remains with graptolites identified particularly with the species *Monograptus uncinatus* var: *orbatus*, *M. chimaera*, and *M. uncinatus* var *microspora* (sic.) (Elles in Lang & Cookson, 1935). Those specimens were placed in synonymy with *M. thomasi* Jaeger (Jaeger, 1966) and the identification of *M. aequabilis notoaequabilis* Jaeger, Stein & Wolfart (Jaeger, Stein & Wolfart, 1969) add to the graptolites found associated with *Baragwanathia*. Tentaculitids of probably Praguian age associated with these graptolite species that are considered by Jaeger (1970: 175) as high in evolutionary stage of development brought about an upward revision of the age of the *Baragwanathia* flora from Silurian to mid-lower Devonian (Siegenian). Lang & Cookson (1935) would have been relieved with this change because their convictions as to the age of these early plants from Victoria would have been satisfied. Their earliest paper on plant remains from the Melbourne Trough (Lang & Cookson, 1927) described *Hostimella* sp., axes compared with *Thursophyton*, and axes classed as *incertae sedis* (= *Arthrostigma* Dawson; Cookson, 1926) that later were referred to *B. longifolia* (Lang & Cookson, 1935: 424). In their words: "Such an assemblage of plant remains, if looked at in the absence of any information as to their horizon and locality, would suggest the Early Devonian flora and perhaps the Middle rather than the Lower Devonian". Even with those judgements they did not want to give much weight to dating the sediments on the plants alone because their knowledge of the pre-Carboniferous flora was still, as they put it, too imperfect. They left the question of the age of the flora to the determination of the age of the fauna. One can sense a certain amount of exuberance over the availability of faunal evidence for dating the *Baragwanathia* flora, but then, in turn, they questioned the significance of the flora in relationship to the evolution of land vascular plants. Their questions about the "*Baragwanathia* Flora" are still valid following the recent discovery of a second *Baragwanathia* horizon, which on the basis of faunal evidence is considered late Silurian age. I therefore pose these questions in slightly modified form.

Why does the flora comprise kinds of plants known only from the Lower to early Middle Devonian of the Northern Hemisphere?

Was Victoria, Australia the only region favourable for the preservation of large, Silurian, vascular land plants to the exclusion of the smaller forms found elsewhere?

Was there an earlier development of land vascular plants in the Southern Hemisphere than in the Northern Hemisphere?

Where is the evidence of ancestral stock for such a plant as *Baragwanathia* that was so large and so morphologically and anatomically complex if, putatively, there are specimens of the genus that may be dated as Silurian?

Answering these questions requires an emphasis of the fact that since the time of Lang & Cookson (1935), the information available to us has expanded significantly regarding Lower and Middle Devonian floras of the Northern and, to some extent (Edwards, 1973; Hueber & Banks, 1981), the Southern Hemisphere (see bibliographies, Chaloner & Sheerin, 1979; Banks, 1980). Of particular significance is the improved stratigraphic control on many of the floras that results in a more substantial basis for interpreting the levels of evolutionary development attained by land and land vascular plants of the times. The concept of the evolutionary level of development of the "*Baragwanathia* Flora" expressed by Lang & Cookson (1935) is essentially unchanged. The flora is suggestive of the Lower and early Middle Devonian floras of the Northern Hemisphere. *Baragwanathia*, with its robust, cauline xylem strand, leaves supplied with a vein composed of tracheids and extending to the tip of the leaf, and sporangia borne possibly in adaxial position to the leaves, reflects, in its complexity, a level of evolution that is also typical of species of the lycopod genus *Drepanophycus*. No lycopods with such a complex structure are known in well-dated strata older than Siegenian (e.g. *D. spinaeformis*, Kräusel & Weyland, 1930; 1935). *Baragwanathia longifolia* characterizes both the "Lower Plant Assemblage" and the "Upper Plant Assemblage" (composed of plants described by Lang & Cookson) in the distribution outlined by Garratt (1979). Morphologically there is no difference between the plants from the two assemblages and the presence of the spores of

B. longifolia in the "Lower Plant Assemblage" (Douglas & Lejal-Nicol, 1981) supports the identification of the species. I feel that any differences are wholly intraspecific, if real differences should be proposed.

The "articulated stem of *Baragwanathia* flora" illustrated by Garratt (1978: fig. 4B), in my opinion, is *Baragwanathia*. I have seen the specimen in the collections of the Geological Survey of Victoria, and it is not articulated. The nodes or joints are artifacts of preservation and are not the condition of the stem during the life of the plant. The jointed appearance is the result of infilling of broad cracks in the stem by the surrounding sediment. Similar preservation is illustrated and described in branches of *Callixylon* (Arnold, 1934) found in the marine, Upper Devonian Antrim, New Albany and Ohio black shales of the United States. Dawson (1862) described a specimen from the Genesee shale of central New York State that exhibited the same type of preservation. He named it *Calamites inornatus* because of the apparent jointing of the stem that reminded him of the Carboniferous arthrophyte genus *Calamites*. The species was transferred to *Pseudobornia* (White, in Prosser, 1912: 522–523) but has been ignored by subsequent workers or thought, but not proved, to be another specimen of *Callixylon newberryi* of the kind described by Arnold (1931).

Two other genera belonging to the classic "*Baragwanathia* Flora" are listed as occurring in the "Lower Plant Assemblage" as well as the "Upper Plant Assemblage". They are *Hedeia* Cookson, 1935 (Douglas & Lejal-Nicol, 1981) and *Yarravia* Lang & Cookson, 1935 (Garratt, 1979). *Hedeia* is a fertile branch system of sufficient complexity to suggest an affinity with the Trimerophytina rather than the Rhyniophytina (Banks, 1968a). Unfortunately, the specimens of *Hedeia* are only fertile, short, terminal lengths of stems that give no indication of having been attached to larger stems; a criterion of the Trimerophytina. The genus would certainly represent the most complex member of the Rhyniophytina whose oldest member, *Cooksonia* Lang (Lang, 1937; Edwards & Davies, 1976; Edwards, Bassett & Rogerson, 1979; Banks, 1972, 1975a, b, c) is the oldest and simplest land vascular plant described so far. That *Hedeia* is structurally highly advanced over the most primitive member of the Rhyniophytina suggests that the genus is younger, probably of early Devonian age. *Yarravia*, in my opinion, after examining the type specimens and subsequent newly collected specimens, represents a poorly preserved form of *Hedeia* rather than a synangium, the form of fertile organ that typifies *Yarravia*. Should it be a synangium, its level of complexity is such that the similarity to Carboniferous spore-bearing organs is suggested over any Devonian genus known, even to Lang & Cookson (1935: 440), and that conclusion is still correct today.

The three remaining plants described as members of the "*Baragwanathia* Flora" are *Zosterophyllum australianum* Lang & Cookson (Lang & Cookson, 1930), *Pachytheca* sp., and *Hostimella* sp. (=*Hostinella* sp.). They belong with the "Upper Plant Assemblage" (Douglas & Lejal-Nicol, 1981) as they are not listed as present in the "Lower Plant Assemblage". The genus *Zosterophyllum* has a stratigraphic range from the Gedinnian through Emsian (Banks, 1980; Banks, Chaloner & Lacey, 1967; Edwards, 1969a, b, 1975). The Zosterophyllaceae ranges into the Frasnian with the genus *Sawdonia* Hueber (Hueber, 1971) (*Psilophyton princeps* of Hueber & Grierson, 1961). Thus, in the Northern Hemisphere the family is not known from any horizons outside of the Devonian. That the Zosterophyllaceae is represented by *Hedeia* and *Salopella* Edwards & Richardson (Edwards &

Richardson, 1974) in the "Lower Plant Assemblage" appears to be a mistake in the discussion by Douglas & Lejal-Nicol (1981: 686) as those genera are questionable members of the Rhyniophytina (Banks, 1975d) and certainly are not zosterophylls. These genera could only be placed in the Zosterophyllaceae if one ignores the identifying and unifying characters of the form and position of the sporangia. *Zosterophyllum australianum*, with its almost strobilar spike of sporangia, if eventually reported from the "Lower Plant Assemblage", would be difficult to explain as a precursor to the other less complex members of the genus and family. *Pachytheca* Hooker (Salter, 1861) is a fossil usually assigned to the algae but with unknown affinities within that group of plants. Its stratigraphic range is from the Wenlockian to Eifelian. Neither its stratigraphic range nor its enigmatic place in the evolution of the plant kingdom make it a particularly good reference fossil. In my own experience in collecting the genus in the Emsian deposits of New Brunswick, Canada, it occurs only in narrow horizons of coarsely comminuted plant debris. There has been no possibility of finding its relationship to other plant parts. The last genus in this discussion of the "*Baragwanathia* Flora" is *Hostinella* Stur (Stur 1882). It is a genus that is used to identify fragments of Silurian and Devonian plant stems that are naked and that branch dichotomously. Banks (1968b) has demonstrated the anatomy in some Devonian axes referred to this genus and one can trust the identification of those specimens with the Trimerophytina. One may only speculate that the Silurian plant fragments referable to *Hostinella* represent trimerophytes or their precursors. However, fragments of plants in that subdivision can be anticipated with certainty in strata from Siegenien through Emsian time. Although I must admit to abbreviating the possible points of explanation that might serve to answer the questions of Lang & Cookson (1935), I believe the essence is: the flora is more surely of Devonian age than it is of Silurian age. That is why the flora contains the plants one sees in it. The positive faunal evidence supports this. The effort to extend the age into the Silurian on the basis of extremely poorly-preserved graptolites which are in turn accompanied by a shelly fauna (Garratt, 1979, 1981) that may be related to other Devonian faunas seems more subjective than to say that the flora throughout its occurrence is of early Devonian age. The basis of this conclusion is that the level of evolutionary development of the flora is the same as compared with Devonian floras in other areas of the world where the stratigraphic control is well-documented. The flora was certainly one that could have grown in the marshlands associated with shorelines of rivers or lakes as suggested by Douglas & Lejal-Nicol (1981) and fragmentary evidence of the flora subsequently could have been washed into the great marine basin represented by the Melbourne Trough. I do not feel that any species of *Baragwanathia* could be interpreted as an aquatic plant merely on the basis of their stems appearing flexuous and being covered with long, filiform leaves (Douglas & Lejal-Nicol, 1981). Certainly, the occurrence of the flora in marine sediments is not to be taken as anything unusual, and neither should the plants be interpreted as marine forms. The marine sediments of Devonian age in western and central New York State, for example, contain fossilized plant remains rafted there originally from the ancient shorelines some 175–225 miles to the east; from the "Catskill Delta". For an indication of the variety of plants found similarly in the marine and continental sediments of New York State see Banks (1966) for a partial list. The number has increased since then but has not been newly summarized.

The answers to these questions are difficult to resolve because documented evidence of Silurian land or land vascular plants is available from very few sources. Rather than repeat the details of the occurrences, I refer the reader to the summaries given by Pratt, Phillips & Dennison (1978) and Banks (1980). In all instances, except for that reported by Pratt *et al.* (1978), the sediments containing the plant fragments are of marine origin and the plant material has drifted into the sites of deposition. If *Baragwanathia* or a plant of similar size were contemporaneous with the *Cooksonia* Zone of Banks (1980) one could certainly anticipate finding some evidence of it among the detritus in other localities. I do not feel that the Melbourne Trough was unique as a site for the preservation of plants of the *Baragwanathia*-kind nor do I feel that the northern and southern floras would have been at so widely different levels of evolution.

The plant remains described by Pratt *et al.* (1978) are found in fluviatile deposits of Llandoverian age. They are the best-documented forms of the oldest array of spores, sheets of cells, plant cuticle and tubes with internal thickenings that are available for comparison and discussion of the origins of land floras. Gray & Boucot (1977) did not document their observations but I have seen photographs of some of their isolates from Silurian sediments besides the trilete spores. The specimens comprised sheets of cells interpreted as cuticle, but more comparable to *Litostroma* Mamay (Mamay, 1959) and fragments of pellicles of arthropods interpreted as plant cuticle, one with a seta base interpreted as a pore. This comment is meant to serve as a note of caution, and it is not new (Banks, 1975a, b). Documentation of all records of possible land plant remains is important for the development of some understanding of the sequence in which the individual characteristics of trilete spores, cuticle, stomata and tracheidal vascularization arose. There is no argument that there has been a long period of pre-Devonian evolution of land and land vascular plants. The problem, at present, is that we have more evidence for the presence of land plants in the Silurian than we have for land vascular plants. No single, macrofossil plant fragment from the Silurian combines all of the characteristics of a land vascular plant. The oldest one closest to realizing all of that is the sterile *Cooksonia*-like axes described by Edwards & Davies, 1976. No plant of the magnitude and complexity of *Baragwanathia* is found in association with the *Cooksonia*-like stems or with *Cooksonia* in any of its occurrences. The primitive rhyniophyte genus *Salopella* (Edwards & Richardson, 1974) is reported as occurring with *Baragwanathia*, *Hedeia*, and *Yarravia* in the "Lower Plant Assemblage" of Garratt (1979). The originators of *Salopella* suggested that a compressed *Rhynia* may resemble *Salopella;* hence there is no problem in associating the genus with the other evolutionarily advanced genera. The rhynias were quite complex rhyniophytes. I would suggest that evidence for the development of land vascular plants earlier in the Southern Hemisphere than in the Northern Hemisphere is not yet available. However, on the basis of well-documented, objectively treated evidence, land vascular plants are well on their way in Upper Ludlow time in the Northern Hemisphere (Edwards & Davies, 1976).

ACKNOWLEDGEMENTS

I wish to thank the Director of the Geological Survey of Canada for the opportunity to report my observations on the fossils described herein; the Smithsonian Research Foundation, the Roland Brown Fund and Secretary

Ripley's Fluid Research Fund for financial support in Australia; John Talent for field assistance in Australia; Thomas Darragh and Thomas Rich of the National Museum of Victoria; Jack Douglas of the Mines Department, Geological Survey of Victoria; Jackie Tims for discussions; Susann Braden, Mary-Jacque Mann and James P. Ferrigno for operation of the SEM and photography.

REFERENCES

ARNOLD, C. A., 1931. On *Callixylon Newberryi* (Dawson) Elkins and Wieland. *Contributions of the Museum of Palaeontology, University of Michigan, 3:* 207–232.

ARNOLD, C. A., 1934. The so-called branch impressions of *Callixylon Newberryi* (Dn.) Elkins and Weyland and the conditions of their preservation. *Journal of Geology, 42:* 71–76.

BANKS, H. P., 1966. Devonian flora of New York State. *The Empire State Geogram, Triannual Newsletter, Geological Survey, (New York), 4:* 10–23.

BANKS, H. P., 1968a. The early history of land plants. In E. Drake, *Evolution and Environment: a symposium presented on the one hundredth anniversary of the foundation of the Peabody Museum of Natural History at Yale University:* 73–107. New Haven & London.

BANKS, H. P., 1968b. Anatomy and affinities of a Devonian *Hostinella*. *Phytomorphology, 17:* 321–330.

BANKS, H. P., 1972. The stratigraphic occurrence of early land plants. *Palaeontology, 15:* 365–377.

BANKS, H. P., 1975a. The oldest vascular land plants: a note of caution. *Review of Palaeobotany and Palynology, 20:* 13–25.

BANKS, H. P., 1975b. Early vascular land plants: proof and conjecture. *Bioscience, 25:* 730–737.

BANKS, H. P., 1975c. Palaeogeographic implications of some Silurian-early Devonian floras. In K.S.W. Campbell (Ed.), *Gondwana Geology:* 71–97. Canberra: Australian National University Press.

BANKS, H. P., 1975d. Reclassification of Psilophyta. *Taxon, 24:* 401–413.

BANKS, H. P., 1980. Floral assemblages in the Siluro-Devonian. In D.L. Dilcher & T.N. Taylor (Eds), *Biostratigraphy of Fossil Plants:* 1–24. Stroudburg, U.S.A.: Dowden, Hutchinson & Ross, Inc.

BANKS, H. P. & GRIERSON, J. D., 1968. *Drepanophycus spinaeformis* Göppert in the early Upper Devonian of New York State. *Palaeontographica, B, 123:* 113–120.

BANKS, H. P., CHALONER, W. G. & LACEY, W. S., 1967. *The Fossil Record,* 219–231. London; Geological Society.

CHALONER, W. G., 1967. Lycophyta. In E. Boureau, W. G. Chaloner, O. A. Hoeg & S. Jovet-Ast (Eds), *Traité de Paléobotanique, 2, Bryophyta, Psilophyta, Lycophyta:* Paris: Masson et Cie.

CHALONER, W. G. & SHEERIN, A., 1979. Devonian macrofloras. In M. R. House. C. T. Scrutton & M. G. Bassett (Eds), *The Devonian System, Special Papers in Palaeontology, 23:* 145–161. London: Palaeontological Association.

COQKSON, I. C., 1926. On the occurrence of the Devonian genus *Arthrostigma* in Victoria. *Proceedings of the Royal Society of Victoria, 38:* 65–68.

COOKSON, I. C., 1935. On plant-remains from the Silurian of Victoria, Australia, that extend and connect floras hitherto described. *Philosophical Transactions of the Royal Society, London B, 225:* 127–148.

DAWSON, J. W., 1862. On the flora of the Devonian period in northeastern America. *Quarterly Journal of the Geological Society, London, 18:* 296–230.

DOUGLAS, J. G. & LEJAL-NICOL, A., 1981. Sur les premières flores vasculaires terrestres dateés du Silurien: Une comparaison entre la "Flore à *Baragwanathia*" d'Australie et la "Flore à Psilophytes et Lycophytes" d'Afrique du Nord. *Comptes rendus hebdomadaires des Séances de l'Académie des sciences, Paris, Série D, 292:* 685–688.

EDWARDS, D., 1969a. *Zosterophyllum* from the Lower Old Red Sandstone of South Wales. *New Phytologist, 68:* 923–931.

EDWARDS, D., 1969b. Further observations on *Zosterophyllum llanoveranum* from the Lower Devonian of South Wales. *American Journal of Botany, 56:* 201–210.

EDWARDS, D., 1973. Devonian Floras. In A. Hallam (Ed.), *Atlas of Palaeobiogeography:* 105–115. Amsterdam, London & New York: Elsevier.

EDWARDS, D., 1975. Some observations on the fertile parts of *Zosterophyllum myretonianum* Penhallow from the Lower Old Red Sandstone of Scotland. *Transactions of the Royal Society of Edinburgh, 69:* 251–265.

EDWARDS, D. & DAVIES, E. C. W., 1976. Oldest recorded *in situ* tracheids. *Nature, London, 263:* 494–495.

EDWARDS, D. & RICHARDSON, J. B., 1974. Lower Devonian (Dittonian) plants from the Welsh Borderland. *Palaeontology, 17:* 311–324.

EDWARDS, D., BASSETT, M. G. & ROGERSON, E. C. W., 1979. The earliest vascular land plants: continuing the search for proof. *Lethaia, 12:* 313–324.

FAIRON-DEMARET, M. 1971. Quelques caractères anatomiques du *Drepanophycus spinaeformis* Göppert. *Comptes rendus hebdomadaires des Séances de l'Académie des sciences, Paris, Série D, 273:* 933–935.

FAIRON-DEMARET, M., 1978. Observations nouvelles sur les axes végétatifs de *Drepanophycus spinaeformis* Göppert de l'Emsien Inférieur des "Nouvelles Carrières" de Dave, Belgique. *Review of Palaeobotany and Palynology, 26:* 9–20.

GARRATT, M., 1979. New evidence for a Silurian (Ludlow) age for the earliest *Baragwanathia* flora. *Alcheringa*, 2: 217–224.

GARRATT, M. J., 1981. The earliest vascular land plants: comment on the age of the oldest *Baragwanathia* flora. *Lethaia, 14:* 8.

GÖPPERT, H. R., 1852. Fossil Flora des Ubergangsgebirges. *Verhandlungen der Kaiserlichen Leopoldinisch-carolinischen Akademie der Naturforscher, Supplement des Band 14:* 1–299.

GRAY, J. & BOUCOT, A. J., 1977. Early vascular land plants: proof and conjecture. *Lethaia, 10:* 145–174.

GRIERSON, J. D. & BANKS, H. P., 1963. Lycopods of the Devonian of New York State. *Palaeontographica Americana, 4:* 217–295.

GRIERSON, J. D., & BANKS, H. P., 1983. A new genus of lycopods from the Devonian of New York State. *Botanical Journal of the Linnean Society, 86:* 57–79.

GRIERSON, J. D. & HUEBER, F. M., 1967. Devonian lycopods from northern New Brunswick. In D. H. Oswald (Ed.), *International Symposium on the Devonian System, Calgary, 2:* 823–826. Calgary, Canada: Alberta Society of Petroleum Geologists.

HARRIS, T. M., 1926. Note on a new method for the investigation of fossil plants. *New Phytologist, 25:* 58–60.

HARTMAN, C. R., 1981. The effect of pyrite on the tracheid structure of *Drepanophycus spinaeformis*, a long-ranging Devonian lycopod. *Review of Palaeobotany and Palynology, 32:* 239–255.

HUEBER, F. M., 1971. *Sawdonia ornata:* a new name for *Psilophyton princeps*, var. *ornatum. Taxon, 20:* 641–642.

HUEBER, F. M. & BANKS, H. P., 1981. Devonian floras of Gondwanaland. *Abstracts: Thirteenth International Botanical Congress*, Sydney, Australia: 202.

HUEBER, F. M. & GRIERSON, J. D., 1961. On the occurrence of *Psilophyton princeps* in the early Upper Devonian of New York State. *American Journal of Botany, 48:* 473–479.

HUNDT, R., 1952. *Von Den Ältesten Landpflanzen Die Neue Brehm-Bücheri*. Leipzig.

JAEGER, H., 1966. Two late *Monograptus* species from Victoria, Australia, and their significance for dating the *Baragwanathia* flora. *Proceedings of the Royal Society, Victoria, 79:* 393–413.

JAEGER, H., 1970. Remarks on the stratigraphy and morphology of Praguian and probably younger monograptids. *Lethaia, 3:* 173–182.

JAEGER, H., STEIN, V. & WOLFART, A., 1969. Fauna (Graptolithen, Brachiopoden) der unterdevonischen Schwarschiefer Nord-Thailands. *Neues Jahrbuch für Geologie und Palaeontologie, Abhandlungen 133 (2):* 171–190.

KRÄUSEL, R. & WEYLAND, H., 1930. Die Flora des deutschen Unterdevons. *Abhandlungen der Preussischen Geologischen Landesanstalt, N.F, 131:* 1–92.

KRÄUSEL, R. & WEYLAND, H., 1933. Die Flora des böhmischen Mitteldevons (Stufe Hh₁ Barrande=h Kettner-Kodym). *Palaeontographica, B, 78:* 1–46.

KRÄUSEL, R. & WEYLAND, H., 1935. Neue Pflanzenfunde im rheinischen Unterdevon. *Palaeontographica, B, 80:* 171–190.

KRÄUSEL, R. & WEYLAND, H., 1948. Pflanzenreste aus dem Devon. XIII. Die Devon-Floren Belgiens und des Rheinlandes, nebst Bemerkungen zu einigen ihrer Arten. *Senckenbergiana, 29:* 77–99.

KRÄUSEL, R. & WEYLAND, H., 1949. *Gilboaphyton* und die Protolepidophytales. *Senckenbergiana, 30(1/3):* 129–152.

LANG, W. H., 1937. On the plant-remains from the Downtonian of England and Wales. *Philosophical Transactions of the Royal Society, London, B, 227:* 245–291.

LANG, W. H., & COOKSON, I. C., 1927. On some early Paleozoic plants from Victoria, Australia. *Memoirs of the Manchester Literary and Philosophical Society, 71:* 45–51.

LANG, W. H. & COOKSON, I. C., 1930. Some fossil plants of early Devonian type from the Walhalla Series, Victoria, Australia. *Philosophical Transactions of the Royal Society London, B, 219:* 133–163.

LANG, W. H. & COOKSON, I. C., 1935. On a flora, including vascular land plants, associated with *Monograptus*, in rocks of Silurian age, from Victoria, Australia. *Philosophical Transactions of the Royal Society, London, B, 224:* 421–449.

LECLERCQ, S., 1960. Refendage d'une roche fossilifère et dégagement de ses fossiles sous binoculaire. *Senckenbergiana Lethaea, 41:* 483–487.

MAMAY, S. H., 1959. *Litostroma*, a new genus of problematical algae from the Pennsylvanian of Oklahoma. *American Journal of Botany, 46:* 283–292.

MCGREGOR, D. C., 1979. Devonian miospores of North America. *Palynology, 3:* 31–52.

MCGREGOR, D. C. & CAMFIELD, M., 1976. Upper Silurian (?) to Middle Devonian spores of the Moose River Basin, Ontario. *Geological Survey of Canada, Bulletin, 263:* 63 p.

OBRHEL, J., 1962. Die Flora der Pridoli-Schichten (Budnany-Stufe) des Mittel-böhmischen Silurs. *Geologie, 11:* 83–97.

PRATT, L. M., PHILLIPS, T. L. & DENNISON, J. M., 1978. Evidence of non-vascular land plants from the early Silurian (Llandoverian) of Virginia, U.S.A. *Review of Palaeobotany and Palynology, 25:* 121–149.

PROSSER, C. S., 1912. The Devonian and Mississippian formations of northeastern Ohio. *Geological Survey of Ohio, Series 4, 15:* 1–574.

ROSELT, G., 1962. Uber die ältesten Landpflanzen und eine mögliche Landpflanzen aus dem Ludlow Sachsens. *Geologie, 11:* 320–333.

SALTER, J. W., 1861. Note on the fossils found in the Worcester and Hereford Railway cuttings. *Geological Society of London, Quarterly Journal, 17:* 161–162.

STUBBLEFIELD, S. & BANKS, H. P., 1978. The cuticle of *Drepanophycus spinaeformis*, a long-ranging Devonian lycopod from New York and eastern Canada. *American Journal of Botany*, *65:* 110–118.

STUR, D., 1882. Die Silur-Flora der Etage H-h₁ in Böhmen. *Sitzungsberichte der kaiserlichen Akademie der Wissenschaften in Wien, 84:* 330–391.

WALTON, J., 1925. Carboniferous Bryophyta. I. Hepaticae. *Annals of Botany, 39:* 563–572.

Botanical Journal of the Linnean Society (1983), *86:* 81–101. With 22 figures

A new genus of lycopods from the Devonian of New York State

JAMES D. GRIERSON AND HARLAN P. BANKS

Department of Biological Sciences, State University of New York,
Binghamton, New York 13901 and
Division of Biological Sciences, Cornell University,
Ithaca, New York 14853, U.S.A.

Received January 1982, accepted for publication June 1982

Drepanophycus colophyllus Grierson & Banks is transferred to *Haskinsia* Grierson & Banks on the basis of the morphology of its leaves and its anatomical structure. Slender, dichotomizing axes bear helically arranged leaves that are falcate, petiolate, simple, lanceolate and entire. A solid cylinder of metaxylem is surrounded by ridges of protoxylem. Pitting on tracheid walls is annular, helical, reticulate, scalariform to circular-bordered. This lycopod extends from early Givetian into early Frasnian time in the Catskill strata of eastern New York State. The effect of cleavage of fossiliferous rock on patterns observed on the exposed fossils and the bearing of recent research on the fossil history of lycopods are discussed.

KEY WORDS:—Catskill strata – Devonian – lycopods – New York State – Protolepidodendrales.

CONTENTS

INTRODUCTION

We have collected specimens of an herbaceous lycopod over a period of years. All specimens come from sediments of the Middle to Upper Devonian Catskill clastic wedge of New York State. All the remains are easily recognizable as lycopods, but can be superficially quite dissimilar (compare Figs 1–5, 8–11, 15, 17). However, we are convinced that our material can be placed in a single taxon, defined by its simple, adaxially recurved (falcate), petiolate, laminar leaves, combined with newly acquired anatomical and morphological data.

Earlier we described some of these specimens as *Drepanophycus colophyllus*

Grierson & Banks (Grierson & Banks, 1963) on the basis of their short, rigid, falcate leaves. The size of the axes, the surface pattern of the stem, and the number and arrangement of the leaves resembled *Protolepidodendron*. The simple, falcate leaf definitely excluded it from that genus which is characterized by divaricate and bifid leaves. (For the concept of that genus see Kräusel & Weyland, 1940; Grierson & Banks, 1963; Fairon-Demaret, 1980.) At that time we were unwilling to erect a new genus on the basis of limited material. Instead, we chose to use the falcate leaf to assign the plant to the heteromorphic genus *Drepanophycus* which had come to include widely disparate species, some probably closely related, others totally unrelated (Grierson & Banks, 1963: 234, 271; Grierson & Hueber, 1967: 824). We suspected that the relationships of our new plant, when more fully known, would prove to lie with the Protolepidodendraceae rather than with the Drepanophycaceae (Grierson & Banks, 1963: 271).

The discovery of new specimens of this plant with more complete leaf morphology and well-preserved pyritic cellular permineralization of the stem requires us to remove it from the genus *Drepanophycus*. We here add materially to the description of the plant. The marked variability in morphological appearance exhibited by our axes (Figs 1–11) requires that we demonstrate that the variability is confined to a single taxon.

TECHNIQUES

Standard palaeobotanical methods were employed for sectioning and polishing the petrified specimens. The following more specialized techniques were utilized for some preparations: dégagement (Leclercq, 1960); bioplastic embedding (Grierson & Banks, 1963); serial bioplastic transfer method (Bonamo, 1977); the long acid etch technique for the study of pyritized vascular tissue studied with the SEM (Grierson, 1976). A Leitz ortholux with Orthomat and Ultra Pak was used for epi-illuminated photomicrography; a Polaroid MP-4 and a Leitz Aristophot were used for macrophotography. An ETEC Bioscan electron Microscope (SEM) was utilized for study and photography.

DESCRIPTION

All axes are slender, aerial, vegetative, leafy branches (Figs 1–11, 15, 17), 5.57–9.14 mm ($\bar{x} = 6.52$ mm) in width, exclusive of leaves. Branching is dichotomous (Fig. 4; Grierson & Banks, 1963: pl. 32, fig. 1). The phyllotaxy is helical with the number of orthostiches varying from six to twelve (Figs 1–11).

Figures 1–4. *Haskinsia colophylla*; specimens photographed immersed in xylene; all figures to scale as in Fig. 1, scale bar equivalent to 10 mm. Fig. 1. Holotype; most leaves appear spine-like, but those that were uncovered demonstrate the leaf base (arrow a), petiole (arrow b), and laminar blade (arrow c); DBPC Type Catalog No. 57; locality: Walton Farm. Fig. 2. A cleaved compression surface (Fig. 22B) showing apparently spine-like leaves along the margins and two nearly complete uncovered leaves at arrows; DBPC Type Catalog No. 58; locality: Walton Farm. Fig. 3. A single axis revealing in the upper part an abaxial view of complete leaves, constituting a leaf laminae compression fracture plane. (Fig. 22, D2); at the lower right edge of the axis the fracture plane broke downward to the surface of the stem, revealing a cleavage compression surface. (Fig. 22, C2); PB-SUNY-B TYPE NO. 53; Locality: Blenheim-Gilboa. Fig. 4. A dichotomously branched stem with falcate leaves seen along the margins of the stem; the stem fracture surface is a cleavage compression (Fig. 22, C2); DBPC Type Catalog No. 271; Locality: Summit Road Cut.

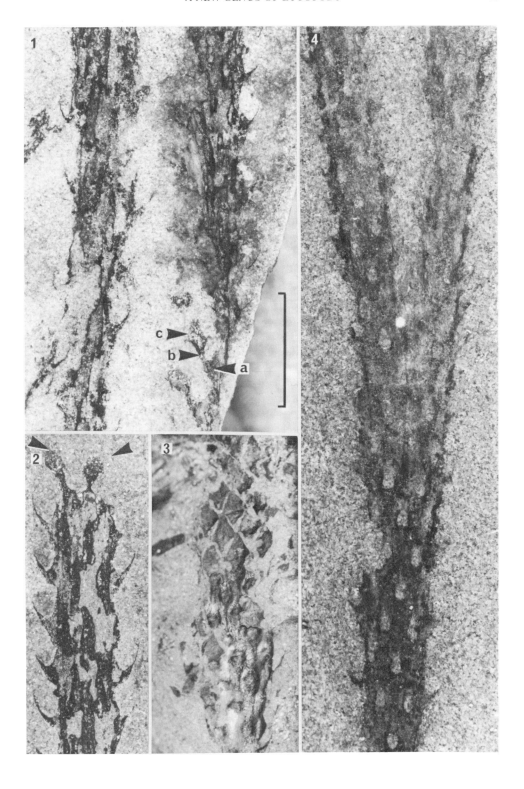

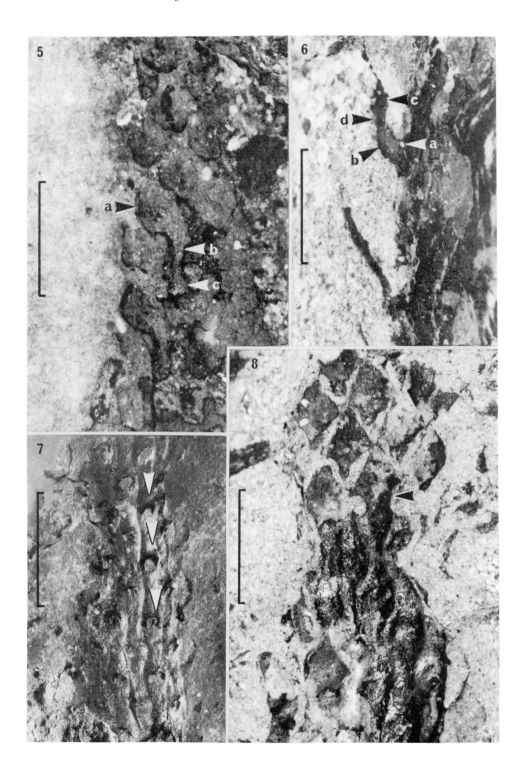

The longest specimen (not illustrated) is 202 mm (with a dichotomy 37 mm) from its proximal end. Just below this dichotomy the stem is 9 mm wide; just above 5 mm wide. Distally the stem broadens gradually to 8 mm in a distance of 165 mm. This widening implies that another dichotomy may have occurred. The dichotomizing specimen in Fig. 4 is 136 mm long, 8 mm wide below the dichotomy, and each branch is 5 mm wide above the dichotomy. All other branching specimens show only dichotomies.

The leaf

The leaves are simple and persistent (e.g. Figs 1, 4, 10, 11). When seen only in side view (Figs 1, 2), the leaves are markedly falcate, the abaxial surface forming a sickle-like convex arc with the axis. The leaves were probably quite rigid in life since they are rarely found to deviate from this typical falcate attitude.

Each leaf consists of a thickened proximal base which is flattened laterally so that its point of attachment to the stem is slightly elliptical in cross-section. From this base the petiole curves upward to the attachment of the lanceolate blade (Figs 1, 2). The lamina of the blade first widens slightly from its junction with the petiole, then tapers to an acute tip (Figs 3, 5, 6, 8). The blade, too, recurves upward following the curve of the petiole. The margin of the leaf is entire (Figs 3, 5, 8–11). Leaf bases that are petrified show a single, mesarch trace when seen in cross-section (Fig. 12).

The more delicate leaf blade failed to survive intact the acid transfer treatment in all transfers prepared for our earlier paper (Grierson & Banks, 1963). The bioplastic serial transfer technique (Banks, Bonamo & Grierson, 1972) applied to our newer, finer grained materials permits us to control the removal of sediment and has provided successful transfers of both leaves (Fig. 6) and stems (Fig. 7).

The serial leaf transfer (Fig. 6) reveals a nearly complete attached leaf illustrating in more detail the enlarged leaf base (arrow a), the thickened petiole (arrow b), and the junction between the blade and petiole (arrow d). Only a portion of the blade remains, a break having occurred from lower right towards upper left of the lamina.

Figures 5–8. *Haskinsia colophylla;* Figs 5 & 8 from specimens photographed immersed in xylene. Fig. 5. An enlargement of a distal portion of the axis in Fig. 10. The fracture plane is a leaf laminae cleavage compression (Fig. 22, D2). A few leaves near the edge of the axis have been displaced laterally (see arrows a–c) revealing more accurately the proportions of the blade (arrow a), petiole (arrow b) and combined leaf base (arrow c); PB-SUNY-B Type No. 54; Locality: Blenheim-Gilboa; scale bar equivalent to 5 mm. Fig. 6. An enlarged portion of an axis following treatment with the bioplastic serial transfer technique; the acid has removed enough of the sandstone matrix that parts of two leaves can be seen; at left the remaining matrix supports portions of the two leaves; the upper leaf shows the leaf base (arrow a), the thickened petiole (arrow b), and its point of attachment to the blade (arrow d); the thin blade (arrow c), is not completely intact, weakened as the supporting matrix was removed, and a portion of the blade from left to upper right was lost; PB-SUNY-B Type No. 55; Locality: Blenheim-Gilboa; scale bar equivalent to 2 mm. Fig. 7. A bioplastic transfer of a cleavage compression stem surface; falcate leaves are visible on the margins of the stem, and the proximal parts of coaly, raised leaf bases are aligned on vertical ridges (arrows); PB-SUNY-B Type No. 56; Locality: Blenheim-Gilboa; scale bar equivalent to 10 mm. Fig. 8. An enlargement of Fig. 3 shows two stem surface patterns; a leaf laminae cleavage compression (Fig. 22, D2), revealing leaves in abaxial view at the upper part of the figure and at the bottom left, and a cleavage compression surface (Fig. 22, C2) on the lower right. Juxtaposition of the two surfaces reveals the extent and form of a complete leaf (arrow) with the attachment of the petiole to the leaf base; PB-SUNY-B Type No. 53; Locality: Blenheim-Gilboa; scale bar equivalent to 5 mm.

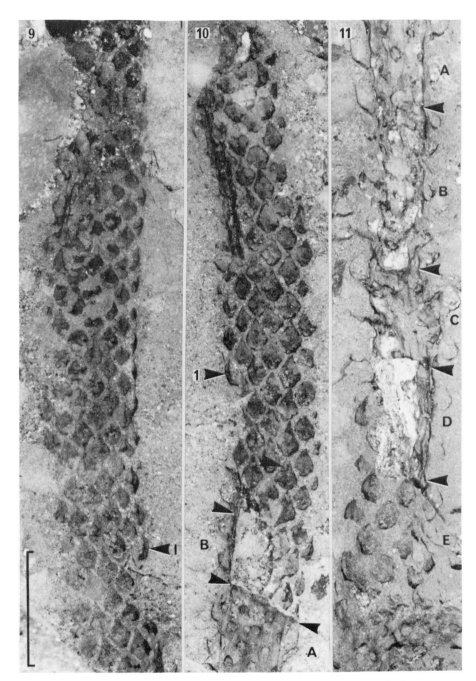

Figures 9–11. *Haskinsia colophylla;* specimens photographed immersed in xylene; all figures to scale as in Fig. 9, scale bar equivalent to 10 mm. Figs 9 & 10. Part and counterpart of an axis showing an especially well-preserved length of leaf laminae cleavage impression (Fig. 9) and compression (Fig. 10); the same leaf (arrow 1) on both figures indicates the corresponding levels of part and counterpart; in Fig. 10, two other fracture patterns are seen on the proximal portions of the axis; a cleavage impression Area A, (Fig. 22, Cl), and cleavage compression Area B, (Fig. 22, C2); PB-SUNY-B Type No. 54; Locality: Blenheim-Gilboa. Fig. 11.This specimen reveals several surface fracture planes; the distal portion of the axis, Area A, shows a cleavage impression surface (Fig. 22, Cl); in Area B, just below Area A, the fracture plane rises, with regards to the axis, to the level of a cleaved compression (Fig. 22B) and exposes a portion of the pyritized vascular strand. Area C is again a cleavage impression; Area D is a cleaved compression split through the vascular column; in Area E, the fracture surface again rises with regards to the axis to the level of a leaf laminae cleavage compression plane; PB-SUNY-B Type No. 57; Locality: Blenheim-Gilboa.

We have also been able to return to some of the type specimens of our earlier study and to demonstrate, by uncovering, that they have the same leaf morphology and attitude as our newer material. (Figs 1, 2.). Some leaves, uncovered by dégagement (Figs 1, 2, arrows), show the abaxial surface of the basal petiole and laminar, falcate blade, features shared by the leaves of all newer specimens.

Stem morphology

The diverse morphological appearances of our axes are attributable to interactions of biological and physical factors during fossilization and to their subsequent planes of fracture. In recent years several workers on Devonian material have re-emphasized problems of interpreting lycopod remains that exhibit a variety of stem surface and subsurface patterns (e.g., Grierson & Banks, 1963; Menéndez, 1965a, b; Chaloner, 1967; Chaloner & Collinson, 1975; Chaloner et al., 1980). As a result of these investigations, it is easier to distinguish between and to interpret the variety of surfaces and levels exposed on lycopod compressions and impressions.

The specimen in Fig. 2 shows a fracture plane exposed by a split through the middle of the axis. The resulting specimen reveals the inside of an axis with leaves present only along the margins. This kind of a split and fracture plane are illustrated diagrammatically on Fig. 22B. This plane of fracture is comparable to the "cleaved compression" terminology of Chaloner & Collinson (1975) of leaf impression and compression fossils and the stems illustrated in Chaloner et al., (1980).

In lycopods with numerous closely-set leaves, a more common fracture pattern occurs when the plane of cleavage passes along the interface between the stem surface and the leaves (Fig. 22C). Such a break results in a cleavage compression part (Fig. 22, C2; Chaloner & Collinson, 1975; fig. 1B2) bearing most of the compression fossil, and a cleavage impression counterpart (Fig. 22, C1; Chaloner and Collinson, 1975; fig. 1B1). The cleavage impression bears the impression of the stem, and, in its matrix, the leaves that were borne on that surface of the fossil.

Specimens showing the cleavage compression pattern (Fig. 22, C2) bear leaves seen in side view along the margins of the fossils, as well as broken attachments of leaves on the exposed stem surface (Fig. 4, lower portion). Our material commonly shows this cleavage compression pattern, but often fractures near the weak junction of the petiole and blade. The stem surface thus revealed bears raised bosses consisting of the leaf base and thickened, basal portions of the petiole (Figs 3, 8; see also Grierson & Banks, 1963: pl. 32, fig. 7). This pattern contributed to our previous conception of the leaf as non-laminar.

The specimen illustrated in Fig. 7 is a bioplastic transfer of an axis with a natural cleavage impression surface (compare with Fig. 11, upper portion). The concave surface of the specimen was first filled with liquid bioplastic, and then transferred by the bioplastic/hydrofluoric acid technique. The result is a cast of the cleavage impression surface as diagrammed in cross-section in Fig. 22, C1. Leaves can still be seen attached to the margins of the axis. The proximal coaly portions of the leaves that had been buried in the rock matrix, now protrude, outlining the leaf bases (Fig. 7, arrow). They are arranged in orthostichies and parastichies; the leaves of a single orthostichy are borne on a slightly raised ridge, represented in Fig. 22A as a ridge on the outer surface. These vertical ridges are separated by sinuous vertical

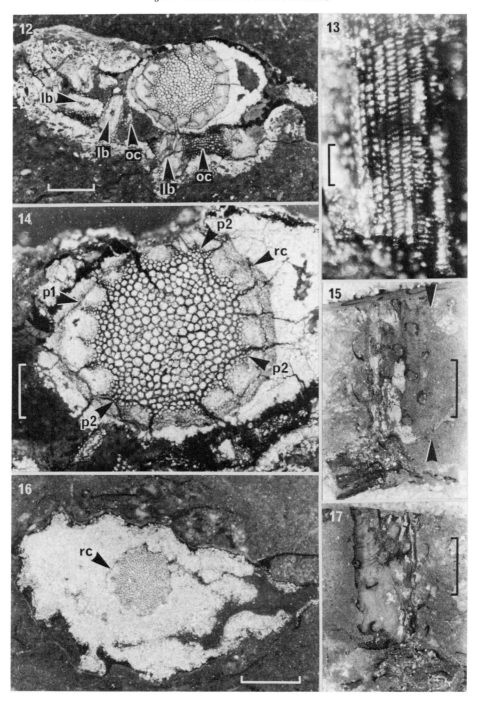

Figures 12–17. *Haskinsia colophylla*. Fig. 12. A cross-section of a petrified axis with a nearly intact xylem column, petrified leaf bases (arrows lb), and occasionally, petrifaction of the thick-walled outer cortical strengthening cells adjacent to the leaf bases (arrows oc); DBPC Type Catalog No. 272; Locality: summit Road Cut; scale bar equivalent to 1 mm. Fig. 13. A radial split through a protoxylem ridge belonging to the same axis seen in Figs 12 and 14; this figure shows the maturation sequence of the protoxylem tracheids increasing, from left to right, in diameter of the cells, in more complex wall patterns, and in rigidity; DBPC Type Catalog No. 272; Locality: Summit Road Cut;

depressions (depicted in Fig. 22A as rounded grooves) and seen in Figs 3, 7, 8, lower portion, between adjacent orthostichies. Anatomically the groove is the site of vertically elongate, thick-walled cells of the outer cortex just below the epidermis (Fig. 12, arrow oc). The slight lateral compression of the axis illustrated in Fig. 7 accentuates the depth and sharpness of these linear depressions.

Specimens illustrating the cleavage impression pattern (Fig. 22, Cl) are shown in Fig. 10, lower portion, and Fig. 11, upper portion. The persistent leaves are visible on the margins of the axis, but the stem surface here revealed is the inner surface of the cortical area on the opposite side of the the axis. What appear to be leaf bases are, in reality, depressions representing the departure of the leaf trace and accompanying tissues into the bases of the leaves protruding downwards into the rock matrix.

A third fracture pattern has been discussed for some species of *Archaeosigillaria* by Menéndez (1965a) and Chaloner *et al.* (1980). This pattern seems to be the result of a combination of biological factors including the close proximity of leaves on the stem, their rigidity, and their laminar, recurved blades that lie more or less parallel to the surface of the axis. In our material, this external, closely fitting, sheath-like mass of laminae is especially important. With this pattern, a fracture plane may follow the erect surfaces of the leaf laminae, 'jumping across' the narrow band of matrix between adjacent leaf blades (Fig. 22, D2). The result is a leaf laminae cleavage compression that consists of the entire stem and almost all of the leaves (Figs 3, 8, 10, upper portions; 11, lower portion). The leaf laminae cleavage impression (Fig. 9) bears only the impression of the leaf laminae, or, sometimes, a portion of the abaxial side of the recurved lamina itself (Fig. 23, D1).

The perception of the width of an axis can be affected by the level of the various fracture patterns. These apparent differences are most obvious in Fig. 11, in which five fracture planes have resulted from a split of the block in which the axis was embedded. Areas A and C represent cleavage impression surfaces (Fig. 22, Cl); Areas B and D, cleaved compressions (Fig. 22B); and Area E, a leaf laminae compression surface (Fig. 22, D2).

The axes revealed by these different patterns vary widely in appearance and proportion, but dégagement, transfer techniques and anatomical evidence prove them to be merely different aspects of a single plant.

---

scale bar equivalent to 50 µm. Fig. 14. An enlargement of a different cross-section of the axis shown in Fig. 12 reveals that some protoxylem ridges have a single protoxylem point (arrow p1); others have two or more (arrow p2); exterior to the vascular strand is a ring of cells possibly representing the remnants of the pericycle, endodermis, and sometimes, inner cortex (arrow rc); DBPC Type Catalog No. 272; Locality: Summit Road Cut; scale bar equivalent to 500 µm. Fig. 15. This axis showing a partial cleavage compression surface (Fig. 22, C2) includes the petrified axis seen in Fig. 16; the fracture plane did not completely reach the right margin of the axis, leaving the right edge of the stem buried in the matrix, the true margin on the right side is indicated by arrows; Specimen photographed immersed in xylene; PB-SUNY-B Type No. 58; Locality: Blenheim-Gilboa; scale bar equivalent to 5 mm. Fig. 16. A cross-section of the petrified axis whose split surfaces are shown in Figs 15 & 17; the xylem strand is well preserved here, a few cells are petrified around the perimeter of the stem but, most of the outer tissues are not preserved; the cellular ring of pericycle, endodermis, and perhaps inner cortex has fallen in on some of the protoxylem ridges softening the outline of the xylem strand (arrow rc); PB-SUNY-B Type No. 58; Locality: Blenheim-Gilboa; scale bar equivalent to 1 mm. Fig. 17. The fracture plane surface of this axis is a cleavage impression (Fig. 22, Cl) and is the counterpart of Fig. 15; specimens photographed immersed in xylene; PB-SUNY-B Type No. 58; Locality: Blenheim-Gilboa; scale bar equivalent to 5 mm.

Figures 18–21. *Haskinsia colophylla*. Fig. 18. A portion of a longitudinal section of a pyritized axis in the metaxylem; the surface was ground and etched; bordered pits are predominately uniseriate or biseriate on this face of the tracheid, the brighter areas, the pyrite casts of the lumen and pits; within the wall pit casts of adjacent pit pairs can be seen (arrows); PB-SUNY-B Type No. 59; Locality: Blenheim-Gilboa; scale bar equivalent to 10 μm. Fig. 19. A portion of a split surface of the petrified axis seen in Figs

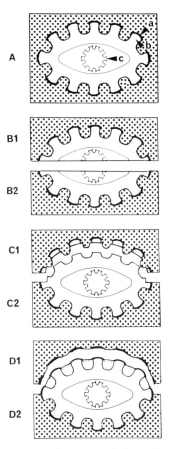

Figure 22. Diagrams illustrating fracture planes or weathering surfaces of compressed lycopod stems in a rock matrix. A. Matrix stippled; arrow a, leaf lamina; arrow b, raised ridge bearing leaves; arrow c, region of petrified vascular tissue. B1, B2. A fracture plane through the middle of the axis reveals essentially similar surfaces (cleaved compression). C. The plane of fracture passed through the interface between the stem surface and the leaves. Cl is a cleavage impression; C2 is a cleavage compression. D. The plane of fracture followed the parallel surfaces of the leaf laminae. This fracture plane results in a leaf laminae cleavage impression, D1, and a leaf laminae cleavage compression, D2.

Anatomy

Only a few well-petrified axes of our material have been found (Figs 12–14, 16, 18–21), although a relatively large number of axes are filled with pyrite showing little or no preservation of cell detail. Even in the petrified axes only a few show cellular detail in the outer regions of the axis and that is usually less distinct than in the xylem strand.

12–14; the pyritized bordered pits are predominantly triseriate and alternate; DBPC Type Catalog No. 272; Locality: Summit Road Cut; scale bar equivalent to 10 μm. Fig. 20. A portion of a split surface seen in Figs 12–14. The pyritized borders show a variety of patterns on a wall; DBPC Type Catalog No. 272; Locality: Summit Road Cut; scale bar equivalent to 10 μm. Fig. 21. A split portion of the metaxylem of a petrified axis; the specimen was prepared by the long acid etch technique, which removed the coaly substances, leaving four (1–4) pyritized lumen casts still bound to each other by their shared pit cavity casts (arrow); PB-SUNY-B Type No. 59; Locality: Blenheim-Gilboa: scale bar equivalent to 10 μm.

The best-preserved specimen (Figs 12–14, 19–20) was embedded in bioplastic. The centre of the stem is occupied by a xylem strand consisting of a solid column of metaxylem tracheids bearing outer vertical ridges made up of smaller tracheids (Fig. 13). There is no intraxylary parenchyma. The flanking ridges are composed of both protoxylem and transitional tracheids (Fig. 13). The complete xylem strand is 1.8 mm in diameter (range 1.7–1.9 mm). The maximum extent of the ridges is 0.2 mm so that the metaxylem core is about 1.4 mm in diameter. On some ridges there are two (Fig. 14) protoxylem areas. At any one level of the stem there is a maximum of 14–16 ridges and 15–22 protoxylem points.

In this specimen, preservation is best in the centre of the xylem strand. Often cracks radiate from the protoxylem ridges toward the exterior of the stem (Fig. 14). These cracks may represent lines of weakness caused by the breakdown of soft tissues accompanying the leaf trace, thereby permitting shifts in position during post depositional compression. Whatever their cause, they make it difficult to determine the exact number of protoxylem points and traces in the serial sections.

The bulk of the xylem strand is composed of large metaxylem cells (Figs 12, 14, 16) that are 26–104 μm in diameter in cross-section. The lumens of the cells are filled with minute pyrite crystals; the cell walls are represented by dark brown, coaly remains (Figs 12, 14, 16, 18–20). The walls of the metaxylem tracheids have conspicuous intertracheary bordered pitting (Figs 18–21). Pits vary in shape from circular-bordered pitted, through oval, to transversely elongate pits that could be called scalariform where they occur in vertical series (Fig. 18). Although there is a continuum in form, oval and circular pits predominate; elongate pits are less frequent. The number of pits across a wall face is not strictly correlated with the width of the cell face. For example, in Figs 18 and 20, one elongate pit occupies the full wall face while just below it two or more oval pits span the same wall face. Depending in part on the degree of crowding, the arrangement of the pits may appear alternate (Fig. 19) or opposite (Fig. 18). Pits are 10–26 μm in width with pit borders $c.$ 1.5–2 μm in thickness.

The specimen in Fig. 21 is a longitudinal split of a portion of a pyritized metaxylem strand. This surface was treated with the long acid etch technique (Grierson, 1976). The result of this treatment is to totally remove the coaly substances of the tracheid walls leaving the pyritized casts of each tracheid intact, and connected to adjacent tracheids by their shared pyritized-bordered pit pair casts. This technique permits comparisons using the great depth of focus and high resolution possible with the SEM to compare the stem vasculature of different areas or between different plants (Grierson, 1976).

The ridges around the metaxylem column are more-or-less equidistant from one another and, as seen in cross-section, taper rather sharply from their bases against the metaxylem to their tips. The cells to the left in Fig. 13 are narrow ($c.$ 5 μm) and the annular rings are distant. These cells represent the early protoxylem and are followed centripetally by cells of larger diameter with increased wall thickenings in a helical and reticulate pattern. The last two cells shown to the right are reticulate, 26 μm in diameter and are transitional between the protoxylem and the central core of metaxylem. Each ridge is composed of about 24 cells (Fig. 14). Those ridges with two protoxylem points (Fig. 14, p 2) consist of more cells, but the number varies from level to level, presumably increasing where leaf traces join the stele. In taking measurements, the inner limit of the ridge was selected on the basis of cell diameter. In those instances where we can examine both cross and longitudinal sections of a portion of the xylem strand, the difference in lateral wall thickenings supports this criterion of size.

Exterior to the xylem column is a ring of cells 3–8 cells in thickness (Figs 12, 14, 16). The width of the ring seems to depend on the degree of preservation at different levels of the stem. As seen in cross-section the cells of this ring are sometimes radially seriate, sometimes alternately arranged. We interpret this cellular zone as the remains of the pericycle, the endodermal area and part of the inner cortex. Occasionally, leaf traces are found among the preserved cells of this ring.

Outside of the inner cortex there is a region of noncellular pyrite that we interpret as the area of the former middle cortex. Presumably the cells comprising this middle cortex were less resistant to decay and compression. The width of this noncellular zone is variable, being narrower at the top and bottom of the cross-section than at the two sides (Figs 12, 14). The stele is nearly adjacent to the top and bottom layers of the stem, indicating that petrifaction of the stele preceded the early decay of the cortex.

Just within the non-cellular coaly perimeter of the stem there are areas of the outer cortex composed of thick, angular cells (Fig. 12, arrows oc). These cells range 18 to 65 μm ($\bar{x} = 41.3\ \mu$m) in diameter. This outer cortical region is discontinuous, being interrupted by the areas of leaf base attachment. It is this region of the cortex between the leaf bases that forms the sinuous pattern that is so prominent on some of the stems that have been compressed or that have lost their epidermis (Figs 7, 8, bottom; Grierson & Banks, 1963: pl. 32, fig. 4). Leaves are situated between these groups of cells, as illustrated by Fig. 12, arrows 1b.

Several more pyritic leaf bases of the same gyre are present at arrows 1b, but are slightly displaced since the outer cortex is not well-preserved. In a few cases, sections of leaves show a single, central, mesarch vascular strand (Fig. 12).

A second petrified axis, (Fig. 16), belongs to the stem whose outer stem morphology is illustrated in Figs 15 & 17. This section is oblique. It demonstrates the pyritized bases of several leaves and a few falcate, coalified leaf blades. The centre of the stem is occupied by the fluted xylem column with approximately 14 peripheral ridges. The xylem strand is well-preserved, but the area between the outer cortex and xylem strand is not cellularly preserved. The partially decomposed ring of pericyclic, endodermal and inner cortical cells is draped over the protoxylem ridges. This gives the xylem column the false appearance of having rounded rather than sharply pointed protoxylem ridges.

One of us, (J.D.G.), is continuing a more detailed anatomical study of this material utilizing the techniques developed earlier for pyritized remains of *Leclercqia complexa* (Grierson, 1976).

SYSTEMATICS AND DIAGNOSES

DIVISION:	Tracheophyta
SUB-DIVISION:	Lycophytina
ORDER:	Protolepidodendrales

Haskinsiaceae Grierson & Banks **fam. nova**

Slender, dichotomizing axes bearing leaves in a low helix; epidermal surface smooth, thick-walled cells of outer cortex arranged in a fusiform pattern around leaf bases; leaves falcate, petiolate, simple, entire; xylem a solid strand with peripheral protoxylem ridges.

TYPE GENUS: *Haskinsia* Grierson & Banks

Haskinsia Grierson & Banks, **gen. novum**

DERIVATION: The genus is named for Vernon Haskins, Curator of the Durham Center Museum, Greene County, New York whose knowledge of the Catskill sediments, and skills as a collector, have been enthusiastically shared with several 'generations' of palaeobotany students.

Slender, herbaceous lycopod, branching dichotomously, and bearing spirally-arranged simple, adaxially recurved falcate leaves. STEM SURFACE smooth between the slightly enlarged leaf bases; compression of axes or loss of epidermis reveals fusiform arrangement of outer cortical strengthening tissues. LEAVES with a short, thickened, basal petiole, widening into a distal, recurved, laminar blade; blade lanceolate with an acute tip and entire margins. PRIMARY XYLEM COLUMN denticulate in cross-section, exarch. Metaxylem tracheids with uniseriate to multiseriate round, oval, and scalariform bordered pits; protoxylem annular to helical; transitional tracheids reticulate.

TYPE: *Haskinsia colophylla* (Grierson & Banks) Grierson & Banks.

Haskinsia colophylla (Grierson & Banks) Grierson & Banks **comb. nova**

BASIONYM: *Drepanophycus colophyllus* Grierson & Banks, *Palaeontographica Americana*, *4* (31): 233 (1963).

Herbaceous lycopod with dichotomous branching. AXES up to 250 mm in length, average width about 6 mm, ranging from 4 to 9 mm, exclusive of leaves. LEAVES spirally-arranged, 6–12 leaves in each gyre; leaves simple, consisting of a short, thickened adaxially recurved basal petiole and a distal, adaxially recurved laminar blade (1.8 mm wide and 2.6 mm long); blade lanceolate, tapering to an acute tip; margins entire; fusiform pattern around leaf bases may be visible upon compression of stem or loss of epidermis; pattern reflects the arrangement of outer cortical, thick-walled cells. PRIMARY XYLEM COLUMN of stem consisting of a metaxylem core bearing lateral longitudinal ridges composed of annular, helical and reticulate protoxylem and transitional xylem; diameter of complete xylem strand up to 1.9 mm; parenchyma absent in the xylem strand; metaxylem consisting of tracheids with circular, oval, or scalariform-bordered pits, uniseriate to multiseriate, opposite or alternate, varying with the degree of crowding; cylindrical region of cells usually lying outside of the tips of the protoxylem ridges. MIDDLE CORTEX not preserved. OUTER CORTEX consisting of thin-walled cells at regions of leaf attachment, thick-walled cells between leaf bases form vertically sinuous, subepidermal ridges as seen in longitudinal view. EPIDERMIS preserved only as a coaly zone without cell preservation. Reproductive structures unknown.

TYPE: Department of Botany Paleobotanical Collection Cornell University (DBPC). Type Catalog Number 57 is the holotype. DBPC Type numbers 58, 59, 60 and 61 are paratypes of the original description, as is Cornell University Paleontology Collection (CUPC) No. 40616. Substantiation for the new combination, based on specimens numbered DBPC Type No. 271, 272 of the Cornell Collection and Type Specimens numbered 53, 54, 55 and 56 of the Paleobotanical Collection, State

University of New York at Binghamton (PB-SUNY-B). Substantiation for the anatomy is based upon specimens DBPC 272 and PB-SUNY-B 57, 58 and 59.

ILLUSTRATIONS: Figs 1–21.

REGIONAL AND STRATIGRAPHIC DISTRIBUTION. Detailed locality and horizon information for all older collections were presented in Grierson & Banks (1963) (see Localities 1, 4–6, 13–16, 28, 32). The holotype, DBPC 57, and paratype DBPC 58, re-illustrated here, came from Locality 4. The additional two specimens from the Cornell collections, DBPC 271 and 272, come from Locality 15 of Grierson & Banks (1963). The specimens of both localities come from strata of the Kiskatom formation, equivalent in age to the marine Moscow formation, Tioughniogan stage, Erian Series, Middle Devonian (equals Givetian).

Our most recent collections were made at the construction site of the Blenheim-Gilboa Pumped Storage Power plant, Schoharie County, New York, in strata of the Panther Mountain formation, Tioughniogan stage, Erian Series, approximately equivalent in age to the Middle Givetian of the European sequence. Additional data on this locality is given in Banks *et al.* (1972).

At present *Haskinsia* is restricted geographically to sediments of the Catskill clastic wedge of New York State, and, stratigraphically, to rocks equivalent in age to the Givetian and Lower Frasnian.

COMPARISON WITH RELATED LYCOPODS

Drepanophycaceae: Grierson & Banks (1963) suggested that the genus *Drepanophycus* included disparate taxa which, when fully understood, would have to be relocated. The one distinguishing feature of the genus was its falcate leaf, the basis on which we placed *D. colophyllus* in the genus. However, our discovery of petiolate, laminar, lanceolate, recurved leaves on this latter species sets it apart from the long, parallel margined, laminar leaves of *D. gaspianus* (Grierson & Hueber, 1967) and also *Baragwanathia*; from the small divaricate, spine-like leaves of *D. eximius* (Menéndez, 1965b), and from the thorn-like leaves of *D. spinaeformis* and *D. spinosus* that taper distally from an oval or vertically elongate base. Anatomically, as well, the cylindrical xylem strand of *Haskinsia* with its regularly arranged, exarch protoxylem ridges is distinct from the slender, armed strand reported for *D. gaspianus* (Grierson & Hueber, 1967; Fairon-Demaret, 1977) and for *D. spinaeformis* (Fairon-Demaret, 1971; Hartman, 1981), and from the lobed-strand of *Baragwanathia* reported by Lang & Cookson (1935). The range of tracheid forms found in *Haskinsia*, annular to multiseriate-bordered pitted, is much broader than that recorded for *D. gaspianus* and *D. spinaeformis* (Grierson & Hueber, 1967; Fairon-Demaret, 1971; Hartman, 1981) where only annular, connected annular, and helical-bordered pitted tracheids are found. The same authors have also described peculiar openings in the walls between the annular rings. Nothing comparable with these openings has been observed in *Haskinsia*. Finally, Grierson & Banks (1963) demonstrated rhomboidal patterns produced by subepidermal strengthening cells on the stem surface of *D. gaspianus*. *Haskinsia* has a fusiform pattern.

Protolepidodendraceae: *Haskinsia* is distinguished by its recurved, petiolate, simple, lanceolate, entire leaf from the divaricate once-forked leaf of *Protolepidodendron*, the five-tipped leaf of *Leclercqia*, and the three-forked leaf of *Colpodexylon*. The first three genera are similar in the morphology of their slender dichotomizing axes, the number of leaves in one gyre, the spacing of the leaves, the fusiform pattern on compressed stems caused by thickened cortical cells, and, where known, the ridged xylem column and bordered pitted tracheids. *Colpodexylon* is basically similar in morphology but has more than twice as many leaves in a gyre and its stems are twice as great in diameter. Its xylem strand is lobed, as seen in transverse section, rather than ridged.

Archaeosigillariaceae: Leaves are known in detail only in *Archaeosigillaria vanuxemii* and *A. kidstonii*. Those of the former are 5–6 mm long, simple, possibly petiolate, with a deltoid lamina that bears three or four opposite pairs of small serrations and tapers to an acute apex tipped by a tiny hair (Senkewich, 1956; Fairon-Demaret & Banks, 1978). Leaves of *A. kidstonii* (*Clwydia decussata* of Lacey, 1962) are 12 mm in length with a rhomboid lamina whose upturned distal portion tapers into a long hair (Lacey, 1962). On younger axes the leaves are opposite decussate. Stems of both species are regarded as characterized by hexagonal patterns produced by subepidermal thick-walled cells that surround the areas of attachment of leaves (Lacey, 1962; Grierson & Banks, 1963).

Haskinsia has simple leaves that are 5–6 mm in length, lanceolate, entire, and that lack hairs. Subepidermal cells produce a fusiform rather than hexagonal pattern when impressed on the surface of the stem. The ridged xylem strand of *Haskinsia* is distinct from the slightly lobed strand of *Archaeosigillaria vanuxemii* (Grierson & Banks, 1963) as seen in transverse section.

Leaf bases have been demonstrated by Chaloner *et al.* (1980) for *Archaeosigillaria de-vriesii*. Their imaginative techniques could permit the interpretation that the leaves were petiolate, with simple laminae, but it will require better preserved specimens to determine if the specimens had broad laminae.

Eleutherophyllaceae: We have seen no original material of *Eleutherophyllum* and can offer only subjective speculation. The genus is found in Carboniferous strata and has been reported to have whorled or pseudo-whorled leaves. We assume the latter to be correct. The leaves illustrated by Remy & Remy (1960a) resemble those described by Grierson & Hueber (1967) for *Drepanophycus gaspianus*. Remy & Remy (1960b) and Chaloner (1967) both refer to the possibility of a relationship to *Drepanophycus*. Leaf cushions are said to be present on the stems. However, we see only typical enlarged leaf bases and patterns caused by cortical strengthening cells, not leaf cushions. Some leaves have been reported to be forked. Remy & Remy doubt they are actually attached to the stems and regard the leaves as simple. If *Eleutherophyllum* is related to *Drepanophycus*, we suggest *D. gaspianus* as the one species to which it might be related because of the size of the axes, number and morphology of the leaves, and the slender xylem strand. *Haskinsia* can be distinguished from *Eleutherophyllum* by the number of leaves per pseudowhorl, leaf attitude (strongly falcate), leaf morphology, and the ridged (denticulate) xylem strand.

Lycopodiaceae: *Lycopodites oosensis* (Kräusel & Weyland, 1937) is an herbaceous lycopod that bears spirally-arranged, spatulate leaves only 1–2 mm in length; strengthening tissues in the outer cortex, when exposed, produce a round to angular pattern around leaf bases; sporophylls borne on a peculiar short, thickened lateral branch. *Haskinsia* differs in the size and outline of its leaves and in the fusiform pattern produced by its thick-walled cortical cells. Anatomy is known only for *Haskinsia*, sporophylls only for *Lycopodites*.

DISCUSSION

Lycopods are among the most frequent, if not the most frequent, of Devonian fossils both in numbers of specimens and of localities around the world. Yet, because of the vagaries of preservation and of cleavage of fossiliferous rocks, one often collects only axes showing evidence of the former attachment of small leaves (often referred to as spines) ranged in a low spiral. These one assumes to be lycopods. The oldest representatives of the group are found in early Devonian time unless one accepts the still controversial reports of *Baragwanathia* in Silurian strata (Garratt, 1978; Douglas & Lejal-Nicol, 1981). Younger specimens occur in the intervening rocks on up through post-Famennian time. The Devonian Period is thus the time of initiation of a group of vascular plants whose fossil record is almost continuous to the present day. It is the longest record of any group of vascular plants.

In view of the preceding statement, our experience with *Haskinsia* and a brief look at some other recent studies of Devonian lycopods will demonstrate the gradual unfolding of a picture of a few lycopodiaceous plants as opposed to masses of engimatic fossil specimens.

The specimens we described earlier as *Drepanophycus colophyllus* (Grierson & Banks, 1963) consisted of axes embedded in sandstones. Such coarse matrices preserve the characteristic falcate attitude of the leaves as they are seen in side view along the margin of the specimen. Their margins are usually abraded and some or all of the laminae lost. The finer-grained siltstones and mudstones of our newer collections have preserved both the laminar leaves (Figs 9–11) and surface patterns (Fig. 7) so completely that we have now been able to describe the plant with some precision. Transfer, uncovering (dégagement) and anatomical techniques have revealed a unique combination of characteristics necessitating the transfer of *D. colophyllus* to *Haskinsia*. Repeated efforts have failed to demonstrate either sporangia or ligules.

It must be emphasized that we had collected a wide range of different-appearing specimens and were confronted by the question: how many of these belonged to a single taxon? We learned something about this problem when we showed (Grierson & Banks, 1963) that patterns on lycopod axes were caused by strengthening tissue in the cortex and not by leaf cushions of the lepidodendrid type. Chaloner & Collinson (1975) made a major advance when they erected a new terminology for the various patterns produced by the fracture of rocks containing lycopods. Chaloner *et al.* (1980) continued the explanation of the formation of various casts and molds. We have here added another pattern of cleavage. The result of these analyses is summarized in Fig. 22 where A illustrates a compression specimen and B illustrates the same specimen cleaved along a diameter. In *Haskinsia* this view shows the typical falcate attitude of the leaf (Fig. 1). However, the split frequently occurs at the interface between the stem and its leaves (Fig. 22C). This produces a cleavage compression specimen bearing most of the fossil and a cleavage impression counterpart bearing an impression of the stem and, in its matrix, the leaves. In *Haskinsia* the cleavage compression varies in appearance from axes with leaves along the margin and broken attachments of leaves on the stem surface (Fig. 4, lower) to a surface with raised bosses consisting of leaf bases and bits of petiole bases (Fig. 3, lower; Fig. 8, lower right; Grierson & Banks, 1963, pl. 32, fig. 7). The frequency of cleavage compression specimens contributed to our earlier interpretation of the leaf as non-laminar. Cleavage impression specimens are seen in Fig. 11, areas A & C; Fig. 17. Here one is viewing the cortical region from the inside and what appear to be leaf bases are in fact depressions through which leaf traces extend into the leaves that are embedded in the matrix.

We have added a third fracture pattern to the terminology of Chaloner & Collinson (1975), the leaf laminae cleavage pattern (Fig. 22D). Similar fractures in some species of *Archaeosigillaria* have been discussed by Menéndez (1965a) and Chaloner *et al.* (1980). This pattern results from the rigidity of the leaves, their close proximity to one another, and the fact that their blades recurve so as to lie parallel to the surface of the stem, forming a sheath-like mass of laminae around the stem. If cleavage occurs along the abaxial surface of the leaves, there results a leaf laminae cleavage compression consisting of stems and most of the leaves (Fig. 22, D2; Fig. 10, upper; Fig. 11, Area E). The leaf laminae cleavage impression (Fig. 9) bears only the impression of leaf laminae, or, occasionally a portion of the abaxial surface of the laminae (Fig. 22, Cl).

There is a need for the collection of a large number of specimens and for the application of various techniques to the diverse surface patterns in order to demonstrate that a single taxon, rather than several species or genera are represented.

Haskinsia appears to be a typical herbaceous lycopod with a trailing stem and ascending, dichotomizing axes bearing simple, spirally-arranged microphylls. Its persistent leaves are attached to a smooth stem with no cushions. Its assignment to a family can be based on several primary characteristics that are now available for the three best-known families of Devonian lycopods, Drepanophycaceae, Protolepidodendraceae, and Archaeosigillariaceae. These are:

> Simple as opposed to divided leaves.
> Ridged xylem strands comparable to those of lepidodendrids as opposed to strands that appear variously lobed or rayed in transverse section.
> Patterns caused in compression by subepidermal strengthening tissue.
> Size of axes and the resulting number of leaves in one gyre.
> The complexity of pitting on metaxylem tracheids.

In gross morphology, *Haskinsia* resembles most closely *Protolepidodendron* (Kräusel & Weyland, 1932, 1940; Fairon-Demaret, 1980) and *Leclercqia* (Banks *et al.*, 1972) from which it differs mainly in its simple leaves. Its axis is much more slender than that of *Colpodexylon* (Banks, 1944) whose xylem strand is lobed rather than ridged. Only simple falcate leaves group *Haskinsia* with *Drepanophycus* (Grierson & Hueber, 1967). In the other four characters, the two differ markedly. Similarly *Haskinsia* shares simple, possibly petiolate leaves with *Archaeosigillaria* but differs in all other respects.

Therefore, we have reluctantly erected a new family, Haskinsiaceae, for *Haskinsia* in the hope that continued careful study will lead to an evaluation of the relative taxonomic significance of at least the five characteristics we have discussed.

The ever-changing and improving understanding of herbaceous Devonian lycopods prompts us to mention briefly some recent research that has obviously coloured our thinking about *Haskinsia*.

First was the discovery of *Leclercqia*, a *Protolepidodendron*-like lycopod whose divided leaves ended in five tips (Banks *et al.*, 1972). Its leaves, leaf traces, veins, stomata, sporophylls, sporangia, spores and anatomy were all preserved. Subsequently, Grierson (1976) published a detailed study of tracheids and their pitting, Bonamo & Grierson (1973) reported briefly on the maturational variation of spores in attached sporangia, and Grierson & Bonamo (1979) recorded the remarkable, and unexpected, discovery of ligules in this apparently homosporous lycopod. *Leclercqia* is now the most completely known of the Devonian lycopods.

The expanding knowledge of *Leclercqia* led to questions about the morphology and anatomy of *Protolepidodendron*. Fairon-Demaret (1980), in a notable paper, demonstrated that the material described by Kräusel & Weyland (1932) as *Protolepidodendron scharianum* (material that rapidly became the archetype of *P. scharianum* Krejči (Krejči, 1880)) is in fact *Leclercqia complexa*. She uncovered its five-tipped leaves and its sporophylls with attached sporangia. She embedded bits of mineralized axis to prove that it has a ridged xylem strand. She proposed that all *P. scharianum* reported since 1932 is, like the material of Kräusel & Weyland, probably *Leclercqia*. One example is her unequivocal demonstration (Fairon-Demaret, 1974) that specimens called *P. scharianum* from Queensland, Australia are *Leclercqia*. She proposed that the one acceptable specimen of *P. scharianum* is its holotype in the Narodni Museum, Prague and that the description (Stur, 1882) of leaves that bifurcated at their midpoint with the segments departing at an obtuse angle, be accepted as the concept of *P. scharianum*. She observed such a leaf on the holotype. In contrast, leaves of *Leclercqia* appear to fork closer to their apices. She accepted *P. gilboense* (Grierson & Banks, 1963) whose xylem strand is a solid, ridged cylinder, as a valid species on the ground that its leaves fork distal to the midpoint and fork only once. Fairon-Demaret (1972, 1979) has also demonstrated that many, if not all, specimens assigned to *P. wahnbachense* should be reassigned to her new taxon, *Estinnophyton*, a possible precursor of sphenophylls. One important result of all this work is the recognition that the much-reproduced reconstruction (Kräusel & Weyland, 1940) labelled *P. scharianum* is probably based largely on specimens of *Leclercqia complexa*.

Mustafa (1975) suggested that the triangular xylem strand described by Kräusel & Weyland (1932) for *Protolepidodendron scharianum* (shown by Fairon-Demaret, 1980, to be *Leclercqia*) belonged instead to *Protopteridium thomsoni* (sic). He also claims (Mustafa, 1975; 121) to have new specimens of *Protolepidodendron scharianum* with anatomy like that of *P. gilboense* (Grierson & Banks, 1963). These, in his opinion, invalidate *P. gilboense*, which was established solely because its anatomy differed so sharply from that described by Kräusel & Weyland (1932) for *P. scharianum*. We are unable to accept his conclusions because (a) he has not illustrated his *P. scharianum* nor shown it to have only once-forked leaves, and (b) he has not illustrated the xylem strand, nor demonstrated that it comes from an axis whose leaves fork once only.

Banks & Grierson (1968) reported the occurrence of *Drepanophycus spinaeformis* in Frasnian strata in eastern New York State. Subsequently, Stubblefield & Banks (1978) described stomata of *D. spinaeformis* from Lower Emsian (Gaspé), Middle Givetian (New York), and Upper Frasnian (New York) strata. All stomata were paracytic. Lang (1932) described stomata of the same species from Siegenian strata in Britain. Stubblefield & Banks (1978) interpret Lang's stomata to be paracytic as well. Thus this long-ranging lycopod provides a good example of evolutionary stasis among early vascular plants.

As data such as these continue to accumulate, the conclusion seems inescapable that lycopods arose relatively early in the conquest of the land and quickly reached the evolutionary level that herbaceous forms have maintained throughout the intervening years up to the present.

ACKNOWLEDGEMENTS

We wish to acknowledge Herman Paikoff for help and advice with our photographs and Stanley Kauffman for the drawings in Fig. 22.

REFERENCES

BANKS, H. P., 1944. A new Devonian lycopod genus from southeastern New York. *American Journal of Botany, 31:* 649–659.

BANKS, H. P. & GRIERSON, J. D., 1968. *Drepanophycus spinaeformis* Göppert in the early Upper Devonian of New York State. *Paleontographica B, 123:* 113–120.

BANKS, H. P., BONAMO, P. M., & GRIERSON, J. D., 1972, *Leclercqia complexa* gen. et. sp. nov., a new lycopod from the Middle Devonian of eastern New York. *Review of Palaeobotany and Palynology, 14:* 19–40.

BONAMO, P. M., 1977. *Rellimia thomsonii* (Progymnospermopsida) from the Middle Devonian of New York State. *American Journal of Botany, 64:* 1272–1285.

BONAMO, P. M. & GRIERSON, J. D., 1973. Sporophylls, sporangia and spores of *Leclercqia complexa. American Journal of Botany, 60 (4 supp.);* 16.

CHALONER, W. G. 1967. Lycophyta: 434–802. In E. Boureau, W. G. Chaloner, O. A. Hoeg & S. Jovet-Ast (Eds), *Traité de Paléobotanique, 2: Bryophyta, Psilophyta, Lycophyta.* Paris: Masson et Cie.

CHALONER, W. G. & COLLINSON, M. E., 1975. Application of SEM to a sigillarian impression fossil. *Review of Palaeobotany and Palynology, 20:* 85–101.

CHALONER, W. G., FOREY, P. L., GARDINER, B. G., HILL, A. J., & YOUNG, V. T., 1980. Devonian fish and plants from the Bokkeveld Series of South Africa. *The Annals of the South African Museum, 81:* 127–157.

DOUGLAS, J. G. & LEJAL-NICOL, A., 1981. Sur les premières flores vasculaires terrestres datées du Silurien: Une comparaison entre la "Flore à *Baragwanathia*" d'Australie et la "Flore à Psilophytes et Lycophytes" d'Afrique du Nord. *Comptes rendus hebdomadaires des Séances de l'Académie des sciences, Paris, Série D, 292:* 685–688.

FAIRON-DEMARET, M. 1971. Quelques caractères anatomiques du *Drepanophycus spinaeformis* Göppert. *Comptes rendus hebdomadaires des Séances de l'Académie des sciences, Paris, Série D, 273:* 933–935.

FAIRON-DEMARET, M. 1972. Les feuilles fertiles de *Protolepidodendron wahnbachense* Kräusel, et Weyland, 1932 du Dévonian Inférieur de Belgique. *Bulletin de l'Institut royal des sciences naturelles de Belgique (Sciences de la Terre), 48:* 1–9.

FAIRON-DEMARET, M., 1974. Nouveaux spécimens du genre *Leclercqia* Banks, H. P., Bonamo, P. M. et Grierson, J. D. 1972, Du Givetien (?) du Queensland (Australie). *Bulletin de l'Institut royal des sciences naturelles de Belgique (Sciences de la Terre), 50:* 1–4.

FAIRON-DEMARET, M., 1977. À propos de certains spécimens de *Drepanophycus gaspianus* (Dawson) Stockmans, F., 1939, du Dévonien Inférieur de Belgique. *Bulletin de l'Académie royale de Belgique (Classe des Sciences), 68:* 781–790.

FAIRON-DEMARET, M., 1979. *Estinnophyton wahnbachense* (Kräusel et Weyland) comb. nov., une plante remarquable du Siegenien D'Allemagne. *Review of Palaeobotany and Palynology, 28:* 145–160.

FAIRON-DEMARET, M., 1980. À propos des spécimens déterminés *Protolepidodendron scharianum* par Kräusel et Weyland. 1932. *Review of Palaeobotany and Palynology, 29:* 201–220.

FAIRON-DEMARET, M. & BANKS, H. P., 1978. Leaves of *Archaeosigillaria vanuxemii*, a Devonian lycopod from New York. *American Journal of Botany, 65:* 246–249.

GARRATT, M. J., 1978. New evidence for a Silurian (Ludlow) age for the earliest *Baragwanathia* flora. *Alcheringa, 2:* 217–224.

GRIERSON, J. D., 1976. *Leclercqia complexa* (Lycopsida, Middle Devonian): Its anatomy, and the interpretation of pyrite petrifactions. *American Journal of Botany, 63:* 1184–1202.

GRIERSON, J. D. & BANKS, H, P., 1963. Lycopods of the Devonian of New York State. *Palaeontographica Americana, 4:* 220–278.

GRIERSON, J. D. & BONAMO, P. M., 1979. *Leclercqia complexa:* earliest ligulate lycopod (Middle Devonian). *American Journal of Botany, 66:* 474–476.

GRIERSON, J. D. & HUEBER, F. M., 1967. Devonian lycopods from northern New Brunswick. In D. H. Oswald (Ed,), *International Symposium on the Devonian System Calgary, 2: 823–836*. Canada: Alberta Society of Petroleum Geologists.

HARTMAN, C. M., 1981. The effect of pyrite on the tracheid structure of *Drepanophycus spinaeformis*, a long-ranging Devonian lycopod. *Review of Palaeobotany and Palynology, 32:* 239–255.

KRÄUSEL, R. & WEYLAND, H., 1932. Pflanzenreste aus dem Devon, II. *Senckenbergiana, 14:* 185–190.

KRÄUSEL, R. & WEYLAND, H., 1937. Pflanzenreste aus dem Devon. X. Zwei Pflanzenfunde im Oberdevon der Eifel. *Senckenbergiana 19:* 338–355.

KRÄUSEL, R. & WEYLAND, H., 1940. Die Gattung *Protolepidodendron* Krejči. *Senckenbergiana Lethaea, 22:* 6–16.

KREJČI, J., 1880. Notiz über die Reste von Landpflanzen in der Böhmischen Silurformation. *Königliche Böhmische Gesellschaft der Wissenschaften in Prague, sitzungberichte Jahrgang, 1879:* 201–204.

LACEY, W. S., 1962. Welsh Lower Carboniferous plants. I. The flora of the lower Brown Limestone in the Vale of Clwyd, North Wales. *Palaeontographica B, 111:* 126–160.

LANG, W. H. 1932. Contributions to the study of the Old Red Sandstone Flora of Scotland. VIII. On *Arthrostigma, Psilophyton* and some associated plant-remains from the Strathmore Beds of the Caledonian Lower Old Red Sandstone. *Transactions of the Royal Society of Edinburgh, 57:* 491–521.

LANG, W. H. & COOKSON, I. C., 1935. On a flora, including vascular land plants, associated with *Monograptus*, in rocks of Silurian age, from Victoria, Australia. *Philosophical Transactions of the Royal Society, London, Series B, 224:* 421–449.

LECLERCQ, S., 1960. Réfendage d'une roche fossilfère et dégagement de ses fossiles sous binoculaire. *Senckenbergiana Lethaea, 41:* 483–487.

MENÉNDEZ, C. A., 1965a. *Archaeosigillaria conferta* (Frenguelli) nov. comb. del Devònico de la Quebrada de la Chavela, San Juan. *Révista de la Asociación Paleontológica Argentina. Ameghiniana, 4:* 67–68.

MENÉNDEZ, C. A., 1965b. *Drepanophycus eximius* (Frenguelli) nov. comb. del Devónico de la Quebrada de la Chavela, San Juan. *Révista de la Asociación Paleontológica Argentina. Ameghiniana, 4:* 139–140.

MUSTAFA, H., 1975. Beiträge zur Devonflora I. *Argumenta Palaeobotanica, 4:* 101–133.

REMY, R. & REMY, W., 1960a. *Eleutherophyllum drepanophyciforme* N. Sp. (aus dem Namur A von Niederschlesien). *Senckenbergiana Lethaea, 41:* 89–100.

REMY, W. & REMY, R., 1960b. *Eleutherophyllum waldenburgense* (Stur) Zimmerman. *Monatsberichte der Deutschen Akademie der Wissenschaften zu Berlin, 2:* 54–62.

SENKEWICH, M. A., 1956. New data on the flora of the Middle Devonian of North Kazakhstan. *Doklady Akademii Nauk SSSR,* [*Reports of the Academy of Sciences of the U.S.S.R.,*] *D106:* 342–344.

STUBBLEFIELD, S. & BANKS, H. P., 1978. The cuticle of *Drepanophycus spinaeformis*, a long-ranging Devonian lycopod from New York and eastern Canada. *American Journal of Botany, 65:* 110–118.

STUR, D., 1882. Die Silur-Flora der Etage H-h₁ in Böhmen. *Sitzungsberichte der kaiserlichen Akademie der Wissenschaften in Wien, 84:* 330–391.

Botanical Journal of the Linnean Society (1983), *86:* 103–123. With 29 figures

A probable pteridosperm from the uppermost Devonian near Ballyheigue, Co. Kerry, Ireland

BRUCE I. MAY

Amoco Production Company, 500 Jefferson Building, P.O. Box 3092, Houston, Texas 77001, U.S.A.

AND

LAWRENCE C. MATTEN, F.L.S.

Department of Botany, Southern Illinois University, Carbondale, Illinois 62901, U.S.A.

Received August 1981, accepted for publication May 1982

A new genus of pteridosperms is described from the uppermost Devonian beds from Ballyheigue, Ireland. *Laceya hibernica* May & Matten is represented by stems bearing spirally arranged fronds. The base of the frond is swollen and is about the same size as the stem. Pinnae are borne alternately and apparently in one plane on the rachis. The anatomy of the stem reveals a three-fluted protostele. Secondary xylem consists of rays and tracheids and secondary phloem is present. The inner cortex contains probable secretory and/or sclerotic cells. The outer cortex is of the sparganum-type. Rachial trace formation is described. The U-shaped xylem strand of the rachis lacks secondary tissue. Pinnae traces are V- to C-shaped. A presumed adventitious root has a triarch protostele, a parenchymatous cortex and lacks a 'sparganum' outer zone. *Laceya* is believed to be a member of the Lyginopteridaceae. The divisions of the sympodial protoxylem strand forming the rachial trace is compared among the Aneurophytales, Buteoxylonaceae, Calamopityaceae and Lyginopteridaceae and is shown to be similar.

KEY WORDS:—Carboniferous – Devonian – Ireland – *Laceya* – Lyginopteridales – pteridosperms.

CONTENTS

INTRODUCTION

In a brief report on the uppermost Devonian flora from Ballyheigue, Ireland, reference was made to a new plant showing resemblance to the pteridosperm genera, *Triradioxylon*, *Tristichia* and *Lyginorachis* (Matten *et al.*, 1980b). This same plant was mentioned at the Palaeontological Association's Devonian System

103

0024–4074/83/010103+21$03.00/0

8

Symposium at Bristol in 1978 and the Botanical Society of America's meeting at Blacksburg, Virginia (May, Matten & Lacey, 1978). This appears to be a most appropriate occasion and place to formally name and describe this plant.

Professor William S. Lacey has been a major figure in the study of the uppermost Devonian flora from southern Ireland (Matten, Lacey & Edwards, 1975, 1976; Matten, Lacey & Lucas, 1978, 1980a; May et al., 1978; Matten et al., 1980b; Matten & Lacey, 1981). In 1973, he organized the first exploratory trip to southern Ireland with John Richardson (BMNH) and Lawrence Matten. He made subsequent trips to Ireland in 1974, 1975 (with Matten) and 1976. His contacts with Bridge (then at Belfast), Van Veen (Utrecht) and Matten (Carbondale) led to a joint study on the sedimentation, palynology and palaeobotany at Kerry Head (Bridge, Van Veen & Matten, 1980). The contribution to palaeobotanical techniques by Professor Lacey has changed the trickle of papers on coal ball studies into the flood of contributions on Carboniferous floras. The cellulose acetate peel technique (Joy, Willis & Lacey, 1956) was developed by Lacey and a visiting student, Joy (from Bristol), in Lacey's laboratory in Bangor. This technique has been of major importance to our study of the new Irish plant. In recognition of this, and especially of our Irish studies, we dedicate this paper to Professor William S. Lacey.

LOCALITY

The new plant is found in an outcrop 35 m north of the beach at Ballyheigue, County Kerry (52°23′20″N, 9°50′30″W). The outcrop is covered during the peak of the high tide but is exposed (in part) during the 3 h preceding and following the low tide. The beach at the foot of the carpark at Ballyheigue is composed of sand. Northwards, a red bed is first encountered, underlain by yellowish to green to gray sandstones and siltstones. The plant-bearing layer is approximately 3 m below the red bed.

The sedimentology and palynology of this locality have been examined (Bridge et al., 1980). The plant-bearing layer occurs within a facies association comprised of interbedded fine to very fine sandstones and siltstones, displaying a diversity of sedimentary structures, rapid lateral and vertical facies transitions, and occurring as broad sheets or channel-filling sequences. This association is interpreted as representing crevasse, levée and channel-fill deposits. The Ballyheigue macro- and microflora indicate that major plants grew in areas very close to palaeochannel margins and the age for this flora is late Famennian – lowermost Tournaisian (Fa 2d – Tn 1b; uppermost Devonian). The flora is within the *Retispora lepidophyta* zone of Van der Zwan & Van Veen (1978). Plants described from this flora include: cupulate *Hydrasperma tenuis*, *Buteoxylon gordonianum*, *Lyginorachis* sp., a small herbaceous lycopod or coenopterid fern, wood fragments and the plant to be described here (Matten et al., 1980a, et al., b). Stratigraphically, the petrified plant zone lies within the upper portion of the Coomhola Formation of Gardiner & Horne (1972) which is equivalent to the Transition Formation or Lack Beds of House et al. (1977) or to the Kilmore Sandstone Formation of Kahn (1955).

MATERIAL AND METHODS

The plants are preserved as silicified petrifactions in an arkosic siltstone. The petrifactions are often visible on the surface of the rock and may help show growth form and relationship of parts. Dégagement was used to show organic connection between the stem and petioles of *Laceya*. This technique involved the use of a small hammer, dissecting needles and small chisels to remove the matrix and uncover the plant parts.

Rocks were cut into 2–4 cm thick slabs. Each surface was polished and etched in 24% HF for approximately 13 min. After washing for a minimum of 30 min, the specimens were dried and peeled (Joy *et al.*, 1956) using 3 mil cellulose acetate sheets. This technique was favoured over the preparation of ground thin-sections because of the increased number of sections possible. However, ground thin-sections were occasionally prepared. Etching times varied somewhat with the different degree of cementation and particle size of the matrix. In addition, some specimens appear to be more amenable to the peel technique, yielding fine detail in the peel. Other specimens produce poor peels, but show fine detail while the peel is still on the etched surface. Therefore, many photographs were taken prior to removal of the peels from the rock surface.

Twelve specimens were selected for study and description. Each of the specimens shows an identifiable stem of *Laceya*. The other organs attributable to this new genus are found in organic connection to one or more of these identifiable stems. Isolated petioles, pinnae and rachises, assignable to the form genus, *Lyginorachis*, will be described at a later time.

SYSTEMATICS

Laceya May & Matten **genus novum**

DERIVATION: Named in honour of Professor W. S. Lacey.

SYNONYMY: Plant B, Matten *et al. Review of Palaeobotany and Palynology*, *29:* 243 (1980b).

Probable pteridosperm possessing stems, petioles, pinnae and roots. STEMS bearing compound fronds in a 1/3 phyllotaxis; cambial activity presumed, producing secondary xylem and secondary phloem present in the stem and absent in the frond and root; 'Sparganum' outer cortex present in the stem, rachis and pinna; absent in the root; primary xylem of stem three-fluted with a mesarch protoxylem strand in each of the ridges; secondary xylem manoxylic with rays separated by 1–4 radial rows of tracheids. RACHIAL TRACE INITIATION preceded by repeated divisions (dichotomous or pseudomonopodial) of the protoxylem in the ridge of primary xylem, accompanied by increased fluting of the ridge until in cross-section the end of the primary xylem arm is 6- to 8-lobed with a protoxylem area in each lobe. RACHIAL TRACE U-shaped with the outer, convex surface (abaxial) corrugated and the inner, concave surface (adaxial) smooth; protoxylem strands on the abaxial surface, one in each corrugation. RACHIS swollen (pulvinus-like) at base, bearing alternately arranged pinnae in a distichous manner; xylem strand of the rachis U-shaped with the ends bent adaxially towards one another. PINNA TRACE FORMATION is marginal with xylem segment containing two protoxylems separating from the end of the rachial strand; xylem of pinna V-shaped. ADVENTITIOUS ROOTS, triarch having only primary tissues, present.

Laceya hibernica May & Matten **sp. nova.**

DERIVATION: From the latin for Ireland.

SYNONYMY: Plant B, Matten *et al. Review of Paeaeobotany and Palynology, 29:* 243 (1980b).

Plants compatible with the generic diagnosis. Sclerenchyma strands in outer cortex exhibit infrequent anastomoses. Inner cortex parenchymatous with clusters of dark-filled cells (secretory?). Secondary phloem of thin-walled, radially-aligned cells and rays. Secondary xylem tracheids are 4- to 6-sided in cross-section, tangential diameter 16–32 ($\bar{x}$ = 23) μm radial diameter 26–53 ($\bar{x}$ = 37) μm. Tracheids with 1- to 4-seriate crowded, alternate, circular-bordered pitting on tangential and radial walls. Pits elongate in the horizontal plane, mean dimensions 10×20 μm. Tracheids 0.3–2.3 ($\bar{x} = 1$) mm long. Rays 1- to 6-seriate, *c.* 50×730 μm–160×2700 μm. Metaxylem tracheids rectangular in cross-section, varying in size from 21×37 μm near the protoxylem to 37×96 μm near the secondary xylem. Mean diameter of protoxylem tracheids in cross-section 15 μm; rachises decurrent, often overlapping forming an outer covering of the stem. Swollen rachial bases up to 21 mm in diameter. Clusters of dark-coloured cells surrounding the xylem strand in the rachis. Exarch protostele of adventitious root 0.25 mm wide; entire root 1.2 mm wide.

TYPE: Holotype. Specimen SIPC (Southern Illinois University—Carbondale Palaeobotanical Collection) 666.4. The collection is housed in the Department of Botany. PARATYPES: SIPC 666.3; 666.5; 666.13; 666.14; 666.15; 666.16; 666.19; 666.66; 666.201; 666.238; 666.239.

LOCALITY: Sandstone exposure 35 m north of the beach at Ballyheigue, Co. Kerry, Ireland (52°23′20″N; 9°50′30″W).

HORIZON: Coomhola Formation of Gardiner & Horne (1972) or Transition Formation of House *et al.* (1977), Uppermost Devonian (Fa 2d – lower part Tn 1b).

DESCRIPTION

External morphology: The holotype consists of a stem, 100 mm long and about 12 mm wide, that bears two rachises (Fig. 2). The lower rachis is 155 mm long and the second rachis is 15 mm long (the ends of both being broken off). A pinna is attached 71 mm from the base of the first rachis and is 5 mm long and 5 mm wide. The plant has been secondarily flattened parallel to the bedding plane. Because the plant was exposed on the surface of the rock it was subject to extensive weathering and as a result some of the plant has been destroyed. This is true for an area of the stem below the second rachis and for the pinna on the first rachis.

There is a slight taper of the stem of approximately 1–2 mm over its length. The internode between the first and second rachis is *c.* 43 mm. The rachises form angles with the stem of 72° and 91° respectively. Based on observations of large, *in situ* specimens at Ballyheigue, it appears that the stem (i.e. stems assumed to be *L. hibernica*) forms a zig-zag pattern as the fronds are produced (i.e. there is a deflection of the stem at the node). This pattern is reminiscent of the pattern found in *Tristichia ovensi*, Long (1962). The base of the rachis is swollen into a pulvinus-like structure and may in fact be almost equal to the stem in diameter (Fig. 1). The base of the lower rachis is *c*. 15 mm in diameter. The diameter of this rachis is only 5 mm *c*. 35 mm above its base, indicating a sharp reduction in size. The taper over the remaining 120 mm of the rachis is only 0.5 mm.

The black, outer surface of the stem shows infrequently anastomosing longitudinal ridges (Fig. 3). This is the external expression of the sclerenchyma in the 'sparganum' outer cortex. The low areas between the longitudinal ridges represent areas of parenchyma. This relief is the result of the weathering of the epidermis and is not the true expression of the surface of the plant which we believe to have been smooth or covered by leaf bases. The outer surface of the rachis also possesses the parallel ridges of the 'sparganum-type' of outer cortex.

Stem anatomy: Stems in our collection range in diameter from 8 to 23 mm, those showing little secondary growth are 8–13 mm wide, those with attached petiole bases are up to 23 mm in diameter. Pieces of wood tissue believed to belong to *Laceya* are even broader (over 30 mm).

The primary xylem of the stem is a three-fluted protostele (Figs 4, 5, 7, 24, 28). In cross-section, the mean length of each arm is 1.8 mm but this value will vary with the stage of trace formation. Swelling and lobation of the end of the primary xylem arm occurs during trace formation.

There is a protoxylem area located in each arm (Figs 5, 8, 13, 17, 19, 20). These areas are generally round and situated a little more than midway between the centre of the stele and the end of each arm. The protoxylem tracheids are thin-walled and have a mean diameter of 15 μm. No definite lateral wall thickenings have been identified in longitudinal sections, although there are some indications of spiral or annular thickenings in some poorly-preserved sections.

The metaxylem tracheids are generally thick-walled and as a result are more distinct than the walls of the protoxylem tracheids (Figs 8, 13). The tracheids are generally rectangular in shape, although this may be due to compression. The mean diameter of the elements in cross-section is 28–60 μm. Lateral wall thickenings, tracheid length and pitting have not been determined. There is no indication of any parenchyma in the metaxylem of any of the sections examined.

Secondary xylem is present in all of our specimens, but is well-developed in only two stems identifiable as *L. hibernica*. In the best example of secondary growth, the secondary xylem zone measures, radially, *c*. 9 mm (Fig. 21). Growth rings are absent. The secondary xylem is composed of rays and tracheids. The tracheids are grouped in files 1–6 cells wide and are separated by vascular rays (Fig. 14). The tracheids are rectangular to hexagonal in cross-section.

Pitting on both the radial and the tangential walls is crowded, circular-bordered (Figs 10, 18). Some difficulty was encountered in determining the tracheid length because of the angle of sectioning and preservation. As a result, it is not known if the complete tracheid was measured. Where tracheid end walls were recognizable, they appeared to taper.

Vascular rays are quite well-developed in some cases. In cross-section, the rays are relatively uniform and expand slightly at the periphery of the wood tissue. Most rays are multiseriate but a few uniseriate rays have been seen (Fig. 11). Although often incomplete, the larger rays are over 160 μm wide and over 2700 μm long. Uniseriate rays had mean dimensions of 50 μm × 730 μm. In general, the multiseriate rays taper very gradually at their ends but are uniform in width

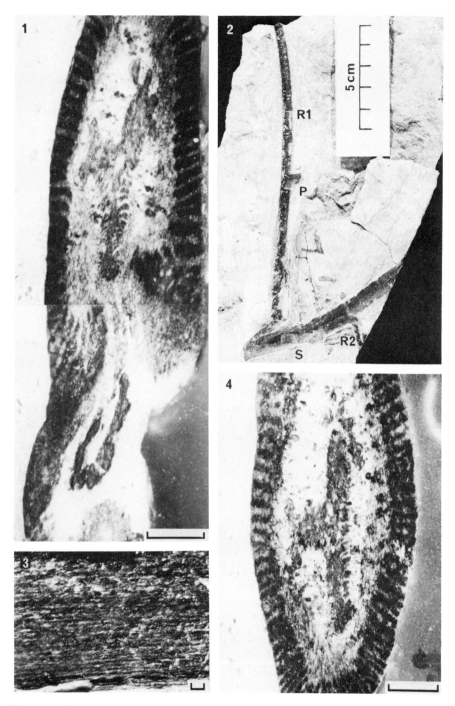

Figures 1–4. *Laceya hibernica*, holotype (SIPC 666.4). Fig. 1. Transverse section of stem with U-shaped rachial trace; section is at level just below the attachment of the second rachis (R2) shown in Fig. 2, etched surface A_b–B_e; scale bar equivalent to 1 mm. Fig. 2. Surface view of specimen showing stem (S), lower rachis (R1) with pinna (P) and upper rachis (R2), scale bar equivalent to 50 mm. Fig. 3. Surface of stem showing parallel striations due to sparganum outer cortex; scale bar equivalent to 1.5 mm. Fig. 4. Transverse section of stem showing bilobed xylem arm (left) and rachial trace (top), etched surface $B_b B_a$; scale bar equivalent to 1 mm.

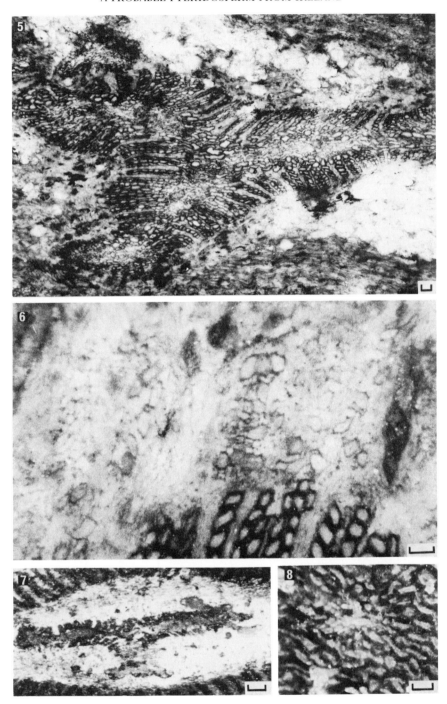

Figures 5–8. *Laceya hibernica*. Fig. 5. Transverse section of stem, SIPC 666.13, peel F–1–A; scale bar equivalent to 100 μm. Fig. 6. Transverse section of stem showing secondary phloem and outermost cells of secondary xylem (bottom, centre), SIPC 666, 14, peel C–3–D; scale bar equivalent to 100 μm. Fig. 7. Transverse section of stem showing ends of xylem arms in various stages of trace formation; the incipient trace on the right has four lobes and six protoxylems, SIPC 666.4, etched surface B_d–B_c; scale. bar equivalent to 300 μm. Fig. 8. Transverse section through protoxylem in arm of stem stele, SIPC 666.4, etched surface B_a–B_b; scale bar equivalent to 50 μm.

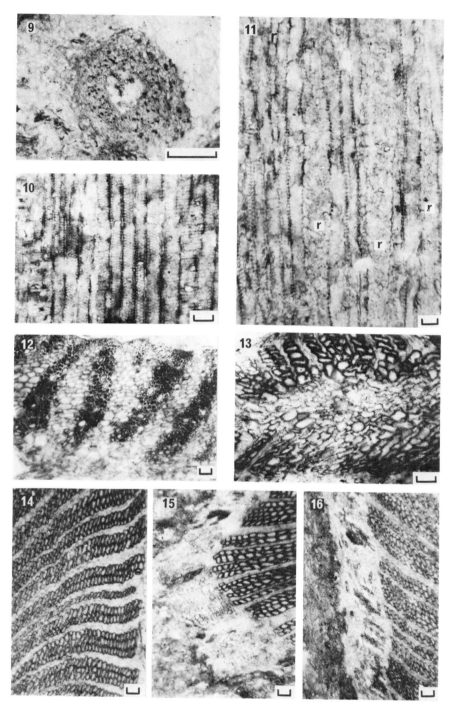

Figures 9–16. *Laceya hibernica*. Fig. 9. Transverse section of root, SIPC 666.14, peel C–3–D; scale bar equivalent to 500 μm. Fig. 10. Radial section of wood, SIPC 666.204, scale bar equivalent to 1 mm. Fig. 11. Tangential section showing rays (r), SIPC 666.204; scale bar equivalent to 500 μm. Fig. 12. Sparganum outer cortex of stem, SIPC 666.14, peel C–3–D; scale bar equivalent to 1 mm. Fig. 13. Transverse section of stem showing primary xylem with a mesarch protoxylem area and secondary xylem, SIPC 666.13, peel F–1–A; scale bar equivalent to 1 mm. Figs 14–16. All to same magnification as Fig. 12. Fig. 14. Transverse section through secondary xylem of stem, SIPC 666.14, peel C–3–D. Fig. 15. Transverse section through secondary xylem and secondary phloem of stem, SIPC 666.14, peel C–3–D, Fig. 16. Transverse section of stem showing secondary xylem, secondary phloem and inner cortex, SIPC 666.14, peel C–3–D.

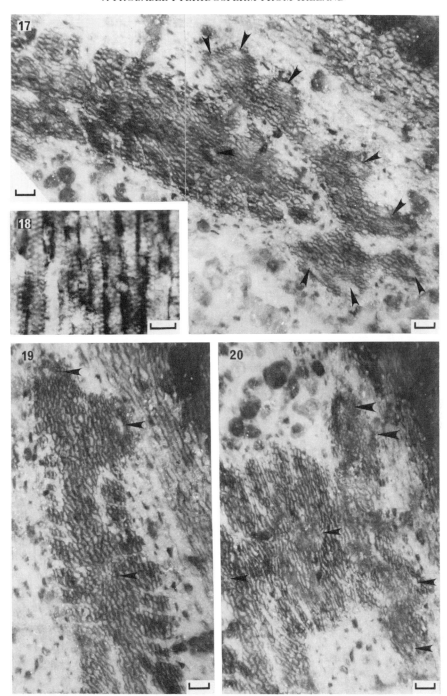

Figures 17–20. *Laceya hibernica*; scale bars equivalent to 100 μm except in Fig. 18. Fig. 17. Transverse section of stem showing rachial trace adjacent to xylem arm. Eight protoxylem areas (arrows) are present in the trace, SIPC 666.4, etched surface B_b–B_a. Fig. 18. Radial section of secondary xylem showing multiseriate pitting, SIPC 666.204; scale bar equivalent to 50 μm. Fig. 19. Transverse section of stem showing bilobed xylem arm with two protoxylem areas (arrows), SIPC 666.4, etched surface B_b–B_a. Fig. 20. Transverse section of stem showing bilobed xylem arm with four protoxylem areas (arrows), SIPC 666.4, etched surface B_b–B_a.

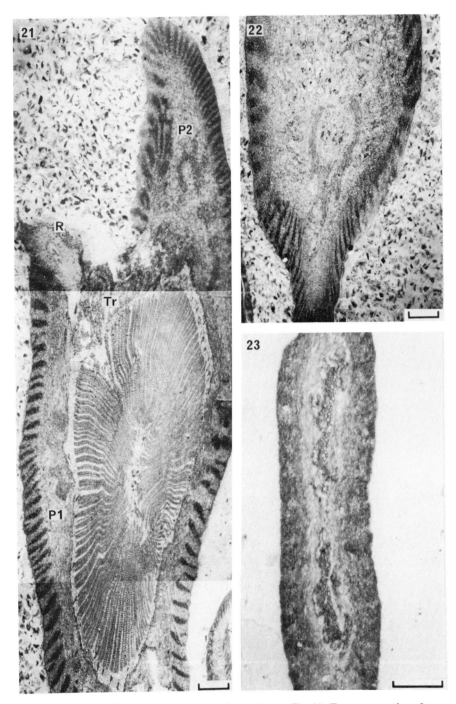

Figures 21–23. *Laceya hibernica*; scale bars equivalent to 1 mm. Fig. 21. Transverse section of stem showing extensive area of secondary xylem, two attached rachial traces (P1, P2), root (R) and a trace (Tr) in the cortex, SIPC 666.14, peel D–1–C. Fig. 22. Transverse section through swollen base of a rachis, SIPC 666.14, peel B–1–A. Fig. 23. Transverse section of isolated rachis showing divided xylem strand. This specimen is thought to probably belong to *Laceya hibernica* and provides evidence for a forking of the rachis, SIPC 666.6.

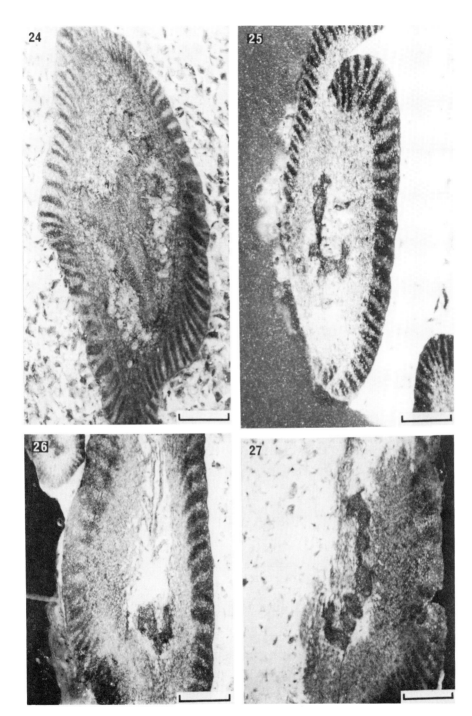

Figures 24–27. *Laceya hibernica*; all to same magnification, scale bars equivalent to 1 mm. Fig. 24. Transverse section of stem, SIPC 666.13, peel F–1–A. Fig. 25. Transverse section of isolated rachis, SIPC 666.5, peel G–1–F. Fig. 26. Transverse section of rachis, SIPC 666.4, etched surface B_c–B_g. Fig. 27. Transverse section of rachis, SIPC 666.4, etched surface B_g–Sur.

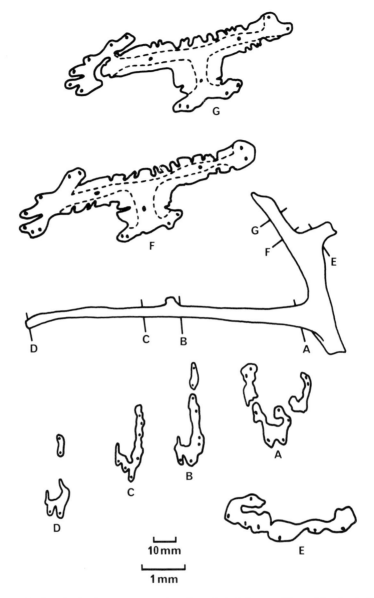

Figure 28. Outline of xylem in transverse sections through the holotype of *Laceya hibernica*. A–D Sections of the primary xylem through the lower rachis. E. An oblique section through the upper rachial base. F & G. Sections through the stem; the black areas (points) represent the position of the protoxylem within the primary xylem; the dotted lines represent contact of the secondary xylem with the primary xylem. Axis outline top scale bar, sections bottom scale bar.

throughout the central portion. In radial section, the ray cells are rectangular in shape and procumbant (Fig. 10). The mean dimensions of the cells are 14 μm wide and 29 μm long. Occasionally, in longitudinal section, ray cells with dark contents are seen. These cells may be secretory in nature or the dark colour may be artifact from the petrifying mineral.

The development of the secondary xylem is greatest opposite the sinuses of the protostele and least opposite the ridges of the protostele (Fig. 5). This indicates a developmental history similar to that commonly seen in roots. That is, the vascular cambium first appears along the flanks and in the sinuses of the protostele. This is followed by the development of a continuous vascular cambium and differential growth rates until a cylindrical cambium is achieved.

Some specimens reveal the presence of light-coloured, thin-walled cells that are radially aligned in the position of the secondary phloem (Figs 6, 15, 16). A narrow zone of small cells separates the secondary phloem from the secondary xylem. These cells, again based on position, may represent the remnants of the vascular cambium.

The inner cortex consists of parenchyma cells which are variable in size. The largest parenchyma cells tend to occur toward the inner portion of the cortex, which is often poorly preserved. Parenchyma cells in cross-section have mean dimensions of $23 \times 33 \mu m$. Differential preservation and crystal growth often give the impression that sclerotic nests or clusters of secretory cells are present. Close examination of some of the cells show that some walls of a cell may be light and other walls may be dark-coloured. The fact that the cells are clustered and their walls are slightly thicker than the surrounding cells leads us to interpret them as sclerotic. Some isolated cells appear to have darker contents than the surrounding cells. This dark amber colour may indicate a secretory function for the cell. But it should be noted that several dark, brown-coloured crystals can be observed in various parts of the cortex. If the interpretation of secretory cells in the rays of the secondary xylem is correct, it seems likely that secretory cells also occur in the cortex.

The outer cortex typically shows a 'sparganum'-type of arrangement of parenchyma and sclerenchyma (Figs 1, 4, 12, 21, 24). The parenchyma cells of the outer cortex are generally smaller than those parenchyma cells found in the inner cortex and they are often radially aligned. The parenchyma cells have mean dimensions in cross-section of $25 \times 15 \mu m$. The parenchyma cells are probably deformed by compression from an isodiametric form.

The sclerenchyma cells appear to be elongated longitudinally, but it is difficult to determine their full length. In cross-section, the mean dimensions of the sclerenchyma cells are $19 \times 25 \mu m$. There seems to be little variation in the diameter of the sclerenchyma cells but there is a great deal of variation in the size of the sclerenchyma bands in the outer cortex. The width of the sclerenchyma bands is $100–200 \mu m$ and the radial dimension is $500–900 \mu m$. Although no definite trend in length or width of the sclerenchyma bands can be seen, the number of bands seems to decrease as one progresses towards the apex. As many as 96 sclerenchyma bands have been counted in cross-section.

Rachial trace formation: The method of trace formation in *L. hibernica* is a unique character. The sequence of stages was assembled from numerous sections of many specimens. However, most of the stages can be seen in any of the specimens (Figs 4, 7, 24). The sequence begins with a tangential division of the protoxylem in the arm of the protostele (producing two radially-oriented poles; for convenience, the description will be based on the appearance of the various parts as seen in cross-section). The distal protoxylem divides again radially producing two tangentially-orientated poles (Fig. 19). The two new protoxylems assume an almost bipolar appearance at the tip of the primary xylem arm. At the next level, the primary xylem arm is bilobed with a protoxylem in each lobe. Above this, the two protoxylems divide again radially so that each lobe at the end of the xylem arm has two protoxylems (Figs 7, 20). This stage is followed by gradual lobation separating each protoxylem strand into its own lobe. At this level there are four lobes at the end of the xylem arm (Fig. 7). Then the protoxylems in the two outer (peripheral) lobes divide again. The protoxylems of the two interior lobes do not divide and remain unchanged. At this stage of trace formation, there are six protoxylem strands in a four-lobed primary xylem arm. Further lobation continues to produce a six-lobed end of a primary xylem arm (Fig. 17). The two outermost protoxylems divide once more making a total of eight protoxylems in a six-lobed tip of a primary xylem arm. At this stage, the trace separates from the arm of the primary xylem (Fig. 4). Departure of the trace is simultaneous, and as a result, the trace immediately assumes a U-shape (Fig. 1).

A subsequent division of the inner protoxylem strand in the arm of the protostele starts the process again. Several specimens have been observed showing each of the three xylem arms in a different stage of trace formation. In addition, several specimens show overlapping bases of rachises still attached to the stem, one opposite each of the three xylem arms. Thus it seems likely that the traces are given off in a 1/3 phyllotaxy. In addition, it should be noted that in some specimens the individual rachises are separated by considerable internodal lengths while in others the rachises overlap. These differences are thought to be environmentally induced.

Some specimens show the rachial trace in the cortex or rachis base as being composed of two or more distinct and separate xylem units (Fig. 28). In most cases this is interpreted as being caused by compression and is a resultant preservation artifact. However, in other cases it appears that the separation of the trace into smaller units is not artifact. Such an occurrence is to be expected. It occurs in several Lower Carboniferous genera such as *Pitus* (Long, 1963), *Calathopteris* (Long, 1976) and *Kalymma* (Galtier, 1974).

Rachis anatomy: The base of the rachis is inflated and approaches or equals the diameter of the stem (Figs 1, 21, 22, 27). Four measured rachis bases range from 10 to 21 mm in diameter. In each case the size of the rachis base was within 1 to 2 mm of the stem diameter. Above their swollen bases, the rachises are *c*. 5 mm wide. The outer cortex is of the 'sparganum'-type. The inner cortex possesses parenchyma cells having mean dimensions 25×69 μm. Some clusters of dark cells have been seen, mostly paralleling the concave surface of the xylem strand.

The convex (abaxial) surface of the U-shaped vascular bundle (Figs 22, 27, 28) is corrugated with a mesarch protoxylem in each corrugation. As the vascular bundle becomes more distal in the rachis it appears to become more angular (Fig. 26). The ends are bent inwards towards each other (Fig. 25). The vascular bundle decreases in size as it becomes more distal, but seems to contain about the same number of protoxylem areas throughout its observed length. In the holotype, the length of the U-shaped bundle in cross-section decreased from 3.4 mm near the base to 2.4 mm below the first pinna (a distance of 70 mm). The metaxylem tracheids vary greatly in dimensions (from 11×11 to 43×73 μm). The mean dimensions of the metaxylem tracheids, in cross-section, is 20×34 μm. No parenchyma is present in the primary xylem.

Bifurcation of the rachis is a key character belonging to many lyginopterid seed ferns. Bifurcating rachises have not been found yet in connection to identifiable stems of *L. hibernica*. However, compressions and petrifactions of dividing rachises are common in our collection. One such forking rachis is illustrated (Fig. 23). The forking of the rachis in *Laceya* (if it did occur) must have occurred at a considerable distance from its insertion and above the level of the first few pinnae. This is commonly the case in *Sphenopteris*.

Pinna trace formation: Pinna trace formation is the result of the elongation and separation of a terminal portion of the xylem strand of the rachis (Fig. 28). There does not appear to be any division of the protoxylem strands prior to trace formation. The trace appears to have two mesarch protoxylem strands and is broadly U-shaped in outline. It does not have the ends of the vascular bundle bent inwards as is the case in the vascular bundle of the rachis. This type of trace departure is termed marginal (Calder, 1935). After separation of the trace, the xylem strand of the rachis is somewhat J-shaped. It retains this configuration until a second pinna trace is formed from the longer arc of xylem. The arrangement of at least the first two pinnae is alternate and distichous.

Pinna anatomy: As noted in the description of the morphology of the holotype, a pinna is attached to the rachis. This pinna is terete in cross-section and has a U-shaped xylem strand with two protoxylem areas located at each end of the strand. The dimensions of the primary xylem elements are similar to that of the rachis elements. The shape and size of the pinna indicates that it was probably not the ultimate unit of the frond but rather bore secondary pinnae and/or pinnules. Numerous detached axes having the same xylem morphology and overall size are present in our collection. Some of these axes bear laterals. The frond of *L. hibernica* was probably twice or thrice compound.

Root: One of the specimens shows a root (Figs 9, 21). It first appears in the inner cortex of the stem adjacent to a rachial trace. It leaves the stem at a level *c.* 5 mm above its first appearance. Outside the stem it appears to divide. At this point the root disappears from our sections. The root is 1.2 mm wide and has a triarch, exarch primary xylem strand that is 0.25 mm wide. Surrounding the primary xylem is a clear area (*c.* 0.4 mm in diameter) that represents the position of the remaining tissues of the protostele (i.e. primary phloem and pericycle). The cortex is homogeneous being composed of parenchyma cells only. A 'sparganum' outer cortex is absent.

Some of the outer tissues of the stem (epidermis and outer cortex) near the root are torn and/or missing. This creates a question about the provenance of the root. The completeness of preservation of the specimen (including such soft tissues as cambial region, secondary phloem, inner cortex) indicates relatively rapid preservation and permineralization. Such rapid processes would tend to act against the occurrence of invading roots from other plants. We prefer to interpret the root as being adventitious and part of *L. hibernica*. As such, it should be noted that it occurs just below a node which is a common point of origin in the pteridosperms and ferns.

DISCUSSION

Laceya hibernica may be compared to a number of Devonian and Lower Carboniferous genera having three-fluted protosteles in the stem. Included in this group are: *Aneurophyton*, *Protopteridium*, *Cairoa* and *Reimannia* from the Middle Devonian; *Triloboxylon* and *Proteokalon* from the Upper Devonian; and *Tristichia*,

Triradioxylon and *Stenomyelon* from the Lower Carboniferous. Each of these genera have in common with *Laceya* the following:

Several, mesarch, protoxylem strands within the protostele.

In cross-section, lobation of the end of a primary xylem arm with at least two protoxylems during ultimate branch or rachial trace formation.

Development of a sparganum or dictyoxylon type of outer cortex.

Multiseriate, circular bordered pitting on the lateral walls of the secondary xylem tracheids.

Aneurophyton germanicum Kräusel & Weyland (Kräusel & Weyland, 1926; Serlin & Banks, 1978) has a three-ridged protostele with four protoxylem strands in the penultimate branch. Its appearance is very much like *Laceya*. However, the ultimate branches also have three-fluted protosteles, lack secondary xylem and bear naked, one to three times dichotomizing (in more than one plane) ultimate appendages. The ultimate branch trace is oval to lobed. Prior to ultimate branch trace formation, the xylem arm of the penultimate branch stele is lobed with two to three protoxylem areas. The ultimate branches and ultimate appendages are spirally arranged. If the ultimate branches and ultimate appendages of *Aneurophyton* and other members of the Aneurophytales (*Cairoa*, *Reimannia*, *Triloboxylon*, *Proteokalon*) are considered to be the morphological equivalent to the frond of *Laceya*, then *Laceya* differs from *Aneurophyton* in trace formation; size, shape and planation of frond; shape of xylem strand within the rachis (i.e. ultimate branch of *Aneurophyton*); and having two protoxylems in the pinna trace (*Aneurophyton* reportedly has no vascular tissue in the ultimate appendage).

For other members of the Aneurophytales we will use the morphological equivalents of penultimate branch for stem, ultimate branch for rachis, ultimate appendage for pinna (realizing that these are arbitrary designations and used for convenience).

Reports of the anatomy of *Protopteridium thomsonii* Mustafa vary (Mustafa, 1975; Bonamo, 1977). Mustafa's specimens come from the Middle Devonian of Germany. The anatomy of most of the specimens seems to be that of a penultimate branch. In cross section, the axes have a three-armed actinostele with six protoxylem areas. Secondary xylem surrounds the protostele and is composed of tracheids with multiseriate, bordered pitting on tangential and radial walls and tall, uniseriate rays. *Laceya hibernica* stems differ in number of protoxylem strands and having multiseriate rays. Bonamo's specimens are from the Middle Devonian of New York State and show the anatomy of a penultimate (stem) and an ultimate branch (rachis). In cross-section, the three-armed protostele of the penultimate branch contains numerous protoxylems in a continuous median band of tracheids. The pattern is similar to that of *Triloboxylon ashlandicum* Matten & Banks (Matten & Banks, 1966). The xylem of the ultimate branch is also three-armed in cross-section with a median strand of narrow tracheids. A single terete ultimate appendage trace (pinna trace) was illustrated by Bonamo for *Protopteridium*. The number of protoxylem strands in the penultimate branch, the median band of narrow tracheids in the penultimate and ultimate branches, the single, terete, ultimate appendage-trace and the apparent spiral arrangement of parts of *Protopteridium* contrasts with the three protoxylems in the stele of the stem, the two protoxylems in the pinna trace, the U-shaped pinna trace, rachial trace and xylem of the rachis, and the distichous arrangement and planation of fronds in *Laceya*.

Cairoa lamanekii Matten (Matten, 1973, 1975) has a penultimate branch (stem) containing, in cross-section, a three-armed protostele with up to nine mesarch protoxylem areas and having an outer cortex with probable periderm. The spirally arranged ultimate branches (rachises) have oval to rhomboidal primary xylem strands in cross-section. The terete, ultimate appendage-traces are thought to be arranged decussately. *Cairoa* is placed with *Proteokalon* (Scheckler & Banks, 1971) in the Proteokalonaceae of Barnard & Long (1975) of the Aneurophytales because of the change in its primary xylem configuration from penultimate to ultimate branch. *Laceya* differs from *Cairoa* in lacking periderm, having three protoxylems in the stem stele, complex U-shaped rachial traces, distichous arrangement and planation of fronds and U-shaped pinna traces with two protoxylems.

Reimannia aldenense Arnold (Arnold, 1935) was described from a Middle Devonian specimen from western New York State. Re-examination of the holotype showed that it consists of a penultimate branch, an ultimate branch and ultimate appendages (Stein, 1978). The penultimate branch (stem) has a three-armed protostele in cross-section, with several mesarch protoxylems aligned in each arm. The ultimate branches (rachises) were apparently helically arranged and bore terete, ultimate appendage traces (pinna traces). The ultimate branch trace is elliptical to rhomboid but becomes three-lobed with four protoxylem strands in the ultimate branch. *Laceya* differs from *Reimannia* in number of protoxylems in the stem stele, origin and shape of the rachial trace, shape of the vascular strand and number of protoxylems in the rachis, shape of the pinna trace and the distichous arrangement and planation of the frond.

Proteokalon petryi Scheckler & Banks (Scheckler & Banks, 1971) from the Upper Devonian of New York State is characterized by having a penultimate branch (stem) with a skewed, four-fluted actinostele with numerous, mesarch protoxylem strands. Xylem parenchyma is present. Secondary xylem is characterized by tall, uniseriate rays with some ray tracheids. Periderm is also present. The ultimate branches (rachises) are borne decussately and contain a three-ribbed actinostele with 4–7 protoxylem strands. The ultimate appendages (pinnae) are arranged in three orthostichies; in a pair at one node alternating with a single unit at the intervening nodes. The ultimate appendages are supplied by a single, terete trace. *Laceya* differs from *Proteokalon* in shape of stele and number of protoxylems in the stem; lack of parenchyma in the stem stele; absence of periderm; multiseriate rays lacking ray tracheids; origin, shape and number of protoxylems in the rachial trace; shape of xylem strand in the rachis; distichous arrangement of rachises on the stem; planation of the frond; and the U-shaped, bipolar pinna traces.

The last three genera to be compared with *Laceya* are considered to belong to the Lyginopteridales of the Lower Carboniferous; *Tristichia* in the Lyginopteridaceae, *Triradioxylon* in the Buteoxylonaceae and *Stenomyelon* in the Calamopityaceae.

Tristichia consists of two species. *Tristichia ovensi* Long (Long, 1962, 1963) is considered to be the ovule-bearing axis of *Lyginorachis papilio* Kidston, the frond of *Pitus*, arising at the bifurcation of the rachis. Galtier (1977) proposed maintaining the genus for those preserved axes whose attachment to *Pitus* has not been demonstrated and named a new species from the Lower Carboniferous of France, *Tristichia longii* Galtier. We agree with Galtier's proposal and will compare *L. hibernica* with *T. longii*.

The stems of *T. longii* are 6–13 mm wide and have (in cross-section) a three-armed protostele with six to twelve mesarch protoxylems disposed along the arms. The metaxylem is composed of tracheids (15–90 μm wide) with multiseriate pits and parenchyma, especially in the centre. The small amount of secondary xylem is composed of multiseriate pitted tracheids (c. 85 μm in diameter) and one- to six-seriate rays (more than 20 cells high). The inner cortex is parenchymatous and contains sclerotic nests and secretory cells. The outer cortex is of the 'sparganum'-type. Rachial traces begin with a tangential division of a distal protoxylem strand. This is followed by enlargement and tangential elongation of the end of the xylem arm. The two distal protoxylems divide radially, supplying each lobe of the xylem arm with two protoxylems. The detached rachial trace is single or double and papillionoid. The petioles correspond to the genus *Lyginorachis*.

Galtier distinguishes *T. longii* from *T. ovensi* on the basis of size of stems (6–13 mm compared with 1.5–8 mm), presence of parenchyma in the protostele (*T. ovensi* lacks this), and the number of protoxylem poles in the xylem of the stem (four in *T. ovensi*) and especially in the xylem arms (only one in *T. ovensi*). Lastly, the stem of *T. ovensi* is reportedly the ovuliferous branch of a rachis borne on *Pitus* while *T. longii*, because of its size, may be a true stem. A similar argument can be made for *L. hibernica*. Its size is such that it is best to assume that it is a stem rather than part of a frond producing a helical arrangement of planated parts.

There are a number of differences between *L. hibernica* and *T. longii*. The mean dimensions of the secondary xylem tracheids are larger in *T. longii*; the rays of *T. longii* are narrower; the number of protoxylem poles in *T. longii* is greater, and the shape of the rachial trace is different (papillionoid compared with U-shaped).

Triradioxylon primaevum Barnard & Long (Barnard & Long, 1975) is characterized by stems with a three-armed protostele (in cross-section) having a central protoxylem and one or two protoxylems in each of the arms. The inner cortex has sclerotic nests and the vascular rays are one- to six-seriate having a slight expansion near the periphery of the secondary xylem. The rachis is swollen at the base and has a T-shaped (in cross-section) vascular strand with two protoxylem groups near the ends of the arms. The rachial (used here interchangeably with petiole) trace often has secondary xylem on the abaxial side. Triarch pinna traces are produced in an alternate and planated manner. *Laceya hibernica* is similar to *Triradioxylon primaevum* in many anatomical features of the stem. It would be extremely difficult to separate stems of the two genera if the stems lacked such characters as traces or attached appendages. However, there is no central protoxylem in the stem stele of *Laceya*. The anatomy of the fronds of the two genera is strikingly different. The U-shaped trace in *Laceya* is very different from the T-shaped trace in *Triradioxylon*. The pinna trace of *Triradioxylon* is apparently triarch while it is bipolar in *Laceya*.

Stenomyelon is characterized by stems with three-armed protosteles in cross-section. All species, but one (*S. primaevum* Long), have xylem parenchyma in the protostele. *Stenomyelon tuedianum* Kidston is thought to have an exarch protoxylem at the tip of each arm, the other species have a mesarch protoxylem area at or near the tip of each arm. Secondary xylem is manoxylic with tracheids having multiseriate pitting on the radial walls only and tall, multiseriate rays. Leaf traces have secondary xylem, at first. They are bipolar and terete and divide repeatedly in the cortex so the base of the rachis has at least eight discrete vascular strands. The petioles are of the *Kalymma*-type. The species with a stem most similar to *Laceya* is *S. primaevum* (Long, 1964). *Stenomyelon* differs from *Laceya* in secondary xylem tracheary pitting, rachial trace formation and configuration of the vascular strand in the rachis.

Tristichia longii, *Triradioxylon primaevum* and *Stenomyelon primaevum* belong to three

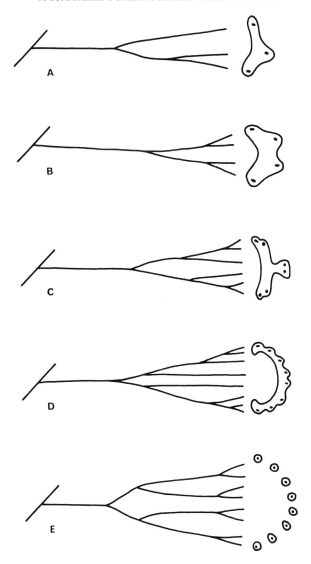

Figure 29. Branching patterns of the stem sympodial protoxylem strand forming the rachial trace. The sympodial protoxylem strand in each pattern divides tangentially, the left (inner) half remaining as part of the stem stele; the right (outer) half dividing to form the characteristic number of protoxylem strands in the trace; the outline of the trace showing the metaxylem around the protoxylem is shown to the right. A. *Aneurophyton germanicum*, *Triloboxylon ashlandicum* or *Proteokalon petryi* of the Aneurophytales. B. *Tristichia longii* (Lyginopteridaceae). C. *Triradioxylon primaevum* (Buteoxylonaceae). D. *Laceya hibernica*. E. *Stenomyelon primaevum* (Calamopityaceae).

different Lower Carboniferous families of the Lyginopteridales. Their stems are very similar in cross-sectional configuration. Each has a three-armed protostele, a relatively small amount of secondary xylem development, an inner, parenchymatous cortex with sclerotic and/or secretory cells and a 'sparganum' outer cortex. The similarity of stem features supports a possible common ancestry. However the phylogenetic relationship appears to be somewhat distant when the rachial traces and the vascular anatomy of the rachises are compared. The

Lyginopteridaceae has papillionoid traces that become U- to V-shaped in the rachis. The Buteoxylonaceae has T-shaped rachial traces and a T-shaped vascular strand in the rachis. The Calamopityaceae has a terete trace that repeatedly divides in the cortex and ultimately forms a *Kalymma*-type rachis. In addition, there is a major forking of the main rachis into two secondary rachises in the Lyginopteridaceae and Calamopityaceae. There is no evidence to date of a major division of the rachis in the Buteoxylonaceae. The rachial anatomy of *Laceya hibernica* is very similar to *Lyginorachis*, the organ genus for the rachis of the Lyginopteridaceae, and we tentatively assign *Laceya* to that family.

The differences in petiolar anatomy among the Lower Carboniferous lyginopteridalean families are not as great as they appear. A comparison of the development and divisions of the protoxylem strands during trace ontogeny reveals a certain amount of uniformity (Fig. 29). Trace formation in the Aneurophytales, Lyginopteridaceae, Buteoxylonaceae, Calamopityaceae and *Laceya* begins with a tangential division of a single, sympodial protoxylem strand of the stem xylem arm into a cauline (reparatory) and a trace strand. The lack of a true leaf gap is the pattern described for gymnosperms (Namboodiri & Beck, 1968a, b, c; Basinger, Rothwell & Stewart, 1974; Rothwell, 1976). The trace protoxylem in these lyginopteridalean and aneurophytalean plants divides one or more times producing 3 (*Aneurophyton*), 4 (*Tristichia*), 6 (*Triradioxylon*) or 8 (*Laceya, Stenomyelon*) protoxylem strands. The subsequent development of metaxylem produces the distinctive *Kalymma*, *Laceya*, *Triradioxylon*, *Tristichia* and *Aneurophyton* configurations.

The similarity in divisional pattern of the protoxylem during trace formation supports the inclusion of the Buteoxylonaceae, Lyginopteridaceae and Calamopityaceae within the Lyginopteridales. The similarity of aneurophytalean genera to primitive members of the Lyginopteridales strengthens the hypothesis of an aneurophytalean ancestry for lyginopterid seed ferns.

ACKNOWLEDGEMENTS

We are grateul to Mr John A. Richardson and Mrs Karen Schmitt for help with the illustrations. The research was supported by grants to the junior author from the Office of Research Development and Administration, Southern Illinois University and the National Science Foundation (DEB 81-08383).

REFERENCES

ARNOLD, C. A., 1935. Some new forms and new occurrences of fossil plants from the Middle and Upper Devonian of New York State. *Buffalo Society of Natural Science, Bulletin, 17:* 1–12.
BARNARD, P. D. W. & LONG, A. G., 1975 *Triradioxylon*—a new genus of Lower Carboniferous petrified stems and petioles together with a review of the classification of early Pterophytina. *Transactions of the Royal Society of Edinburgh, 69:* 231–250.
BASINGER, J. F., ROTHWELL, G. W. & STEWART, W. N., 1974. Cauline vasculature and leaf trace production in medullosan Pteridosperms. *American Journal of Botany, 61:* 1002–1015.
BONAMO, P. M., 1977. *Rellimia thomsonii* (Progymnospermopsida) from the Middle Devonian of New York State. *American Journal of Botany, 64:* 1272–1285.
BRIDGE, J. S., VAN VEEN, P. M. & MATTEN, L. C., 1980. Aspects of the sedimentology, palynology and palaeobotany of the Upper Devonian of southern Kerry Head, Co. Kerry, Ireland. *Geological Journal, 15:* 143–170.
CALDER, M. G., 1935. Further observations on the genus *Lyginorachis* Kidston. *Transactions of the Royal Society of Edinburgh, 58:* 549–559.

GALTIER, J., 1974. Sur l'organization de la fronde des *Calamopitys*, Ptéridospermales probables du Carbonifère inférieur. *Comptes rendus hebdomadaires des Séances de l'Académie des sciences, Paris, Série D, 279:* 975–978.

GALTIER, J., 1977. *Tristichia longii*, nouvelle Ptéridospermale probable du Carbonifère de la Montagne Noire. *Comptes rendus hebdomadaires des Séances de l'Académie des sciences, Paris, Série D, 284:* 2215–2218.

GARDINER, P. R. R. & HORNE, R. R., 1972. The Devonian and Lower Carboniferous clastic correlatives of southern Ireland. *Geological Survey of Ireland, Bulletin, 1:* 335–366.

HOUSE, M. R., RICHARDSON, J. B., CHALONER, W. G., ALLEN, J. R. L., HOLLAND, C. H. & WESTOLL, T. S., 1977. A correlation of the Devonian rocks in the British Isles. *Geological Society of London, Special Report, No. 8:* 1–110.

JOY, K. W., WILLIS, A. J. & LACEY, W. S., 1956. A rapid cellulose peel technique in paleobotany. *Annals of Botany, 20:* 635–637.

KAHN, M. F. H., 1955. The Old Red Sandstone of the Kerry Head anticline, County Kerry. *Proceedings of the Royal Irish Academy, 57:* 71–78.

KRÄUSEL, R. & WEYLAND, H., 1926. Beiträge zur Kenntnis der Devonflora II. *Senckenbergische naturforschende gesellschaft, Abhandlungen, 40:* 115–155.

LONG, A. G., 1962. *Tristichia ovensi* gen. et sp. nov., a protostelic Lower Carboniferous pteridosperm from Berwickshire and East Lothian, with an account of some associated seeds and cupules. *Transactions of the Royal Society of Edinburgh, 64:* 477–489.

LONG, A. G., 1963. Some specimens of *Lyginorachis papilio* Kidston associated with some stems of *Pitys*. *Transactions of the Royal Society of Edinburgh, 66:* 35–46.

LONG, A. G., 1964. Some specimens of *Stenomyelon* and *Kalymma* from the Calciferous Sandstone series of Berwickshire. *Transactions of the Royal Society of Edinburgh, 65:* 435–447.

LONG, A. G., 1976. *Calathopteris heterophylla* gen. et sp. nov., a Lower Carboniferous pteridosperm bearing two kinds of petioles. *Transactions of the Royal Society of Edinburgh, 69:* 327–336.

MATTEN, L. C., 1973. The Cairo flora (Givetian) from eastern New York. I. *Reimannia*, terete axes, and *Cairoa lamanekii* gen. et sp. n. *American Journal of Botany, 60:* 619–630.

MATTEN, L. C., 1975. Additions to the Givetian Cairo flora from eastern New York. *Bulletin of the Torrey Botanical Club, 192:* 45–52.

MATTEN, L. C. & BANKS, H. P., 1966. *Triloboxylon ashlandicum* gen. et sp. n. from the Upper Devonian of New York. *American Journal of Botany, 53:* 1020–1028.

MATTEN, L. C. & LACEY, W. S., 1981. Cupule organization in early seed plants. In: R. C. Romans (Ed.), *Geobotany II*. New York: Plenum Press.

MATTEN, L. C., LACEY, W. S. & EDWARDS, D., 1975. Discovery of one of the oldest gymnosperm floras containing cupulate seeds. *Phytologia, 32:* 299–303.

MATTEN, L. C., LACEY, W. S. & EDWARDS, D., 1976. An Upper Devonian/Lower Carboniferous transition flora from southwest Eire. *Courier Forschungsinstitut Senckenberg, 17:* 87.

MATTEN, L. C., LACEY, W. S. & LUCAS, R. C., 1978. Cupulate seeds of *Hydrasperma* from Kerry Head, Ireland. *Botanical Society of America, Miscellaneous Series, Publication 156:* 3.

MATTEN, L. C., LACEY, W. S. & LUCAS, R. C., 1980a. Studies on the cupulate seed genus *Hydrasperma* Long from Berwickshire and East Lothian in Scotland and County Kerry in Ireland. *Botanical Journal of the Linnean Society, 81:* 249–273.

MATTEN, L. C., LACEY, W. S., MAY, B. I. & LUCAS, R. C., 1980b. A megafossil flora from the uppermost Devonian near Ballyheigue, Co. Kerry, Ireland. *Review of Palaeobotany and Palynology, 29:* 241–251.

MAY, B. I., MATTEN, L. C. & LACEY, W. S., 1978. A protostelic stem bearing *Lyginorachis*-like petioles from the Devonian-Carboniferous transition beds of southern Ireland. *Botanical Society of America, Miscellaneous Series, 156:* 30.

MUSTAFA, H., 1975. Beiträge zur Devonflora I. *Argumenta Palaeobotanica, 4:* 101–133.

NAMBOODIRI, K. K. & BECK, C. B., 1968a. A comparative study of the primary vascular system of conifers. I. genera with helical phyllotaxis. *American Journal of Botany, 55:* 447–457.

NAMBOODIRI, K. K. & BECK, C. B., 1968b. A comparative study of the primary vascular system of conifers. II. genera with opposite and whorled phyllotaxis. *American Journal of Botany, 55:* 458–463.

NAMBOODIRI, K. K. & BECK, C. B., 1968c. A comparative study of the primary vascular system of conifers. III. stelar evolution in gymnosperms. *American Journal of Botany, 55:* 464–472.

ROTHWELL, G. W., 1976. Primary vasculature and gymnosperm systematics. *Review of Palaeobotany and Palynology, 22:* 193–206.

SCHECKLER, S. E. & BANKS, H. P., 1971. *Proteokalon* a new genus of progymnosperms from the Devonian of New York State and its bearing on phylogenetic trends in the group. *American Journal of Botany, 58:* 874–884.

SERLIN, B. S. & BANKS, H. P., 1978. Morphology and anatomy of *Aneurophyton*, a progymnosperm from the late Devonian of New York. *Palaeontographica Americana, 8:* 343–359.

STEIN, W. E. 1978. Additional information on *Reimannia aldenense* Arnold from the Middle Devonian of Erie County, New York, *Botanical Society of America, Miscellaneous Series, Publication, 156:* 2.

VAN DER ZWAN, C. J. & VAN VEEN, P. M., 1978. The Devonian-Carboniferous transition sequence in southern Ireland: integration of palaeogeography and palynology. *Palinologia, numéro extraordinaire, 1:* 469–479.

Botanical Journal of the Linnean Society (1983) 86: 125–133. With 15 figures

A re-examination of *Lepidostrobus* Brongniart

SHEILA D. BRACK-HANES

Eckerd College, St Petersburg, Florida 33733, U.S.A.

AND

BARRY A. THOMAS

University of London, Goldsmiths' College, New Cross, London SE14 6NW

Received December 1981, accepted for publication May 1982

Lepidostrobus is a genus that was established by Brongniart for Palaeozoic lycopod cones. Since then the genus has been used for a variety of cones that have similar characters. There is now compelling evidence that the genus represents a heterogeneous group of monosporangiate and bisporangiate cone species and that it should be divided. *Lepidostrobus* is rediagnosed as a genus of microsporangiate cones. *Flemingites* Carruthers is rediagnosed as a genus of those bisporangiate cones formerly included in *Lepidostrobus*.

KEY WORDS:—*Flemingites* – fossil cones – *Lepidostrobus*.

CONTENTS

INTRODUCTION

Lepidostrobus is a loosely defined genus that includes distinctive petrifactions as well as a rather ambiguous collection of casts or compression cones that may or may not be *Lepidostrobus*, *Lepidocarpon*, *Achlamydocarpon* or even some other kind of fructification. A re-examination of the genus, *Lepidostrobus*, is in order.

Brongniart (1828–1838) named and briefly described the cone figured by Parkinson (1804) as the holotype of *Lepidostrobus* (Fig. 1). For the next 150 years, all cones that even vaguely fitted Brongniart's criteria were included. By the end of the nineteenth century, much of the anatomy was known from structurally preserved lepidostroboid specimens examined by Hooker (1848), Carruthers (1865), Schimper (1869–1874), Williamson (1872, 1893), Maslen (1899) and a

© 1983 The Linnean Society of London

genus revision had even taken place (Kidston, 1897). The cone was shown to have a variable size and shape; a three-zoned cortex around the vascular cylinder; one vascular bundle per sporophyll; sporophylls mainly spirally arranged; one sporangium attached by nearly its entire length to the adaxial surface of the sporophyll; a sporangial wall of a single layer of columnar cells; a ligule on the same surface of the sporophyll as the sporangium and distal to it, and trilete spores. Some cones were microsporangiate and some were bisporangiate with microspores (c. 20 μm diameter) in the apical sporangia and megaspores (up to 1.0 mm diameter) in the basal sporangia. There was also, by then, an awareness that features of the cones were rather uniform and reflected a narrow range of variation. This, of course, made "species distinction a problem" (Arber, 1922). The same problem has persisted.

Lepidostrobus is generally accepted to be a heterogenous group. For taxonomic purposes, sporophyll morphology was first considered the most important character and Abbott (1963) attempted to establish genera based primarily on sporophyll features. She was especially concerned with the difficulty of correlating compressions with petrifactions. However, it has become obvious that, wherever possible, the sporangial contents need to be thoroughly investigated for an accurate determination of these fructifications. Chaloner (1953) and Felix (1954) first showed the real potential of using spores to characterize cone species. Spores have the advantage over all other cone features in that they are often preserved with minimal damage and, therefore, may be used with greater success in correlating cones with each other and also in correlating compressions with petrifactions. Compression cones have a greater chance of being found attached to stems, thereby increasing our knowledge of the plants as a whole. For stratigraphic purposes, there is the need to make precise comparisons of dispersed spores with those from fructifications. This is especially true for the ubiquitous arborescent lycophytes, producing cones like *Lepidostrobus*, where *in situ* spores are regularly compared with dispersed spore taxa. Earlier taxonomic efforts emphasized the characteristics of the megaspores from these cones rather than those of the microspores. However, the SEM now permits a more critical examination of the microspore. Work by Balbach (1966), Courvoisier & Phillips (1975) and others is clearly in support of work on the dispersed spore genus, *Lycospora* (Potonié & Kremp, 1956; Smith & Butterworth, 1967). Although microspores from *Lepidostrobus* can be correlated with the genus *Lycospora*, it is still difficult, and sometimes impossible, to relate such spores from fructifications in terms of species of dispersed spores. Apart from ontogenetic considerations (Brack & Taylor, 1972), there is increasing evidence to suggest that a larger part of the difficulty may be because of the occurrence of at least two distinctive groups within *Lycospora*. It has already been suggested that the microspores extracted from microsporangiate species of *Lepidostrobus* have much broader equatorial flanges than those of bisporangiate cones (Thomas, 1970, 1978; Thomas & Dytko, 1980). In addition, a reinvestigation of microspores recovered from lepidostroboid cones shows that distinctive proximal ornamentation occurs on spores of microsporangiate cones, whereas the proximal surfaces of microspores recovered from bisporangiate cones are smooth (Brack-Hanes & Thomas, 1981).

It is widely assumed that microsporangiate cones are the counterparts of megasporangiate cones such as *Lepidocarpon*, *Achlamydocarpon*, or *Caudatocarpus* (Brack-Hanes, 1981). Therefore, it is unsuitable to include both microsporangiate

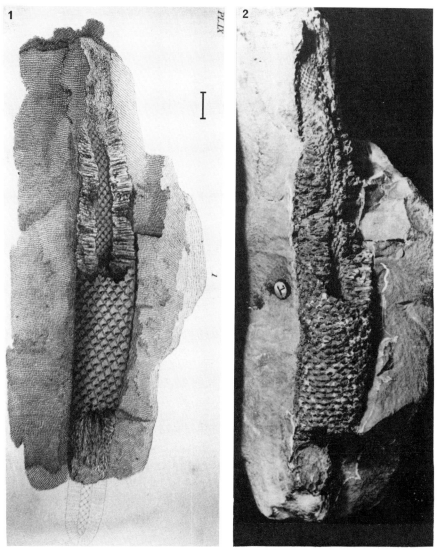

Figures 1 & 2. *Lepidostrobus ornatus* (type for *Lepidostrobus* Brongniart). Fig. 1. Parkinson's illustration. Fig. 2. Cone in nodule (British Museum (Nat. History), V. 16440). Scale bar equivalent to 7.5 mm.

and bisporangiate cones within the same cone genus. However, in order to limit the use of *Lepidostrobus* to one kind of cone, it is necessary to determine the spore content of the type species of *Lepidostrobus*.

DESCRIPTION OF THE TYPE SPECIMEN OF *LEPIDOSTROBUS*

The first species that Brongniart (1828–1838) included within *Lepidostrobus* was *L. ornatus*, thereby naming a specimen that had been previously figured but not named by Parkinson (1804) (Fig. 1). This specimen is now in the collections of the British Museum of Natural History (V16440) and is labelled as coming from the Coal Measures of Derbyshire. The cone is partially preserved in a clay ironstone

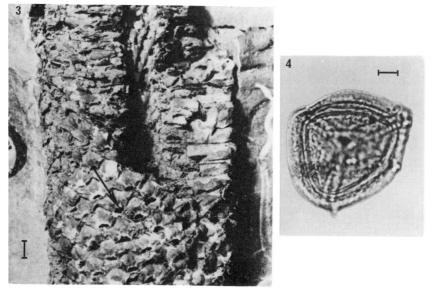

Figures 3 & 4. *Lepidostrobus ornatus* (type for *Lepidostrobus* Brongniart). Fig. 3. Cone enlarged to show ends of sporangia, arrow to vascular bundle ridge and hollow pedicel centre beneath sporangium. Scale bar equivalent to 2 mm. Fig. 4. Microspore proximal trilete mark. Scale bar equivalent to 5 μm.

nodule. It is 140 mm long but broken at both ends and is 22 mm broad. Where the cone is split longitudinally, its axis is 6 mm broad showing the remains of the pedicel bases in low spirals.

Sporangia can be seen in longitudinal section where the cone is split and in end view where the cone is still entire. They are 8 mm long, 2 mm broad and 3 mm high. The sporophylls have not been fully preserved. The centres of the pedicels appear as hollows below those sporangia seen in end view (Fig. 2). We interpret the ridges running along the abaxial side of the pedicel in the base of the hollow as remains of the vascular bundle. Microspores were recovered in large numbers from intact sporangia near the basal end of this cone. Some of the microspores were mounted in 'Cover bond' (from Harleco: piccolyte in xylene) for observation with the light microscope, others were prepared for SEM (Figs 5 & 6). They range in size from 24 to 32 μm and are roundly triangular in polar view. The arms of the trilete

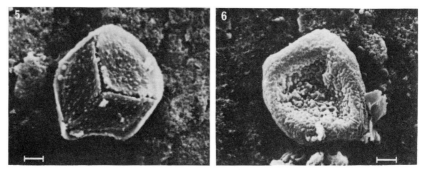

Figures 5 & 6. *Lepidostrobus ornatus* (type for *Lepidostrobus* Brongniart). Fig. 5. Microspore SEM of proximal surface. Scale bar equivalent to 0.005 mm. Fig. 6. Microspore SEM of distall surface. Scale bar equivalent to 5 μm.

mark extended to the inner edges of the broad equatorial flange. The proximal surface has numerous small papillae while the distal surface is granulose. The equatorial flange is 3–5 μm wide with an entire margin. There is sometimes difficulty in distinguishing the flange when the spores are slightly collapsed by desiccation. The effect is to dimple or even implode the thinner distal wall, thereby rolling the equatorial zone and flange distally. This is especially evident in those spores which have imploded during evacuation prior to coating for the SEM. These spores, if found dispersed would clearly be included within the genus, *Lycospora*. These spores most closely resemble *L. granulata*.

DISCUSSION

From our knowledge of bisporangiate, lepidostroboid cones, we know that megaspores usually occur in the base and microspores in the apex. That portion of Parkinson's cone that yielded microspores is well below the region of its maximum width and is therefore most probably basal. This strongly suggests that there were no megaspores in this cone. In addition, broad-flanged, papillate microspores that have been recovered from *L. ornatus* are comparable with those microspores that occur in microsporangiate cones (Thomas, 1970, 1978; Brack-Hanes & Thomas, 1981). Therefore, *L. ornatus*, which is the type species for *Lepidostrobus*, is most likely microsporangiate. Ideally, we believe that the genus *Lepidostrobus* should be reserved for microsporangiate cones. However, at this stage we propose only to remove the bisporangiate species, leaving in *Lepidostrobus* those cones for which no data on spores is available, such as *L. gracilis* (Newberry, 1853; Schmalhausen, 1877).

Carruthers (1865) described a lycopod cone which he named *Flemingites gracilis* rather than as a species of *Lepidostrobus*. He established a new genus for the cone because he mistakenly thought that the megaspores were sporangia. He believed that each sporophyll had two rows of rounded sporangia on its upper surface. Carruthers' type specimen of *F. gracilis* is in the British Museum of Natural History (V6041) (Fig. 7). It is a compressed megasporangiate fragment of cone showing little detail except size. Chaloner (1953) examined the megaspores from this cone (Figs 8, 9) and believed them to be identical to those of *Lepidostrobus dubius* Binney (Binney, 1871), Fig. 10. We have re-examined the megaspores and agree with his interpretation (Figs 11, 12). Although Carruthers described *F. gracilis* before Binney described *L. dubius*, the combination *L. gracilis* could not be used because that combination had been previously used for other and presumably different cones (Newberry, 1953; Schmalhausen, 1877). *Lepidostrobus dubius* Binney *sensu* Chaloner is a bisporangiate cone. Chaloner (1953) described the microspores from *L. dubius* as *Lycospora* spores with smooth proximal surfaces; distal surfaces that were ornamented with low papillae; and slight zonal flanges (Figs 13–15). Such microspores are of the form found in other bisporangiate cones that are included in *Lepidostrobus* (Felix, 1954; Brack, 1970). We accept Chaloner's view that realistic taxonomic treatment must take into account cone fragments, and believe that Carruthers' specimen was bisporangiate.

Lepidostrobus, as we have outlined, has as its type the microsporangiate *L. ornatus*, so it should not include the bisporangiate cones of the lepidostroboid kind. As *Flemingites* is the first valid generic name given to these bisporangiate cones we propose that this genus be used for such cones that were formerly included in

Lepidostrobus. Carruthers' specimen is the type for the genus and should retain the name *Flemingites gracilis*. It has preference over *L. dubius* (Binney, 1871) although that specimen is regarded as the same species and is a more complete cone than that of Carruthers.

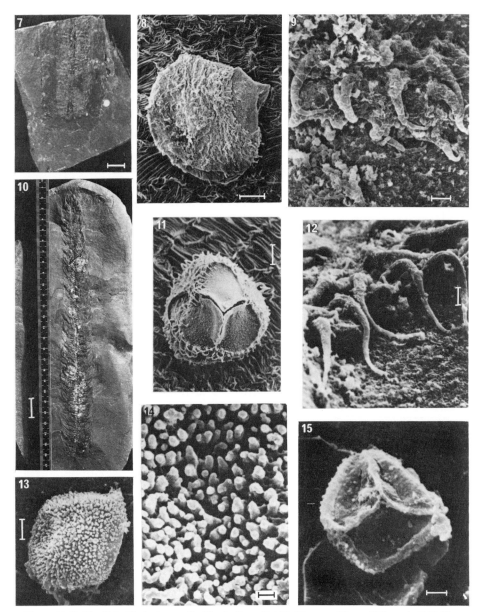

Figures 7–15. Figs 7–9. *Flemingites gracilis*. Fig. 7. Cone (British Museum (Natural History), V 6041). Scale bar equivalent to 10 mm. Fig. 8. Megaspore SEM. Scale bar equivalent to 0.2 mm. Fig. 9. SEM enlargement of spines on distal surface at equatorial zone. Scale bar equivalent to 0.02 mm. Figures 10–15. *Lepidostrobus dubius* (type for *L. dubius* forma *hermaphrodibus*). Fig. 10. Cone in nodule (British Museum (Natural History), V 1244). Scale bar equivalent to 20 mm. Fig. 11. Megaspore SEM. Scale bar equivalent to 0.2 mm. Fig. 12. Megaspore SEM of spines on distal surface at equatorial zone. Scale bar uivalent to 0.02 mm. Fig. 13. Microspore SEM distal surface. Scale bar equivalent to 5 μm. Fig. 14. Microspore SEM of papillae on distal surface. Scale bar equivalent to 0.001 mm. Fig. 15. Microspore SEM proximal surface. Scale bar equivalent to 5 μm.

SYSTEMATIC DESCRIPTIONS

Lepidostrobus Brongniart, Brongniart, *Histoire des végétaux fossiles ou recherches botanique et géologique sur les végétaux renfermés dans les diverses couches du globe* (1828–1838).
EMENDED DIAGNOSIS: Sporophylls in spirals on the cone axis. Axis with exarch vascular bundle surrounded by cortical zones. Sporangium with narrow attachment along its length to adaxial surface of sporophyll pedicel. Ligule on adaxial surface of pedicel distal to sporangium. Lateral laminae extending beneath sporangium. Abaxial keel along length of pedicel. Pedicel extended distally to upturned lamina and downturned heel. Cones microsporangiate. Microspores of *Lycospora*-type with broad equitorial flange and proximal ornamentation, distal surface ornament variable.

TYPE SPECIES: *Lepidostrobus ornatus* Brongniart.

Lepidostrobus ornatus Brongniart, Brongniart, *Histoire des végétaux fossiles ou recherches botanique et géologique sur les végétaux renfermés dan les diverses couches du globe* (1828–1838).

ILLUSTRATION: Brongniart (1828 : plate 1, figs 1–6).

LOCALITY AND HORIZON: Derbyshire, Coal Measures.

SPECIMEN EXAMINED: Holotype, British Museum (Natural History) No. V16440.

Flemingites Carruthers, 1865, Carruthers, *Geological Magazine, 2:* 433–438 (1865).
EMENDED DIAGNOSIS: Sporophylls in spirals on the cone axis. Axis with exarch vascular bundle surrounded by cortical zones. Sporangium with narrow attachment along its length to adaxial surface of sporophyll pedicel. Ligule on adaxial surface of pedicel distal to sporangium. Lateral laminae extending beneath sporangium. Abaxial keel along length of pedicel. Pedicel extended distally to upturned lamina and downturned heel. Cones bisporangiate with apical microsporangia and basal megasporangia. Megaspores either *Lagenicula*- or *Lagenioisporites*-type. Microspores of *Lycospora*-type with narrow equitorial flange and usually smooth proximal surface, distal surface variable in ornament.

TYPE SPECIES: *Flemingites gracilis* Carruthers.

Flemingites gracilis Carruthers, Carruthers, *Geological Magazine, 2:* 433–438 (1865).

SYNONYM: *Lepidostrobus dubius* Binney *sensu* Chaloner. Chaloner *Annals of Botany, 17:* 264 (1953).

LOCALITY AND HORIZON. Airdrie, Lanark., Carboniferous.

SPECIMEN EXAMINED: Holotype. British Museum (Natural History) No. V6041.

Other species to be included within *Flemingites* are:

Flemingites allantonensis (Chaloner) Brack-Hanes & Thomas **comb. nova**
SYNONYM: *Lepidostrobus allantonensis* Chaloner, Chaloner, *Annals of Botany, 17:* 2, pl. 8. (1953).
Flemingites bartletti (Arnold) Brack-Hanes & Thomas **comb. nova**
SYNONYM: *Lepidostrobus bartletti* Arnold, Arnold, *American Journal of Botany, 17:* 1028–1032 (1930).

Flemingites brownii (Unger) Brack-Hanes & Thomas **comb. nova**

SYNONYM: *Triplosporites brownii* Unger, *Genera of Species plantarum fossilium:* 270 (1850).

    Lepidostrobus brownii (Unger) Schimper, *Traité de paléontologie végétale ou la flore du monde Primitif:* 67, pl. 62, figs 3–33 (1870).

Flemingites diversus (Felix) Brack-Hanes & Thomas **comb. nova**

SYNONYM: *Lepidostrobus diversus* Felix, Felix, *Annals of the Missouri Botanic Garden, 41:* 351–394. 353 (text-figs 1 & 2), pl. 13–15 (figs 16–19) (1954).

Flemingites gallowayi (Arnold) Brack-Hanes & Thomas **comb. nova**

SYNONYM: *Lepidostrobus gallowayi* Arnold, Arnold, *American Journal of Botany, 17:* 51–56 (1935).

Flemingites major (Germar) Brack-Hanes & Thomas **comb. nova**

SYNONYMS: *Volkmannia major* Germar, Germar, *Die Versteinerungen des Steinkohlengebirges von Wettin und Lobejun im Saalkreise:* 92, pl. 32 (figs 5–7) (1851).

    Sigillariostrobus major (Germar) Zeiller, *Études des gîtes minéraux de France:* 172, pl. 54 (fig. 1–1, a–c) (1906).

    Lepidostrobus germari Chaloner, *Traité de Paléobotanique, 2 Lycophyta:* 586 (1967).

Flemingites noei (Mathews) **comb. nova**

SYNONYM: *Lepidostrobus noei* Mathews, *Botanical Gazette, 102:* 35–48 (figs 3–7) (1940).

Flemingites olryi (Zeiller) **comb. nova**

SYNONYM: *Lepidostrobus olryi* Zeiller, *Études des gîtes minéraux de la France:* 502, pl. 77 (fig. 1) (1886–1888).

Flemingites russelianus (Binney) **comb. nova**

SYNONYMS: *Lepidostrobus russelianus* Binney, *Palaeontographical Society:* 51, pl. 9 (fig. 1) (1871).

    Lepidostrobus hibbertianus Binney, *Palaeontographical Society:* 55, pl. 10 (fig. 2) (1871).

Flemingites schopfii (Brack) **comb. nova**

SYNONYM: *Lepidostrobus schopfii* Brack, *American Journal of Botany, 57:* 317–330, figs. 1–32 (1970).

Flemingites scotti (Jongmans) **comb. nova**

SYNONYMS: *Lepidostrobus* Williamson, *Philosophical Transactions of the Royal Society of London, 162:* 294–298, pl. 44 (1872).

    Lepidostrobus veltheimianus Scott, *Studies in Fossil Botany:* 170 (1900).

    Lepidostrobus scotti Jongmans, *Fossilum Catalogus, 2:* 519 (1930).

Flemingites spectabilis (Renault) **comb. nova**

SYNONYM: *Sigillariostrobus spectabilis* Renault, *Bulletin de la Société Histoire Naturelle, 1:* 177 & 180, pl. 3 (figs 1–7), pl. 4 (fig. 1) (1888).

ACKNOWLEDGEMENTS

This study was supported in part by the National Science Foundation grants DEB 7905087 and TFI 8020585 to the first author.

REFERENCES

ABBOTT, M. L., 1963. Lycopod fructifications from the Upper Freeport (No. 7) coal in southeastern Ohio. *Palaeontographica, B, 112:* 93–118.

ARBER, E. A. N., 1922. Critical studies of Coal-measure plant-impressions. 1. A revision of the British Upper Carboniferous species of the genus *Lepidostrobus* Brongn., preserved as incrustations, *Journal of the Linnean Society (Botany)*, *46:* 171–188.

ARNOLD, C. A., 1930. A petrified lepidophyte cone from the Pennsylvanian of Michigan. *American Journal of Botany*, *17:* 1028–1032.

ARNOLD, C. A., 1933. Fossil plants from the Pocono (Oswayo) sandstones of Pennsylvania. *Papers from the Michigan Academy of Science Arts and Letters*, *17:* 51–56.

ARNOLD, C. A., 1935. Notes on some American species of *Lepidostrobus*. *American Journal of Botany*, *22:* 23–25.

BALBACH, M. K., 1966. Microspore variation in *Lepidostrobus* and comparison with *Lycospora*. *Micropalaeontology*, *12:* 334–342.

BINNEY, E. W., 1871. Observations on the structure of fossil plants found in the Carboniferous strata. II. *Lepidostrobus* and some allied cones. *Palaeontographical Society Monograph:* 33–62.

BRACK, S. D., 1970. On a new structurally preserved arborescent lycopsid fructification from the Lower Pennsylvanian of North America. *American Journal of Botany*, *57:* 317–330.

BRACK, S. D. & TAYLOR, T. N., 1972. The ultrastructure and organization of *Endosporites*. *Micropaleontology*, *18:* 101–109.

BRACK-HANES, S. D., 1981. On a lycopsid cone with winged spores. *Botanical Gazette*, *142:* 294–304.

BRACK-HANES, S. D. & THOMAS, B. A. 1981. A re-examination of *Lepidostrobus*. *Association of Southern Biologists Bulletin*, *28:* 51.

BRONGNIART, A. 1828–1838. *Histoire des Végétaux Fossiles ou Recherches Botanique et Géologique sur les Végétaux Renfermés dans les Diverses Couches du Globe.* Vol. I (1828–1836): 1–488; Volume II (1836–1837): 1–72.

CARRUTHERS, W., 1865. On an undescribed cone from the Carboniferous beds of Airdrie, Lanarkshire. *Geological Magazine*, *2:* 433–440.

CHALONER, W. G., 1953. On the Megaspores of four species of *Lepidostrobus*. *Annals of Botany*, *17:* 264–273.

CHALONER, W. G., 1967. E. Boureau, W. G. Chaloner, O. A. Hoeg & S. Jovet-Ast (Eds), *Traité de Paléobotanique*, *2 Lycophyta:* 435–802. Paris.

COURVOISIER, J. M. & PHILLIPS, T. L., 1975. Correlation of spores from Pennsylvanian coal-ball fructifications with dispersed spores. *Micropaleontology*, *21:* 45–59.

FELIX, C. J., 1954. Some American arborescent lycopod fructifications. *Annals of the Missouri Botanical Garden*, *41:* 351–394.

GERMAR, E. F., 1851. *Die Versteinerungen des Steinkohlengebirges von Wettin und Lobejun im Saalkreise:* 1–116. Hale.

HOOKER, J., 1848. Remarks on the structure and affinities of some lepidostrobi. *Geological Survey of the United Kingdom*, *2:* 440–456.

JONGMANS, W. 1930. *Fossilium Catalogus (Plantae)-Ilycopodiales II 2:* Pars 16. Berlin:

KIDSTON, R., 1897. Additional records and notes in the fossil flora of the Potteries Coal Field, North Staffordshire, *Transactions of the North Staffordshire Field Club:* 61–65.

MASLEN, A. J., 1899. The structure of *Lepidostrobus*. *Transactions of the Linnean Society II, (Botany)*, *5:* 357–377.

MATHEWS, G. B., 1940. New lepidostrobi from central United States. *Botanical Gazette*, *102:* 26–49.

NEWBERRY, J. S., 1853. Fossil plants from the Ohio coal basin. *Annals of Science*, *1:* 106–108.

PARKINSON, J., 1804. *Organic Remains of a Former World.* London: Sherwood, Meety & Jones.

POTONIÉ, R. & KREMP, G., 1956. Die *Sporae dispersae* des Ruhrkarbons, ihre Morphographie und Stratigraphie mit Ausblicken auf Arten anderer Gebiete und Zeitabschnitte. 2. *Palaeontographica, B*, *99:* 85–191.

RENAULT, B., 1888. Notices sur les Sigillaries, *Bulletin de le Société d'Histoire Naturelle*, *1:* 121–199.

SCHIMPER, W. P., 1869–1874. *Traité de Paléontologie Végétale ou la Flore du Monde Primitif.* Paris: J. B. Bailliere et fils.

SCHMALHAUSEN, J., 1877. Die Pflanzenreste aus der Ursa-Stufe im Flussgeschiebe des Ogur im Ost-Sibirien. *Bulletin de l'Académie impériale des sciences de St. Pétersbourg*, *21:* 277–291.

SCOTT, D. H., 1900. *Studies in Fossil Botany.* London: A. & C. Black.

SMITH, A. H. V. & BUTTERWORTH, M. A., 1967. Miospores in the coal seams of the Carboniferous of Great Britain. *Palaeontological Association, Special Paper in Palaeontology*, *1:* 1–324.

THOMAS, B. A., 1970. A new specimen of *Lepidostrobus binneyanus* from the Westphalian B of Yorkshire. *Pollen et Spores*, *12:* 217–234.

THOMAS, B. A., 1978. Carboniferous Lepidodendraceae and Lepidocarpaceae *Botanical Review*, *44:* 321–364.

THOMAS, B. A. & DYTKO, A., 1980. *Lepidostrobus haslingdenensis:* A new species from the Lancashire Millstone Grit. *Geological Journal*, *15:* 137–142.

UNGER, F., 1850. *Genera et Species Plantarum Fossilium.* Vienna.

WILLIAMSON, W. C., 1872. On the organisation of the fossil plants of the Coal-Measures. 3, Lycopodiaceae. *Philosophical Transactions of the Royal Society, London*, *162:* 283–318.

WILLIAMSON, W. C., 1893. On the organisation of the fossil plants of the Coal-Measures. 19, *Philosophical Transactions of the Royal Society, London*, *184:* 1–38.

ZEILLER, R., 1886. Bassin houiller de Valenciennes. Description de la flore fossile. *Études des gîtes minéraux de la France, texte 1888:* 1–731; *atlas 1886:* 1–94.

ZEILLER, R., 1906. Bassin houiller et permien de Blanzy et du Creusot. Flore fossile. *Études des gîtes minéraux de la France, texte:* 1–261; *atlas:* pl. 1–51, 1906.

Botanical Journal of the Linnean Society (1983), *86:* 135–148. With 10 figures

Leaf and stem growth in the Lepidodendrales

W. G. CHALONER AND B. MEYER-BERTHAUD

*Department of Botany, Bedford College,
Regent's Park, London NW1 4NS.*

Received May 1982, accepted for publication July 1982

Measurement of leaf length and stem diameter in three Upper Carboniferous *Lepidodendron* species shows an exponential relationship between these parameters. Larger axes bear longer leaves even in material which had apparently completed its primary growth. It is suggested that this is a direct product of the determinate growth pattern of the ontogeny of arborescent lycopods. Measurements of *Cyclostigma* from the Irish Devonian are not explicable in these terms, and it is suggested that some specimens were immature at fossilization. It is recommended that the term leaf cushion be applied to the expanded leaf base in the Lepidodendrales at all growth stages, and not only after leaf abscission. Problems in the terminology of 'young' and 'old' growth stages in these determinate plants are discussed.

KEY WORDS:—*Cyclostigma* – determinate growth – leaf cushion – leaf growth – *Lepidodendron* – ontogeny.

CONTENTS

INTRODUCTION

Our picture of the ontogeny of typical members of the Lepidodendrales has been greatly influenced by three papers, namely Walton (1935), Andrews & Murdy (1958) and Eggert (1961). Walton first drew attention to differences in the anatomy of what were evidently different levels of stems or branches of the same species of *Lepidophloios*, which could not be simply reconciled as 'young' and 'old' stages of the same plant. Andrews & Murdy (1958), working principally with *Lepidophloios*, developed this suggestion and argued that growth in this plant was

0024–4074/83/010135+14 $03.00/0

10

determinate. This explains why the relative proportions of primary and secondary wood and of the pith differ in stems sectioned at successive levels up a plant. Finally, Eggert (1961) has reviewed the evidence from a wide range of petrified axes, and developed a more comprehensive picture of determinate growth in lepidodendrids. Since that time further work on lepidodendrid petrifactions has generally confirmed Eggert's interpretation, or at least been consistent with it.

Recently, DiMichele (1979, 1981) has gone further in trying to circumscribe the concepts of the genera *Lepidodendron* and *Lepidophloios*. He has recognized different growth strategies in these two genera, and has reconstructed the mature appearance of some species of both. Unfortunately, while permineralized preservation has given a sound basis for reconstructing the changing anatomy and leaf cushion development in these plants, it gives no clear indication of leaf length, and so of the outward appearance of different parts of the plant when the leaves were still attached. Indeed, the external appearance of the whole lepidodendrid tree, and in particular the growth (and eventual abscission) of the leaves has received little attention in the interval since Eggert (1961). Early work had generally assumed that the young condition of a lepidodendrid tree trunk must have been that of a distal leafy shoot of the same species as is the case, for example, in most living conifers. However, Andrews & Murdy (1958) remark that "some of the arborescent lycopods bore small leaves of only a few centimetres in length, while in others a leaf length in excess of 75 cm is known". Eggert went much further and suggested that as a general rule the main trunk, before its branching, may have borne very long leaves, while the small terminal branches of the crown, formed during the late (apoxogenetic) stage of determinate growth, bore much smaller leaves. In support of this idea Eggert, and Andrews & Murdy, cited a specimen which was later figured and described by Kosanke (1979). This was of a 100 mm diameter *Lepidodendron obovatum* Sternbg. stem bearing leaves up to 0.78 m in length, interpreted by the author as a young trunk prior to the first dichotomy. It might be noted that Hirmer (1927: fig. 205) figures a similar "long leaved *Lepidodendron*" which he attributes to *L. dichotomum* Sternbg.

In review, despite the firm documentation of determinate growth in lepidodendrids some 20 years ago, our knowledge of the external features of the lepidodendrid tree as it grew and particularly the form of the leaves on different parts of the plant remains very incomplete.

TAXONOMIC PROBLEMS IN LEAFY LEPIDODENDRIDS

A large number of species of *Lepidodendron* has been described since the genus was first recognized. There are certainly in excess of 300 species (see Jongmans, 1929; Jongmans & Dijkstra, 1969) and the number is probably nearer to 350. Most compression/impression species are based either on fragments of large stems lacking leaves (and so showing clear abscission scars) or on generally smaller branches with leaves still attached—only rarely do we have secure evidence of connection between leafy shoots and large (main axis?) fragments.

Thomas (1967, 1970, 1977) has used cuticle characters to strengthen the definition of a number of compression species of *Lepidodendron*, but unfortunately this added information is only rarely available. His cuticle material has generally been from larger stems, with leaves already shed, and so far has not led to the linking of leafy shoots and leafless axes.

The problems of trying to reconcile taxa based on parts of a large axis (with large leaf cushions and abscissed leaves) and smaller leafy shoots have been explored by few workers. Němejc (1947) suggested synonymies between some of the 'large' species, and those based on small shoots (e.g. he regarded *L. dichotomum* as representing the smaller branches of *L. obovatum*, i.e. *L. mannabachense, sensu* Thomas, 1970), and urged that species concepts should be broadened to include both leafy shoot and leafless cushion 'states' within the same species. This problem of naming of specimens bearing leaves still in attachment is aggravated by the circumstantial evidence that some lepidodendrids either had long-persistent leaves, or possibly never abscissed the leaves to form a leaf scar. In the latter case, the leaves were presumably eventually sloughed off, together with the leaf base ('leaf cushion') when periderm formed beneath them, as occurs on main stems of the conifer *Sequoia gigantea* (Lindley) Decne. Such species include *Lepidodendron simile* Kidston and *L. acutum* Presl, both in the sense used by Němejc (1947).

Some authors have favoured including such leafy lepidodendrid shoots, lacking evidence of abscission, in the genus *Ulodendron* (e.g. Renier, Stockmans & Willière; see Thomas, 1967: it should be noted that Jonker, 1976, places a rather different interpretation on the concept of this genus). Thomas (1967) has shown that the type species of *Ulodendron*, *U. majus* Lindley & Hutton has a cuticle with sunken stomata, similar to that of *Lepidodendron*. However, the latter genus was founded on species with clear leaf abscission (*L. obovatum*, *L. dichotomum* and *L. aculeatum* Sternbg. all have some claim to be regarded as types of the genus—see Andrews, 1970; Crookall, 1964; Chaloner, 1967). Thomas (1967) suggests that *Lepidodendron* may be distinguished from *Ulodendron* in having wider leaf cushions. The rather unsatisfactory state of the demarcation between these genera is well summed up in the remark (Thomas, 1967) that "some small leafy shoots can be included in *Ulodendron*, [but] . . . others are undoubtedly the small terminal shoots of other arborescent lycopod genera".

LEAF GROWTH IN LEPIDODENDRIDS

Until we have a fuller understanding of the range of leaf form within a whole lepidodendrid tree throughout its growth, the definition of taxa based on compression/impression fossils of leafy stem fragments lacking cuticle will remain arbitrary and unsatisfactory. (It has also to be acknowledged that more complete information will not necessarily give us a tidy taxonomy, applicable to fragmentary material.) Short of finding a specimen complete enough to give this sort of information directly, two approaches to the problem may be explored:

(1) We can investigate the relationship of leaf size and stem diameter within a single specimen showing some range of branch size.
(2) We can investigate the same parameters in separate specimens of varied size, where we have sound evidence that we are dealing with a single species. (For example, in *Cyclostigma kiltorkense* Haughton, dealt with below, where distinctive characters of leaf shape and attachment show continuous gradation through a range of specimens believed to represent a single species.)

In both situations, guarded extrapolations may then be made beyond the limits of the material available. We consider two such instances here, attempting both approaches with some large Westphalian (Coal Measures) lepidodendrid specimens, and the latter approach alone with the Late Devonian lycopod *Cyclostigma kiltorkense*.

Observations on leafy shoots of Lepidodendron

Three species of *Lepidodendron*, based on leafy shoots, occur commonly in the British Westphalian (Upper Carboniferous). These are *L. simile*, *L. acutum* and *L. ophiurus* Brongniart. They are reviewed in Němejc (1947) who gives a brief synonymy, and also in Chaloner (1967). The characters on which these species may be separated are also summarized in Chaloner & Collinson (1975). We have measured and plotted leaf length and the corresponding diameter of the axis for a number of moderately-sized specimens of these species, in the British Museum (N.H.) and the Institute of Geological Sciences. Some of the specimens used are shown in Figs 1–4. They were deliberately selected to give us as great a range of axis diameter as possible within each species; in some cases (Figs 1, 3) one specimen gave several corresponding length/diameter measurements. Others gave a single measurement (Fig. 2).

Our choice of specimens has been restricted by this concentration on relatively large branching shoots, showing some range of axial diameter. Our sample size has thus been very small for all three species. Nevertheless, we suggest that our results offer the basis for a 'first working hypothesis' on lepidodendrid leaf growth.

The measurement of leaf length seen on a leafy shoot in a compression fossil presents certain difficulties. The extent of exposure of the leaves by the fracture plane which has revealed the fossil is, in part, a function of the angle that the leaf makes with the stem (see Rex & Chaloner (in press)). When the tapering leaf tip is seen intact, lying on a bedding plane above the axis to which it is attached, and its base is also exposed, its full length can be measured with confidence. More commonly, leaves are seen more-or-less in profile at the stem margin (e.g. Fig. 8), and in these cases the leaf apex may still be hidden in the matrix. Measurement of the observed length of such leaves will only be a minimum value. However, where many leaves are seen on a stem of given diameter, the largest leaf observed may be taken to approximate to the true leaf length at that position on the specimen. Accordingly, a number of leaves was measured at each stem diameter and the highest value taken.

The data obtained have been plotted on a log/log basis in Fig. 9. The three species show comparable, approximately linear relationships of leaf length and axis diameter on the log scale plot, although there are discernible differences between the species. A regression line has been plotted for each species, and these have slightly different gradients (ranging from 0.29 for *L. acutum* to 0.43 for *L. simile*). While our data source is very limited, it appears that there is an exponential relationship between the length of leaves and the diameter of the stem bearing them.

We assume that we are observing the distal branches of the crown of the mature tree. As we move back from the finest (smallest) shoots, leaf length increases as a power between 0.2 and 0.5 of the diameter of the axis. In other words, larger axes bear longer leaves, but with the leaf length increasing at rather less than the square root (power of 0.5) of the axis diameter.

Since in all the specimens that we have used the leaf cushions are in contact with one another, we assume that the axial growth (i.e. the cortical tissue) is primary

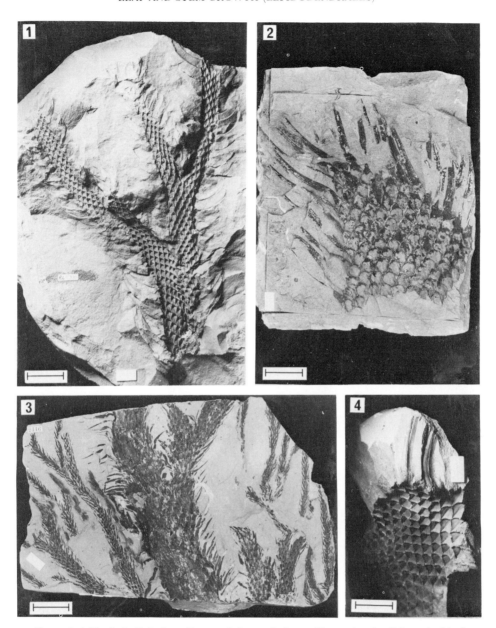

Figures 1–4. Fig. 1. *Lepidodendron ophiurus*, scale bar equivalent to 20 mm. A twice-dichotomized leafy shoot. Note the longer leaves on the larger axis at the base. The arcuate leaves contrast with the more or less sigmoid ones of *L. simile* (see Fig. 3). Fig. 2. *Lepidodendron acutum*, scale bar equivalent to 20 mm. A large stem with relatively long and broad, sigmoid leaves. Fig. 3. *Lepidodendron simile*, scale bar equivalent to 20 mm. Note the difference in the leaf length in different parts of the shoot system. Fig. 4. *Lepidodendron* cf. *acutum*, scale bar equivalent to 14 mm. Impression of a leafy shoot. The axial line of the shoot slopes to the left. Three attached leaves are seen in profile at the side of the axis. The impressions of apparently leafless cushions on the stem surface presumably have leaves still attached, like those seen, but dipping into the body of the matrix behind the specimen. (All the specimens are from the British Westphalian, Upper Carboniferous, and are housed in the Palaeontology collection of the Institute of Geological Sciences, London, with the specimen numbers: Fig. 1, *RC 3208*; Fig. 2, *18232*; Fig. 3, *2446*; Fig. 4, *15209*.) Figs. 2 and 3 were photographed under ethanol.

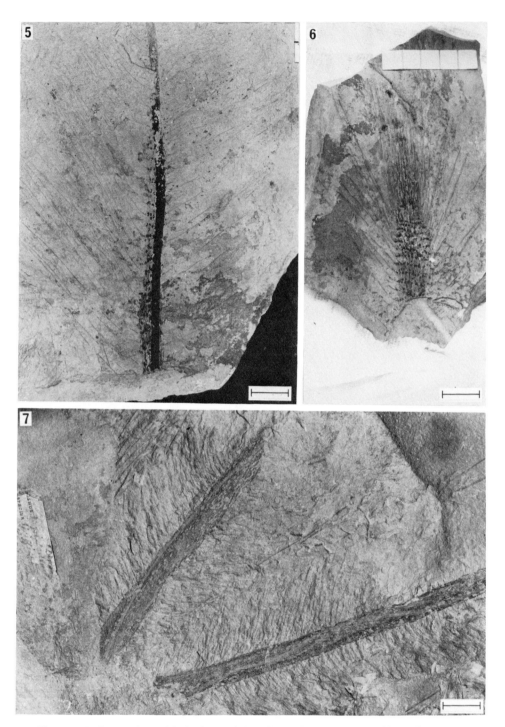

Figures 5–7. Compressions of leafy shoots of *Cyclostigma kiltorkense*, from the Upper Devonian of Kiltorcan, Kilkenny, Ireland. Fig. 5. A long leaved shoot, scale bar equivalent to 20 mm. Note the vascular cylinder as a broad dark band at the right side of the axis, and the narrow leaf attachment. Fig. 6. Apical part of a young shoot with long leaves, scale bar equivalent to 20 mm. Note the difference in spacing between the pseudoverticils of leaves from the base to the apex of the specimen. Fig. 7. Compressions of two short-leaved shoots, scale bar equivalent to 20 mm. (Specimen numbers: Fig. 5, *V 29261*, British Museum (Natural History) London. Fig. 6, unnumbered specimen in the Geological Survey of Ireland, Dublin. Fig. 7, *32*, Irish National Museum Collection, Dublin.) Figs. 5 and 6 were photographed under ethanol.

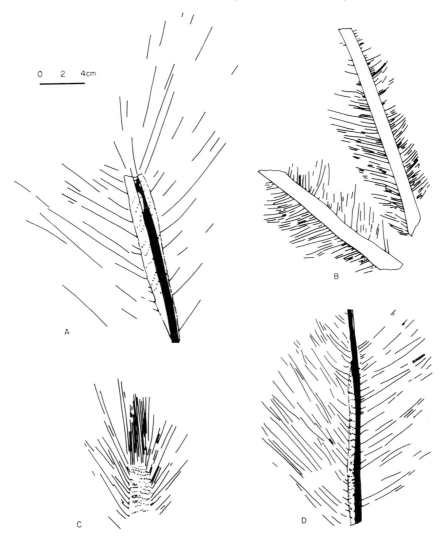

Figure 8. Leafy shoots of *Cyclostigma kiltorkense*. A, C, D. Stems, believed to be immature, bearing nearly mature leaves. B. Mature stems bearing mature but shorter leaves. We suggest that A, C and D would have undergone further increase in girth and particularly in the case of C, length increment, if they had continued their development. (A, B: unnumbered specimen and specimen *no. 32* respectively, Irish National Museum, Dublin. C: unnumbered specimen in the Geological Survey of Ireland, Dublin. D: No. *V 29261*, British Museum (Natural History) London.)

only. We also believe that the leaf growth was more or less complete. This supposition is supported by the fact that the leaves appear to have been robust in their disposition in the matrix, since some evidently protruded into the sediment as it accrued above them.

The size differences between the three species appear to be significant, but we have not enough data to demonstrate this. On the material at our disposal, it seems that the leafy shoots of *L. simile* and *L. ophiurus* differ not only in leaf shape but in relative size, the former having slightly shorter leaves than the latter on stems of corresponding diameter.

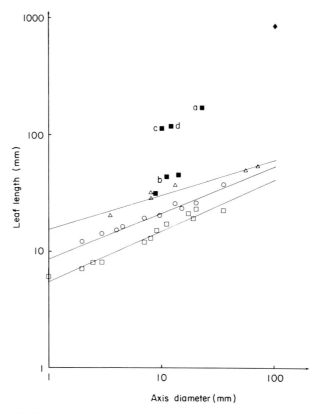

Figure 9. A plot of leaf length against the corresponding stem diameter, for branches of various sizes (1–70 mm diameter) for five species of Lepidodendrales. □: *Lepidodendron simile*, from five specimens; equation of the regression line: log L = 0.43 log D +0.74. ○: *Lepidodendron ophiurus*, from four specimens; equation of the regression line: log L = 0.39 log D+0.94. △: *Lepidodendron acutum*, from five specimens; equation of the regression line: log L = 0.29 log D+1.19. ■: *Cyclostigma kiltorkense*, from four specimens. The letters a, bc, c and d refer to the specimens figured respectively under the letters A, B, C and D of Fig. 8. ◆: *Lepidodendron obovatum* var. *grandifolium* (from Kosanke, 1979). All the *Lepidodendron* specimens come from the Westphalian (Upper Carboniferous) of Britain; the *Cyclostigma*, from the Irish Famennian.

If, as suggested above, we are seeing mature shoots at the periphery of the crown, these were formed in the last stages of growth of the lepidodendrid tree. The implications of determinate growth in these plants has been graphically expressed by Harris (1981): "*Lepidodendron* grew from a *Selaginella*-like megaspore almost to full stature, then it branched with progressively more slender branches, and then it coned and died and its fallen remains . . . became the coal of Europe and North America". We consider that the exponentially diminishing leaf length of the smaller branches of the crown demonstrated in our material is one of the features of the apoxogenetic growth phase of the determinate growth pattern.

One of the major gaps in our knowledge is our ignorance of the nature of the small shoots presumably formed at the crown of a trunk of the *Lepidodendron obovatum*-type. It is possible that such distal leafy shoots pass unrecognized among the 'small leaved' species dealt with here (*L. simile*, *L. ophiurus*, *L. acutum*) despite the very long leaves observed on Kosanke's large specimen of *L. obovatum* var. *grandifolium* Kosanke. It is clear from Fig. 9 that such very long-leaved axes (diamond, upper right) do not lie on an extrapolation of the lines derived from the peripheral shoots of the crown. This is consistent with the general contention of Eggert (1961) that leaves borne on the lepidodendrid trunk are different in kind from those borne on the terminal shoots. Leaf length reflects this difference, as much as does the internal anatomy.

A question related to the present problem is the growth of leaf cushions and scars. Circumstantial evidence suggests that while increase in stem girth on large lepidodendrid axes has caused separation of leaf cushions, with deepening furrows of cortical tissue opening between them, the cushions themselves do not show evidence of prolonged growth. In this they evidently behaved like the decurrent leaf bases of the larger stems of the living conifer, *Sequoia gigantea*. Observations on lepidodendrid leaf scar shapes suggest similarly that there is no lateral expansion of the scars after leaf shedding. In this respect there is a significant difference from the growth of a number of living conifers; for example, in *Araucaria araucana* (Molina) K. Koch, the primary stem surface is retained for many years, showing old leaf scars stretched tangentially far beyond the original leaf width. We do not attempt to consider the problems of lepidodendrid leaf cushion and scar expansion further here, but this subject would surely repay investigation in material showing a suitable range of leaf cushion size.

Leaf length in Cyclostigma

We have plotted leaf length and stem diameter data for this species from a range of specimens housed in the Irish National Museum and other collections in Dublin. All this material came from the 'Old Quarry' on Kiltorcan Hill, mostly collected prior to the First World War. This horizon probably falls within the Famennian but the possibility that it is of early Tournaisian age cannot be ruled out. We do not contest the view of Johnson (1913) that all the lycopod axes from that quarry, with and without leaves, represent a single species. The narrow linear leaves with a circular area of attachment (a leaf scar) lying flush with the stem surface constitutes a distinctive combination of characters. (*Sublepidodendron* cf. *isachseni* (Heer) Schweitzer occurs higher in the section in the 'New Quarry', but this has characteristic fusiform leaf cushions giving it a very different appearance.)

The leaves present on shoots of varied diameter are obviously not of constant length, and in this *Cyclostigma* conforms with our observations on other Lepidodendrales. But the relationship between leaf length and axial diameter in *Cyclostigma kiltorkense* (Fig. 9) does not present as simple a picture as that for the *Lepidodendron* species described above. Rather few of our Irish specimens of *C. kiltorkense* still bore leaves. Most of the material (including the syntype specimens) are of larger leafless axes with clear leaf scars. It may be noted that in the extensive material attributed to *C. kiltorkense* from Spitsbergen by Nathorst (1914) and Schweitzer (1969), only a single leafy shoot is represented, and in this the leaf length is apparently not seen intact. We had only five specimens in which the axial diameter could be clearly seen, and the leaf length measured (points on Fig. 9). The five specimens we have, although all of rather comparable diameter (around 10–30 mm), show considerable variation of leaf length (30–120 mm). These dimensions obviously do not fall within the simple linear relationship of our *Lepidodendron* species (Figs 5–7 & 8A–D). A number of explanations could account for this. One (that we discount) is that we are dealing with more than one species of plant; another (which we favour) is that we are seeing specimens which represent parts of the plant which were still growing in length, girth, or both, at the time of death and fossilization.

All the plant material from the Kiltorcan Old Quarry is apparently 'hypautochthonous'—it has been transported, but probably not far, from its site of growth. None of the plants is in a position of growth, and the several stigmarian bases of *Cyclostigma* known (Johnson 1913, Chaloner, personal observation) lie in the plane of bedding rather than in a growth position, as though uprooted, and then transported and incorporated in sediment deposited nearby. We suggest that our material includes specimens such as that of Fig. 8C in which neither leaf length nor stem length increment has ceased. It is possible that the three long-leaved specimens (upper three points of Fig. 9) are major axes which would have undergone considerable girth increase, while the smaller-leaved axes are distal branches of the crown. (This would mean in effect that the three upper *Cyclostigma* points in Fig. 9 would eventually have matured to a position well to the right of that shown). However, we have no other evidence to support this special pleading beyond a supposition that *Cyclostigma* shared a common growth pattern with other members of the Lepidodendrales.

A model of leaf growth in Lepidodendrales

We attempt to interpret our observations on leaf size and stem diameter in terms of an imaginary growth model, which we suppose is common in general terms to all Lepidodendrales. This is based on five suppositions, any or all of which may be shown to be invalid by further observations. We advance this model simply as a hypothesis, hoping that its validity may be tested by evidence from new material. Our assumptions are:

(1) We suppose that the young leaves elongated rapidly as the stem was growing, soon attaining a dimension close to their mature length. Leaf elongation probably continued at a slower pace after stem elongation had ceased, but we have no evidence of this.

(2) From evidence of stelar and cortical secondary thickening in petrifactions, we presume that stem girth increase continued after leaf growth (and leaf base or 'cushion' growth) ceased.

(3) It is highly probable that the leaves on most of the branches (and trunk) were shed before increase of stem girth was complete; the leaves of the slender terminal branches were probably not deciduous. (Presumably they were still attached when the tree died on completion of its determinate growth pattern.)

(4) On large axes (the trunk, major branches) the leaf cushions were eventually sloughed off by continued girth increase (secondary thickening) of the cortex, leaving a furrowed cortical surface.

(5) The nature of the growth involved in girth increase we assume to have been primary in the early stages, and later (presumably, at least after leaf growth has ceased) by secondary growth of the cortex. We have not attempted to separate these two phases of growth in Fig. 10, acknowledging that in compression material the extent of cortical secondary thickening is not evident until separation of the leaf cushions indicates that it is well advanced. None the less it is probable that the flat parts of the curves of Fig. 10 correspond to

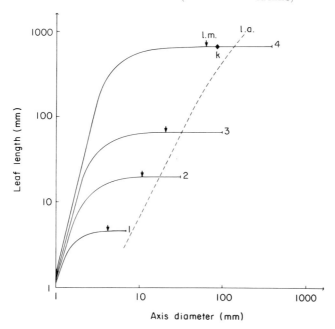

Figure 10. Hypothetical model for the development of four leaves in different positions on an arborescent lycopod. Each line traces the changing leaf length and stem diameter during the ontogeny of that part of the plant. The end of each line represents the condition at the completion of determinate growth. l.m. and arrows: leaf maturity; l.a. and dashed line: leaf abscission. 1: slender shoot, bearing short persistent leaves. 2, 3: distal branches of the crown, shedding their leaves before attaining their maximum diameter (i.e. line extends across l.a., indicating continued axial diameter increase after leaf-shedding). 4: part of the trunk or major branches of the crown. Kosanke's long-leaved specimen of *Lepidodendron obovatum* can be represented by the point k of the curve, prior to reaching the line of leaf abscission (l.a.), but probably with potential for further increase in diameter, had death and fossilization not ensued.

secondary girth increase, after leaf growth (and presumably also stem length extension growth) had ceased.

These assumptions, coupled with the data given in Fig. 9, form the basis of our diagram in Fig. 10. In this figure, each line is the hypothetical pathway of growth of a single leaf. For simplicity, only four leaves are followed, on different parts of the tree. All start as leaf initials of 1 mm length on a 1 mm diameter apex, simply to give a common origin to the curves on a logarithmic scale. (Obviously the 'diameter' of the axis at the point of leaf origin, on a broad dome-shaped apex, is an artificial and arbitrary concept.)

Each curve shows an early rapid increase in length (first, steep part of the curve) followed by leaf growth slowing, while the stem on which it is borne continues to increase in girth. Two 'events' in leaf development are shown: the cessation of length growth by the leaf (arrows) and leaf abscission (the line l.a.). Curve 4 is of growth on a major stem, attaining an eventual diameter of nearly a metre, after leaf abscission. Curves 3 and 2 represent the growth of leaves on smaller branches of the crown (with eventual leaf abscission) while curve 1 is the growth pathway of a leafy shoot attaining only *c.* 10 mm diameter, still retaining its leaves as the tree completes the determinate growth. Obviously the actual figures and gradients would vary from species to species and to a lesser extent within a species. Implicit in this model is the broad observed correlation between leaf length and mature stem diameter, when the leaves were mature, demonstrated above. We show, at 'k', the significant specimen of Kosanke (1979) referred to above, in order to put it in the context of the other data.

A number of living plants, particularly monocotyledons lacking the ability to increase in girth beyond that attained by prolonged primary growth, offer an interesting model for lepidodendrid development. For example, the leaves of the giant bamboo *Dendrocalamus* Nees. are of very different morphology from the photosynthetic leaves of the 'crown' of that tree. The leaves of the 'trunk' are broad-based, more or less scarious, and apparently protect the apex during the early stage of internode elongation. They are shed while the leaves of the crown are still photosynthetically active. Although in other aspects of their growth *Lepidodendron* and *Dendrocalamus* were obviously very different, in their leaf heterophylly between axes of different status, they show this degree of similarity.

Problems of terminology

From what has been said above, two problems of terminology arise in describing *Lepidodendron* and related genera.

'Young' and 'old' growth stages: A good deal of confusion and ambiguity in accounts of *Lepidodendron* growth have resulted from a terminology borrowed from indeterminate growth, characteristic of dicotyledonous trees. Authors have spoken of 'old shoots' and 'young shoots' with the general assumption that the former were derived from a state represented by the latter (as would be the case with, say, an oak tree). Evidently in the Lepidodendrales this was not necessarily true; the long leaves borne on the main axis of *Lepidodendron* seem to be a feature of that early stage of its ontogeny and are better referred to as leaves of the juvenile state. A comparable usage is employed in heterophyllous living conifers, where the juvenile leaves (formed early in the plant's growth) are quite different from those formed later (e.g. in *Pinus*, and in *Cupressus*). If this is correct, the long leaves of Kosanke's specimen are better regarded as the mature leaves of the juvenile state of a main axis or a major branch of the *Lepidodendron obovatum* tree; our observations are consistent with the supposition that young branches of the crown of that species bore much shorter leaves, although this has still to be proved.

Leaf cushions: The characteristic feature of *Lepidodendron* is its expanded pyramidal (or, loosely, 'diamond-shaped') decurrent leaf base. From the early days of palaeobotany this leaf base, particularly after the leaf has been abscissed, has been called a leaf cushion (French, 'coussinet foliaire'; German, 'Blattpolster'). Many authors take the view that the leaf base does not constitute a leaf cushion until the leaf has been shed. Scott (1900: 118) writes of *Lepidodendron*—"when the leaves were shed, their bases remained attached to the surface of the stem, forming the leaf cushions". Taylor (1981: 131) follows essentially the same interpretation in saying "the leaf cushion actually represents the expanded leaf base left behind after the leaf drops off".

A significantly different interpretation of 'leaf cushion' is used in describing conifers, where Harris (1979) writes of the Jurassic genus *Elatides*: "Leaves persistent, . . . arising from decurrent cushion". Thomas (1967) extends this usage to *Ulodendron*, of which he records the "leaf cushions . . . bearing persistent linear leaves". We accordingly use the term here for lepidodendrids more broadly than in the sense of Scott (1900) or Taylor (1981), for an expanded decurrent leaf base, regardless of whether or not there is any evidence for abscission (Fig. 4).

CONCLUSIONS

Within each of the three *Lepidodendron* leafy shoots we have studied, the broadest stems bear the longest leaves. In more exact terms, there proves to be an exponential relationship between leaf length and the diameter of the stem bearing the leaves. This appears to be a feature innate in the determinate growth of these plants, although we have only demonstrated it for the late stage (apoxogenetic phase) of growth. The wide occurrence of ulodendroid scars on several genera of Lepidodendrales suggests that the shedding of mature leafy branches may have been widespread in this group. We do not know whether the leafy shoots on which we have made our measurements represent the distal parts of such deciduous branches, or of the crown of a tree which had reached the end of its determinate growth.

While the measurements recorded here are consistent with a determinate growth pattern this does not in itself resolve the problem of recognizing well-defined species in such fragmentary compression/impression fossils. We still cannot securely link the long-leaved major stems (trunks) of *Lepidodendron* with species based on small distal leafy shoots. It seems that leaf growth followed a different course in the (epidogenetic) growth of the main trunk from that shown in the (apoxogenetic) formation of the crown. A search should be made for specimens showing organic connection between leafy shoots and larger stems from which leaves have been abscissed.

ACKNOWLEDGEMENTS

We record our grateful thanks to the British Council whose award of a Fellowship has made possible a period of research in London by one of us (B.M-B.), during which visit this work was carried out. We also gratefully acknowledge the generous loan of specimens by the Trustees and staff of the British Museum (Natural History), the Institute of Geological Sciences and the Irish National Museum, Dublin.

REFERENCES

ANDREWS, H. N., 1970. *Index of Generic Names of Fossil Plants, 1820–1965. Geological Survey Bulletin, 1300.* Washington: United States Government Printing Office.
ANDREWS, H. N. & MURDY, W. H., 1958. *Lepidophloios* and ontogeny in arborescent lycopods. *American Journal of Botany, 45:* 552–560.
CHALONER, W. G., 1967. Lycophyta *2: Bryophyta, Psilophyta, Lycophyta.* In E. Boureau, W. G. Chaloner, O. A. Hoeg & S. Jovet-Ast (Eds), *Traité de Paléobotanique,* 434–802. Paris: Masson et Cie.
CHALONER, W. G. & COLLINSON, M. E., 1975. An illustrated key to the commoner British Upper Carboniferous plant compression fossils. *Proceedings of the Geologists' Association, 86:* 1–44.
CROOKALL, R., 1964. Fossil Plants of the Carboniferous Rocks of Great Britain, (2nd Section). *Memoirs of the Geological Survey of Great Britain, Palaeontology, 4:* 217–354.
DIMICHELE, W. A., 1979. Arborescent lycopods of Pennsylvanian age coals: *Lepidophloios, Palaentographica, B, 171:* 57–77.

DIMICHELE, W. A., 1981. Arborescent lycopods of Pennsylvanian age coals: *Lepidodendron*, with description of a new species. *Palaeontographica, B, 175:* 85–125.

EGGERT, D. A., 1961. The ontogeny of Carboniferous arborescent Lycopsida. *Palaeontographica, B, 108:* 43–92.

HARRIS, T. M., 1979. *The Yorkshire Jurassic flora—V. Coniferales.* London: British Museum (Natural History).

HARRIS, T. M., 1981. The pteridophytes of the ancient world. *Geography, 11:* 5–10.

HIRMER, M., 1927. *Handbuch der Paläobotanik.* Bd. I. Munich: Oldenbourg.

JOHNSON, T., 1913. On *Bothrodendron (Cyclostigma) kiltorkense. Scientific Proceedings of the Royal Dublin Society,* (*N.S.*), *13:* 500–528.

JONGMANS, W., 1929. *Fossilium Catalogus (Plantae)—Lycopodiales,* II, VII. 's Gravenhage.

JONGMANS, W. J. & DIJKSTRA, S. J., 1969. *Fossilium Catalogus (Plantae)—Lycopodiales,* II, VII. Pars. 72. Ed. Dijkstra, 's Gravenhage.

JONKER, F. P., 1976. The Carboniferous "Genera" *Ulodendron* and *Halonia*—an assessment. *Palaeontographica, B, 157:* 97–111.

KOSANKE, R. M., 1979. A long leaved specimen of *Lepidodendron. Geological Society of America, Bulletin, 1:* 431–434.

NATHORST, A. G., 1914. *Nachträge zur paläozischen Flora Spitzbergens—zur fossilen Flora der Polarländer, 1:* Stockholm.

NĚMEJC, F., 1947. The Lepidodendraceae of the coal districts of Central Bohemia (a preliminary study). *Acta Musei Nationalis Prague, 3B:* 45–87.

REX, G. M. & CHALONER, W. G. (in press). The experimental formation of plant compression fossils. *Palaeontology.*

SCHWEITZER, H. J., 1969. Die Oberdevon-flora der Bäreninsel. 2. Lycopodiinae. *Palaeontographica, B, 126:* 101–137.

SCOTT, D. H., 1900. *Studies in Fossil Botany.* London: A. & C. Black.

TAYLOR, T. N., 1981. *Paleobotany. An introduction to Fossil Plant Biology.* New York: McGraw-Hill.

THOMAS, B. A., 1967. *Ulodendron* Lindley & Hutton, and its cuticle. *Annals of Botany, 31:* 775–782.

THOMAS, B.A., 1970. Epidermal studies in the interpretation of *Lepidodendron* species. *Palaeontology, 13:* 145–173.

THOMAS, B. A., 1977. Epidermal studies in the interpretation of *Lepidophloios* species. *Palaeontology, 20:* 273–293.

WALTON, J., 1935. Scottish Lower Carboniferous plants: the fossil hollow trees of Arran and their branches (*Lepidophloios wünschianus* Carruthers). *Transactions of the Royal Society of Edinburgh, 58, 2:* 313–337.

Botanical Journal of the Linnean Society (1983), 86: 149–159. With 10 figures

The stem of *Pachypteris papillosa* (Thomas & Bose) Harris

TOM M. HARRIS

Geology Department, University of Reading, Reading RG6 2AS

Received December 1981, accepted for publicaton March 1982

Pachypteris papillosa (Thomas & Bose) Harris, leaves of a pteridosperm, are locally abundant in the Middle Jurassic of Yorkshire. Microsporophylls too are known but the stem has been represented only by a few puzzling fragments. Better-preserved stem material is now described. When young the stems were thick (up to 15 mm) and largely composed of soft parenchyma and thus succulent. The leaves are borne sparsely, each at the end of a prominent mamilla. The stem surface is covered by hemispherical emergences with a thickly cutinized epidermis which, when isolated, are beret-shaped. This cuticle is the only robust part of the young stem.

Later the stem formed a thick cylinder of wood and the surface was stretched. A new surface (presumed to be cork), was formed and the 'berets' and mamillae were lost. Branching occurs at distant intervals, when the stem divides nearly equally into widely diverging parts.

Pachypteris papillosa is pictured as a large shrub forming a mangrove-like thicket along tidal rivers.

KEY WORDS:—Mesozoic pteridosperm – *Pachypteris* stem.

CONTENTS

INTRODUCTION

Pachypteris papillosa is one of the Mesozoic pteridosperms. These were recognized and gradually accepted over the last 60 years but we still cannot define them clearly. Most have fern-like leaves, but unlike ordinary fossil fern leaves they can be lifted from the rock with a knife; their strength is in their thick cuticles. *Pachyteris lanceolata* Brongniart (Brongniart, 1838) was the first, and this, like more than half the species, is only known from its leaves. *Pachypteris papillosa* is better known for we have its microsporophyll and information about its stem. Its female organs are undescribed.

In about 1912 Hamshaw Thomas found a leaf coal composed of compacted *P. papillosa* leaves at Little Roseberry, near Great Ayton, N Yorkshire in Middle Jurassic rock (Thomas, 1913, 1915). He also found the same leaf at Roseberry

149

Topping nearby and, with the leaves, what looked like a seed-bearing axis. His study of their cuticles convinced him that the leaves and axis belonged to the same species and this is accepted. However, he identified the leaf as *Thinnfeldia rhomboidalis* Ettingshausen, later transferred to *Pachypteris* by Doludenko (1969, 1974). Gothan (1914: 115) pointed out that the Yorkshire leaf was not *T. rhomboidalis* but it was not named formally until Thomas & Bose (1955) described it as *Pachydermophyllum papillosum* which Harris (1964) transferred to *Pachypteris*. Thomas took the round seed-like bodies on the axis to be true seeds, on which he reported orally at scientific meetings under the name *Sarcostrobus*. This was printed in resumés but without any description and is hence a *nomen nudum*. He did much work on this fossil but was frustrated by being unable to find any microscopic part of a gymnosperm seed, the micropyle, the nucellus or megaspore membrane and published no more on them.

I described these perplexing fossils as stems attributed to *"Pachypteris papillosa"* (Harris, 1964), but they remained puzzling because I could find no attached leaves or leaf scars. I firmly rejected the round bodies as seeds and called them 'berets' from the shape of their cutinized epidermis. Better specimens have since been found which do show their leaf scars borne on prominent cushions or mammillae (and they are now recognized in some of the earlier described specimens). The 'berets' remain peculiar; I take them to be local swellings of photosynthetic tissue, emergences formed by the outermost cortex. The new specimens also show that the stems eventually became woody; the plant must have been a large shrub or a tree and branched in an unusual way. It seems also that the stems were rather succulent when young.

OCCURRENCE

At first Thomas only knew *P. papillosa* from Little Roseberry and Roseberry Topping (Thomas, 1913, 1915) but added others later in Thomas & Bose (1955); Harris (1964) listed 22 localities. The stems are known only from three of the best leaf localities, Roseberry Topping, Hasty Bank and Farndale Hillhouse Nab but isolated 'berets', just like those on the stems, have been obtained by bulk maceration of shales providing *P. papillosa* leaf material from 10 more localities. They have not been found apart from *P. papillosa*.

All 22 leaf localities are at the base of the plant-bearing rocks of the Yorkshire Middle Jurassic and at a short distance above the marine *communis* zone of the Upper Lias. They deserve more study. Unlike most Yorkshire plant beds which have land spores only, these have marine acritarchs as well. The *P. papillosa* beds seem to be part of what Yorkshire geologists have called the Dogger, a varied and partly marine series between the fully marine Upper Lias and the freshwater Bajocian and are placed by some in the Aalenian; to others the Aalenian is the lowest stage of the Bajocian. The English word Dogger has other meanings. Primarily it is a local name for an ironstone nodule. It has been used by some continental geologists for a large part of the Middle Jurassic. Thomas thought that his Roseberry beds were Liassic partly on the basis of his *Thinnfeldia* determination. Probably no bed with *P. papillosa* is truly Liassic, but it is likely that they are slightly older than those basal Yorkshire plant beds without it.

MATERIAL

No stem has an attached leaf and all well-preserved leaves had been shed by abscission. Their expanded petiole bases are clean cut; unfortunately all are flattened so the cut surface is not seen. Leaves are available in huge numbers, and except possibly at Little Roseberry, all had been transported by water; They are often undamaged and form excellent compressions. The leaves have very thick cuticles, extending inwards along the epidermal walls and it is the cuticle which makes the leaf a good fossil. Its interior is a brittle coal which looks featureless apart from darker cells along the veins and even when the leaf is partly cleared by decay no hypodermis or fibres are seen. Since certain conifer leaves preserved along with *P. papillosa* do show both hypodermis and fibres, I think they were probably absent in *P. papillosa*. Certain *P. papillosa* leaves which were folded before compression suggest an original thickness of nearly 1 mm, and if hard internal tissues were absent this would make it fleshy. I imagine it resembled the leaves of some xerophytes where a heavily cutinized epidermis encloses a soft, parenchymatous interior.

Fragments of stems of various sizes are occasional among the numerous leaves at the three localities mentioned but at Hillhouse Nab there was a lenticle about 1 m broad and a few centimetres thick where the stems are numerous and commoner than the leaves. The rock is a weak sandy silt and the size of the blocks limits the length of the stem specimens to pieces up to 20 cm long and broken at both ends. The broader stems are compressed to a crumbly coal layer 2–4 mm thick. These Hillhouse stems are unattractive specimens but are the best of their kind that we have.

All three localities gave stems varying in width from 10 mm upwards. In general, the fragments show no evident taper and only in the two branched stems, which happen to be broad, could the top and base be recognized. The leaves were sparse and each is represented by a scar at the end of a mamilla. These mamillae project 5 mm at the edges of the compressed stem but on the surface are totally flattened by compression, though a slight trench may mark an edge of a mamilla. When the rock is split, the break often passes through the coal substance rather than over the stem surface, and the leaf scars are best seen as imprints on the matrix after all the coal has been removed.

In the following description, I suppose that all the stems were at first similar, but with age became thicker and woody. In the primary state they were 10–15 mm thick (with an additional 5 mm where there is a mamilla). Such stems have a rather firm outer cortex which, when denuded, shows no growth splits. It is mostly, and I think at first, entirely covered with 'berets'. There is little firm tissue inside this cortex and the whole, including cortex and 'berets', compresses to a layer of coal barely 1 mm thick. Often the interior is delayed and occupied by mud. I recall Thomas' manuscript name, *Sarcostrobus*, 'fleshy cone' for these fossils.

None of the stem fragments 10–15 mm wide shows a leaf scar on its surface and I refrained from removing the coal in the hope of exposing the imprint of one on the rock. But one fragment, figured in Harris, 1964, shows a bulge at each end and I now regard these bulges as laterally compressed mamillae. The present narrow fragments give no other fresh information but I repeat what one shows about the interior (p. 154).

Stems 20–40 mm wide still show their mamillae and there are always some

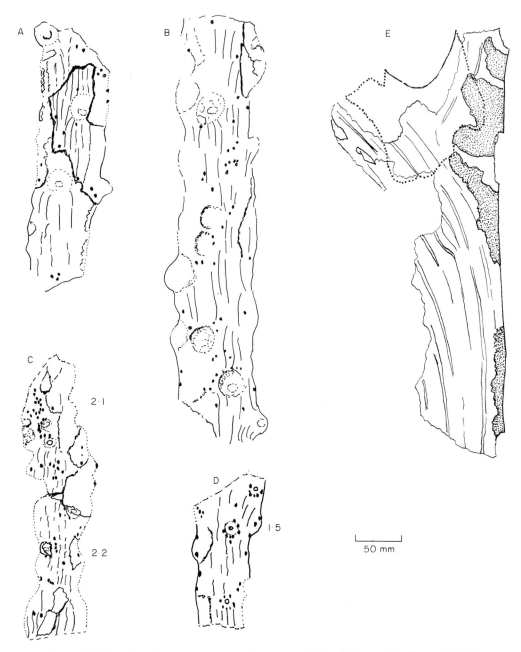

Figure 1. *Pachypteris papillosa* compressed stem fragments, all from Hillhouse Nab (Lower Deltaic). The specimens are rock imprints with some adherent coal 'berets' and imprints of 'berets' are both shown as black spots, bark splits are shown as vertical lines but the less clear ones are omitted. Dotted outlines represent obscurely shown edges or parts lost by breaking of the rock. Scale equivalent to 50 mm. A, counterpart of B, mamilla at top left corresponds to the bulge in B (see also Fig. 5) V. 60659b. B, longest fragment, also figured in Fig. 3. V. 60659a. C, fragment with less prominent mamillae (see also Fig. 6). Regions labelled 2.1 and 2.2 are shown in Figs 3 & 4. V. 60660a. D, fragment with weakly swollen mamillae (see also Figs 4, 7). V. 60661. E, large forked stem, no 'berets' or mamillae remain. V. 606629. One fragment of the counterpart (reversed) V. 60662b is shown in position. All the left edge of this stem was lost in collecting.

'berets', but much of the cortex is denuded. The surface of the outer cortex shows longitudinal splits caused by growth, the largest are up to 15 mm long and 0.5 mm wide in the middle, but most are much smaller. The stem is compressed to a coal layer about 2 mm thick in the inner part, but thinner towards the margins. I regard the thicker coal as compressed secondary wood.

Stems over 40 mm wide show no projecting mamillae and very few 'berets' remain attached. The substance is compressed to a coal 3 mm or more thick and it is thick almost to the edge. I imagine that the primary critical tissues are now lost. The surface is marked with longitudinal cracks which may be up to 2 mm wide. I imagine that the surface is formed by a cork layer. The specimens mentioned below without localities are from Hillhouse Nab.

V. 60660 a–c text Fig. 1C is mainly about 25 mm wide, but the margins are poorly shown in both part and counterpart. The coal layer is nearly 1 mm thick, cortical growth cracks are narrow and short. As originally studied (when most of

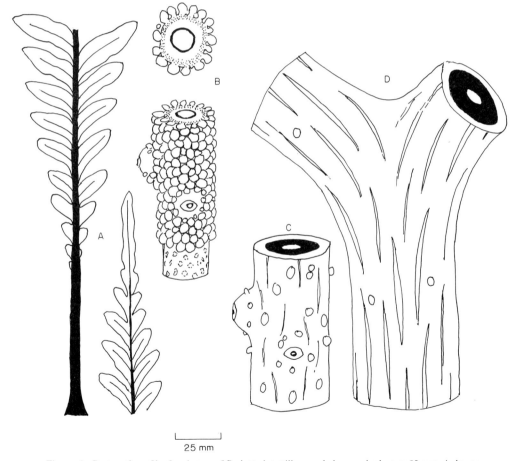

25 mm

Figure 2. Restoration of leaf and stem of *Pachypteris papillosa*, scale bar equivalent to 25 mm. A, lower half and top of leaf of normal size. B, young stem in transverse section and in side view. Xylem cylinder black, firm outer cortex stippled, one leaf base mamilla is in lateral view, another in face view. At the base the 'berets' are removed. C, stem with some secondary growth, these are small splits in the surface and some berets are missing. D, older, forked stem, few 'berets' remain, leaf bases are not seen and the surface splits are wider.

the coal was already lost), the part imprint Fig. 6) was uninformative, though it does retain a good many 'berets'. However, when thin patches of broken coal which stick to the surface at a few points were removed, good leaf scars (Figs 8, 9) were seen surrounded by a compression rim marking the edge of the cushion.

V. 60659 a, b is mainly 35 mm wide but with mamillae (leaf cushions) projecting another 5 mm. Coal layer is about 2 mm thick but thin at the margins. Surface mainly formed by the denuded outer cortex (or cork) but a few 'berets' remain. Growth cracks are conspicuous. On the surface of the part (V. 60659a) there are three clearly marked compressed mamillae and some other obscure ones; there are also some more or less clear laterally projecting mamillae. The best of the surface mamillae show an obscure leaf scar but there are, in addition, less definite marks. If all the obscure bulges on both part and counterpart represent mamillae their number would be at least 12 on 24 cm of stem length (but might be more because the counter part surface is less well seen). The part slab V. 60659a Fig. 3) has an ill-defined bulge without definite features at the top right end but the counterpart V. 60659b (Fig. 5) shows a rounded cushion in the corresponding position. No leaf scar was recognized.

V. 45265; Roseberry Topping, collected by Hamshaw Thomas. This is a stem fragment about 20 × 3 cm; with coal nearly 3 mm thick. The surface shows some 'berets' and some clearly marked leaf scars. This specimen had presumably been examined both by Hamshaw Thomas and by me but its nature was not recognized; if it had been, this axis would have been described much earlier.

V. 60662 a–c (Fig 1E), comprises three stems over 5 cm wide. The width may be greater than 5 cm as one edge is missing. The stem divides into two large branches diverging at about 70°. The coal is almost 4 mm thick and the growth cracks on the surface are up to 2 mm wide. A single 'beret' remained.

V. 60663 (Fig. 10); the width exceeds 12 cm (one edge is missing). The coal is 3 mm thick, growth cracks are strongly marked and the surface is entirely denuded of berets. It divides into two large branches which diverge at about 80°. (The specimen was identified by its association with others, its forked branching and its cracked outer cortex).

Some other large stems with thick coal were collected from Hillhouse Nab. but they broke up, gave no further information and were discarded.

There are also broad fragments of detached outer cortex, torn at all edges and forming a thin layer of coal. Only such fragments as retain some berets were determined.

V. 45470 (figured by Harris, 1964: pl. 6, fig. 7). Roseberry Topping.

V. 45473 (figured by Harris, 1964: pl. 7, fig. 8). Roseberry Topping.

These bark fragments demonstrate that the detached bark is strong enough to be transported. Similar pieces were transferred; their inner surface proved rough, being covered with lumps about 0.25 mm across. The lumps dissolve quickly when macerated.

Interior of the stem: A good many stems are partly denuded after decay and some of the internal parts are visible, particularly the outer surface of the outer cortex (perhaps supplemented by cork) and the surface of the xylem. The most informative specimen is that already described by Harris, 1964: p. 134, a small stem 15 mm

Figures 3–7. *Pachypteris papillosa*. Fig. 3. Longest fragment, see also Fig. 1B, V. 60659a. Counterpart is shown in Figs 1A & 5. Scale bar equivalent to 10 mm. Fig. 4. Fragment, see also Figs 10 & 7, V. 60661. Scale bar equivalent to 10 mm. Fig. 5. Counterpart of Figs 1B & 3, illuminated from bottom right to show the mamilla at the top end clearly. Also figured in Fig. 1A. V. 60659b. Scale bar equivalent to 10 mm. Fig. 6. Fragment, figured in Fig. 1C, 8 & 9, V. 60660a. Scale bar equivalent to 10 mm. Fig. 7. Leaf cushion (mamilla) and leaf scar, position indicated in Fig. 1D. V. 60661. Scale bar equivalent to 2.5 mm.

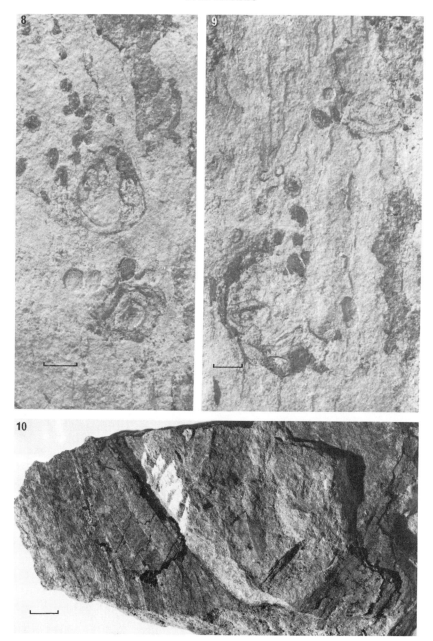

Figures 8–10. *Pachypteris papillosa*. Figs 8, 9. Mamillae (leaf cushions) and 'berets' of specimen in Fig. 1C & 6, V. 60660a. Scale bar equivalent to 2.5 mm. Fig. 10. Fragment from a forked stem (broken in collecting), V. 60663. Scale bar equivalent to 10 mm.

wide is which mud had penetrated between tissues. I give the same data (but expressed as approximate radial thicknesses):

Pith or pith cavity	2.5–3.0 mm
Xylem ring	1.0 mm
Cortex	2.0 mm
'Berets'	1.5 mm

Outside the xylem there is a very thin layer which might be phloem fibres; the cortex consists of a crumbly inner layer and a compact outer layer, apparently composed of hard parenchyma rather than fibres, since there is no sign of long cells.

The outer cortex of small stems is covered with very shallow dimples giving it a shagreened surface. The nature of the dimples is unknown; they are much smaller and more numerous than the 'berets' and can scarcely be related to them. Where the 'berests' were lost before deposition there is nothing to indicate their position and I state explicitly that I saw nothing like a vascular bundle scar to a 'beret'. Only where a 'beret' is pulled off in splitting the rock is there an imprint of its base as a coaly circle corresponding to the inner surface of beret on the counterpart.

DISCUSSION

Pachypteris papillosa had an unusual stem. When young it was fleshy and the prominent leaf bases suggest the mamillae of a cactus. The only other Yorkshire stems known to be of somewhat succulent habit are certain conifers of the form-genus *Pagiophyllum*, particularly *P. kurrii*, but in the Upper Jurassic *Brachyphyllum nepos* looks succulent (Jung, 1974, and in the Lower Cretaceous there is the strongly succulent *Pseudofrenellopsis varians* (see Watson, 1977).

On the young stems, the 'berets' were crowded and their swollen tops touched though their bases were about 1 mm apart. They were very easily detached, at least after death and even the young stems are denuded partly, the older ones almost completely. Where they are crowded the effect was to produce a heavy armour of cuticle at the outside with the stomata limited to the protected sides and opening into the narrow chinks between the 'berets'. I imagine that photosynthetic tissue occupied the whole interior. At the base, the 'beret' cuticle becomes thin and bends outwards and I merely guess that this thinly cutinized epidermis extends to the next 'beret'. I have not seen it. Certainly there was very little tissue of any kind between the 'beret' and the firm outer cortex. As secondary growth enlarged the stem the outer cortex was stretched and splits formed. I imagine that a cork layer now covered the surface, if so the epidermis and 'berets' would become loose and readily exfoliated even in life.

I regard the 'berets' as proliferations of subepidermal photosynthetic tissue; they are a form of 'emergence'. In compression, whether vertical or lateral the contents of the 'beret' forms a thin layer of glassy or crumbly coal showing no cells either on fracture or gentle maceration; soft parenchyma is preserved in this way. The 'beret' yields no resin body when macerated and had there been hard tissue like stone cells I would expect to have seen signs of it. The numerous and well-developed stomata on the sides of the 'berets' suggests a photosynthetic tissue at least as well-developed as in the leaf. A casual survey of organs of living plants showed that parts which were photosynthetic, at least when young had plenty of stomata, whether the green parts were leaves, stems or floral organs; but parts which lack chlorophyll, again without regard to their morphological category, have an epidermis with only a few small stomata.

Two other species of *Pachypteris* are associated with berets:

Pachypteris desmomera (Saporta) Barale 1971: Barale (unpublished) subsequently found detached 'berets' associated with the leaves in the Kimmeridgian of France.

In addition there were two associated stem fragments; one 22 mm wide shows oval prints 4–6 mm wide which may be leaf scars and there are 'berets', sometimes in pairs side by side. The other stem, only 7 mm wide and also with 'berets' shows obscure lateral bulges 3 mm high which possibly represent prominent leaf bases. I am indebted to M. G. Barale for this information.

Pachypteris lanceolata Brongn. This species has been found in Yorkshire in association with detached 'berets', very like those of *P. papillosa* but smaller (Harris 1964: 144). No stem has been seen.

In *P. papillosa*, *P. lanceolata* and *P. desmomera* I imagine that the 'berets' remained as functional photosynthetic organs for some time after the leaves had dropped. But whatever their function they are an item in which the three species agree. How closely similar the whole but unknown plants of leaf-genera may have been is open, but the presence of 'berets' makes me think that these three at least were close.

I have compared the 'berets' with certain emergences formed by proliferated subepidernal tissue. Many plants (roses, brambles) have prickles formed by hard tissue, but these I exclude. Such tissue would show its cells in the fossil and would behave differently in vertical and lateral compression. *Hypericum balearicum* L. has prominent oil glands, but the oil would form a solid body of resin and the epidermis of the bulge has no stomata. The leafless asclepiad *Cynanchium marnieranum* Rauh, a Madagascan xerophyte, has the most similar organs. There are warts on its stem 1–2 mm high and broad and these warts are full of green parenchyma. However, there are differences; the warts are not crowded except locally and much of the stem is without them; they are firmly attached; the whole stem surface is thickly cutinized and has stomata in grooves; stomata occur also on the sides of the warts but not at the top. I am indebted to G. Rowley and R. Rutherford of the Botany Department, University of Reading, for the *Cynanchium* and the *Hypericum*.

The 'berets' look somewhat like the cutinized blisters which cover the rachis of the Mesozoic pteridosperm *Lepidopteris* but there are important differences. *Lepidopteris* blisters are firmly attached, the cuticle between them is thick and strong and they have scarcely any stomata. Townrow (1960) showed there was a trichome on the top of each blister and he compared them with the swellings of the petioles of *Rumex hydrolapathum* Hudson.

A young stem might show how branching starts but the only branched specimens are old and thick. I do not regard the fork branching of the wide stems as true dichotomy, though this is possible. Very likely one or both branches are lateral shoots arising from buds, as in a number of woody dicotyledons. But whatever the origin of the fork, its effect on the form of the plant would be considerable if forking happened throughout growth. A woody plant with forked stems could not produce a main trunk like a pine.

Form and habitat of P. papillosa: The drawings show certain specimens as I imagine they looked before damage and compression. I do not draw the whole shrub or tree for which I have little basis. Nevertheless the sumach, *Rhus typhina* L. offers a possible model in its forked branching, though the young stems of *P. papillosa* are thicker and the intervals between branches longer. This plant is a large shrub, seldom exceeding 6 m. But the leathery leaves of the fossil and particularly the 'berets' after the leaves had fallen would make it look different.

I have suggested (Harris 1964: 131) that *P. papillosa* grew on the banks of tidal river channels and here reaffirm this. Most plant beds contain plant material of land origin only but along with *P. papillosa* there are ordinary land plant leaves and spores and also many acritarchs, spores considered to be of marine origin. Now that we have its stem, we can imagine it as a large mangrove shrub forming a thicket beside the river. This would explain its peculiar frequency and distribution; most Yorkshire plants of wide distribution vary from abundant, through frequent or occasional to very rare, but where *P. papillosa* occurs it is often the commonest species, yet in a large number of localities it is unknown. Leaves of a mangrove might be expected to be abundant in a tidal reach deposit, but absent above the level where high tides brought saltwater.

I make no suggestion about the habitats of the other *Pachypteris* species. *Pachypteris lanceolata* in Yorkshire shows no special link with marine horizons. *Pachypteris desmomera* on the other hand, like all the French Kimmeridgian flora, occurs in a marine lithographic limestone. Clearly its leaves, like those of the other land plants had floated out to sea and then sank and were buried. Its frequency does at least fit a coastal habitat but I believe nothing is known of the land which produced these plants.

REFERENCES

BARALE, G., 1971. *Pachypteris desmomera* (de Saporta) nov. comb., feuillage filicoïde du Kimméridgien de Creys (Isère). *Bulletin de la Société géologique de France* (7) *8:* 174–180.

DOLUDENKO, M. P., 1969. On the relation of the genera *Pachypteris* and *Thinnfeldia. Trudy Geologicheskogo Instituta. Akademiya nauk SSSR, 190:* 14–34 (in Russian).

DOLUDENKO, M. P., 1974. On the relation of the genera *Pachypteris, Thinnfeldia* and *Cycadopteris. Birbal Sahni Institute (Lucknow), Special publication, No. 2:* 8–16.

GOTHAN, W., 1914. Die unter-liassische ("rhätische") Flora der Umgegend von Nürnberg. *Abhandlungen naturhistorischen Gesellschaft zu Nürnberg, 19:* 91–186.

HARRIS, T. M., 1964. *The Yorkshire Jurassic Flora II.* London: British Museum (Natural History).

JUNG, W., 1974. Die Konifere *Brachyphyllum nepos* Saporta aus den Solnhofener Plattenkalken (unteres Untertithon), ein Halophyt. *Mitteilungen der Bayerischen Staatssammlung fuer Palaeontologie und Historische Geologie. 14:* 49–58.

THOMAS, H. H., 1913. The Jurassic plant beds at Roseberry Topping. *The Naturalist, May 1913:* 198–200.

THOMAS, H. H., 1915. The *Thinnfeldia* leaf bed at Roseberry Topping. *The Naturalist, Jan. 1915:* 7–11.

THOMAS, H. H. & BOSE, M. N., 1955. *Pachydermophyllum papillosum* gen. et sp. nov. from the Yorkshire Jurassic. *Annals and Magazine of Natural History, London, 12:* 535–543.

TOWNROW, J. A., 1960. The Peltaspermaceae, a pteridosperm family of Permian and Triassic age. *Palaeontology, 3:* 333–361.

WATSON, J., 1977. Some Lower Cretaceous conifers of the Cheirolepidiaceae from the U.S.A. and England. *Palaeontology, 20:* 715–749.

Botanical Journal of the Linnean Society (1983), *86:* 161–167. With 13 figures

A new species of the conifer *Frenelopsis* from the Cretaceous of Sudan

JOAN WATSON, F.L.S.

Departments of Botany and Geology, The University, Manchester M13 9PL

Received December 1981, accepted for publication April 1982

Conifer fragments from the Lower Cretaceous of Sudan are preserved as internal and external silica moulds. Low viscosity silicone rubber has been used to prepare casts showing fine epidermal details which enable identification of the plants. One is *Pseudofrenelopsis parceramosa* (Fontaine) Watson, the other is described as a new species of *Frenelopsis* Schenk.

KEY WORDS:—Cheirolepidiaceae – conifer – Cretaceous – *Frenelopsis* – Sudan.

CONTENTS

INTRODUCTION

Recent work on conifers of the Cheirolepidiaceae from the Lower Cretaceous of England, U.S.A. and Poland led me to re-examine some silicified material from the Sudan which Edwards (1926) had previously identified as *Frenelopsis hoheneggeri* (Ettingshausen) Schenk. The preservation of this material proved to be rather unusual in that the plants are represented by internal and external silica moulds of the cuticle with no actual plant material remaining. The moulds can be used with low viscosity silicone rubber to produce casts which show very fine epidermal details. The plants were eventually studied almost entirely by SEM of these casts. Details of this exceptional preservation and the techniques employed have been published elsewhere (Watson & Alvin, 1976). The specimens figured by Watson & Alvin (1976) were all referred to as *Frenelopsis* Schenk as the various genera were undergoing revision at that time. It is now clear that there is an intimate mixture of two species, a species of *Frenelopsis* Schenk with leaves in whorls of three and a species of *Pseudofrenelopsis* Nathorst with spirally arranged leaves. On the basis of the external features of the *Pseudofrenelopsis* stomatal apparatus, I am satisfied that

161

0024–4074/83/010161+07$03.00/0

the latter is *Pseudofrenelopsis parceramosa* (Fontaine) Watson. A silicone rubber cast of the stomatal apparatus has been figured side by side with cuticle preparations of English and American *P. parceramosa* (Watson, 1977). However, I cannot match the *Frenelopsis* to any known species and it is therefore described here as a new species. The type species of *Frenelopsis* is *F. hoheneggeri* (Ettingshausen) Schenk (Reymanówna & Watson, 1976).

TAXONOMY

ORDER:	Coniferales
FAMILY:	Cheirolepidiaceae
GENUS:	*Frenelopsis* Schenk

Frenelopsis silfloana Watson **sp. nova**

DERIVATION: Named in honour of 'Silflo' the dental silicone rubber which has proved indispensable in this study.

SYNONYM: *Frenelopsis hoheneggri* (Ettingshausen) Schenk *sensu* Edwards, *Quarterly Journal of the Geological Society* (1926).

Diagnosis based on unbranched shoot lengths of up to 8 internodes. Shoots bearing 3 (probably sometimes 2) leaves per node. Triangular part of leaf up to 1 mm high, apex rounded, margins entire. Stomata tending to be arranged in rows sometimes ill-defined. Stomata with 4 or 5 subsidiary cells, encircling cells common; diameter of stomatal apparatus about 80 μm; guard cells sunken at bottom of shallow pit (orientation unknown). Outer surface of subsidiary cells forming thickened circular rim around stomatal pit; rim lacking papillae. Large rounded papillae present inside stomatal pit, probably one per subsidiary cell. Each ordinary cell bearing on the outer surface a hemispherical papilla. Ordinary epidermal cells rounded, squarish or elongated, arranged roughly in files.

TYPE: The holotype is the shoot on specimen V. 17224 British Museum (Natural History), of which Fig. 7 is a cast.

ILLUSTRATIONS: Figures 1–13.

LOCALITY: Jebel Dirra, about 45 miles east of El Fasher, Eastern Darfur, Sudan.

STRATUM: These fossils, of Nubian Sandstone Age, are presumed to be Lower Cretaceous on the evidence of the presence of *Pseudofrenelopsis parceramosa*. Details have been given in Edwards (1926) and Watson & Alvin (1976).

DESCRIPTION AND DISCUSSION

Specimen numbers refer to the British Museum (Natural History). All the lengths of shoot upon which the diagnosis of *Frenelopsis silfloana* is based are

---

Figures 1–6. Fig. 1. Block showing a mixture of *Frenelopsis* and *Pseudofrenelopsis* shoots. Area A is the shoot of which Fig. 4 is a cast; B the shoot of which Fig. 5 is a cast; C, *Pseudofrenelopsis parceramosa* (Fontaine) of which Watson & Alvin, 1976: pl. 96, fig. 2 is a cast. V. 21708, actual size. Figs 2–6 *Frenelopsis silfloana* to same magnification, scale bar in Fig. 6 is equivalent to 50 mm. Figs 2 & 3. Silica specimens, Fig. 2, with steinkern intact V. 17222; Fig. 3, external mould from which casts can be made V. 17222. Figs 3–6. Silicone rubber casts of shoots. Fig. 4, clearly with three leaves per node. Fig. 5, perhaps with 2 leaves per node. Fig. 6, showing stomata arranged in well-defined rows. Figs 4–6, V. 21708.

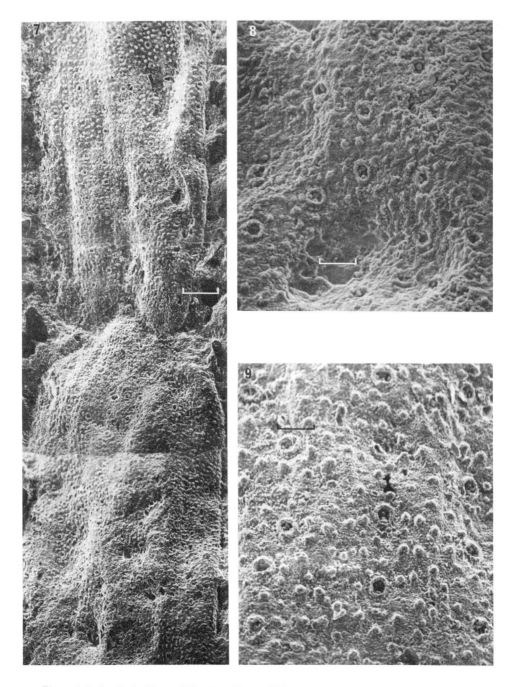

Figures 7–9. *Frenelopsis silfloana*, all V. 17224. Fig. 7. SEM montage of silicone rubber cast of holotype, scale bar (a), equivalent to 200 μm. Figs 8 & 9. Scale bar (b), equivalent to 50 μm. Surface details of holotype cast in SEM showing stomatal rims, papillae inside stomatal pit and hemispherical papillae on ordinary epidermal cells.

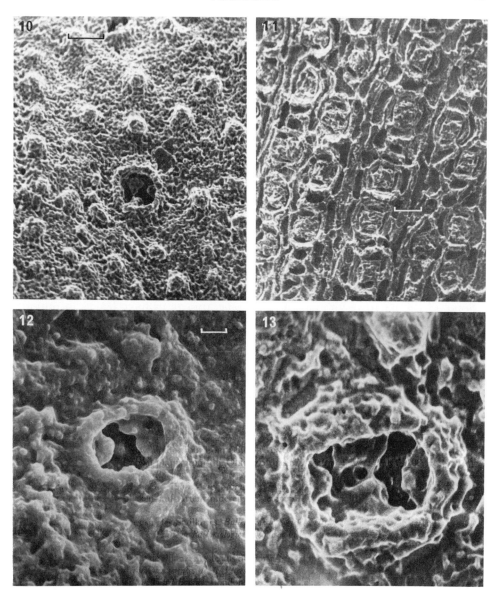

Figures 10–13. *Frenelopsis silfloana*, all silicone rubber casts in SEM. Fig. 10. Outer surface showing particularly well-preserved papillae both on the surface and inside the stomatal pit. V. 17224. Scale bar equivalent to 20 µm. Fig. 11. Internal details of cuticle. V. 21708. Scale bar equivalent to 50 µm. Figs. 12 & 13. External views of 2 stomata. V. 17224. Scale bar equivalent to 5 µm. Imperfections in the silica mould are apparent at high magnification.

unbranched, 1–3 mm wide with segments (see Watson, 1977: 717, text fig. 1 for shoot terminology) up to 10 mm long. All other parts of the plant are unknown though it is not unlikely that some of the associated wood and cones belong to *F. silfloana*.

Figure 2 shows the *Frenelopsis* shoot figured by Edwards (1926: fig. 3). This clearly shows a leaf arrangement of alternating whorls of three. Numerous such short lengths of *F. silfloana* are present on several blocks mixed with shoots of

Pseudofrenelopsis parceramosa (Fontaine) Watson. This association illustrated in Fig. 1 which shows a block bearing several shoots which have been identified by means of 'Silflo' casts and figured in detail. Figure 1C is the area which was used to produce the cast of a branched specimen of *P. parceramosa* which was illustrated by Watson & Alvin (1976: l. 96, fig. 2). The details shown by the silicone rubber casts are dependent upon the state of the silica matrix. The shoots in Fig. 1A B are in the form of fine silica external moulds which were used to produce the 'Silflo' casts shown in Figs 4 & 5 respectively. Figure 3 shows such a mould in detail. Clearly only specimens like this with no steinkern will produce the fine surface details such as are shown in Figs 7–10, 12 & 13. The cast in Fig. 11 is of course produced from the surface of a steinkern. However, many specimens (e.g. Fig. 2) have a rather weathered surface with the steinkern firmly attached to the matrix. Such specimens cannot be used to obtain casts and sometimes not even the leaf arrangement can be seen. It should be mentioned that I have also obtained epidermal casts which I cannot attribute either to *P. parceramosa* or *F. silfloana*. Thus the confidence with which I can identify any given shoot fragment is somewhat variable and indeed I do not rule out the presence of a third cheirolepidiaceous species.

Some of the narrowest shoots of *F. silfloana* appear to have leaves in opposite decussate arrangement rather than the usual alternating whorls of three. This leaf arrangement in *Frenelopsis* has already been described by Alvin & Pais (1978) in *Frenelopsis teixeirae* Alvin & Pais which has opposite decussate leaves on three orders of branching. Figure 5 shows a shoot which I think has leaves in twos, and perhaps Fig. 6 also. These should be compared with Fig. 4 which clearly shows leaves in threes. The exact leaf arrangement in the shoot of which I am uncertain could perhaps be confirmed by picking away the remaining steinkerns and in effect looking at both sides of the shoots.

The figured specimens of Watson & Alvin (1976), except for the cones, can now be correctly identified:— pl. 96, fig. 1: a mixture of *P. parceramosa* and *F. silfloana*; pl. 96, figs 2 & 3: *P. parceramosa*; pl. 97, figs 1–6: *F. silfloana*; pl. 98, figs 1–5: *F. silfloana*; pl. 99, figs 5 & 6: *P. parceramosa*.

There are isolated cones present, probably both male and female (Watson & Alvin, 1976: pl. 99, figs 1–4). Both sorts of cone have very strongly papillate surfaces but I cannot attribute them with any confidence to either of the two species of shoot. However, surface features of the male cone scale do not seem to me to match the male cones of *P. parceramosa* which are known in detail (Alvin, Spicer & Watson, 1978).

COMPARISON

Most known species of *Frenelopsis* have stomata with large papillae situated inside the stomatal pit. *Frenelopsis ramosissima* Fontaine from the Potomac Formation, U.S.A. is a notable exception (Watson, 1977). The thickened rim around the stomatal pit is a feature seen in several species but *F. silfloana* seems to be the only species with a circular opening within the rim. In other species the rim is lobed to give a star-shaped opening (e.g. *Frenelopsis alata* (Feistmantel) and *Frenelopsis oligostomata* Romariz—see Alvin 1977). Species without a thickened rim usually have a star-shaped opening flush with the surface (e.g. *Frenelopsis occidentalis* Heer and *Frenelopsis teixeirae* Alvin & Pais, 1978). A closely similar rimless pit with internal papillae is seen in *Cupressinocladus valdensis* (Seward) Seward from the English Wealden (Watson, 1977). However, it is clearly distinguished from *F. teixeirae* in its possession of conspicuous sutures between adjacent leaf bases. The stoma of *F. ramosissima* is characterized by a thickened rim with small papillae on its inner margin, very similar to the two known species of *Pseudofrenelopsis*, *P. parceramosa* from the Lower Cretaceous of several countries and *P. varians* from the Lower Cretaceous of Texas (Watson, 1977).

ACKNOWLEDGEMENTS

I should like to express my thanks to Mr J. M. Davis of Flexico Developments Ltd., the manufacturers of 'Silflo'; the Department of Textile Technology, U.M.I.S.T. for use of the SEM, and Mr Ian Miller who printed the photographs.

REFERENCES

ALVIN, K. L., 1977. The conifers *Frenelopsis* and *Manica* in the Cretaceous of Portugal. *Palaeontology, 20:* 387–404.
ALVIN, K. L. & PAIS, J. J. C., 1978. A *Frenelopsis* with opposite decussate leaves from the Lower Cretaceous of Portugal. *Palaeontology, 21:* 873–879.
ALVIN, K. L., SPICER, R. A. & WATSON, J., 1978. A *Classopollis*-containing male cone associated with *Pseudofrenelopsis*. *Palaeontology, 21:* 847–856.
EDWARDS, W. N., 1926. Fossil plants from the Nubian Sandstone of Eastern Darfur. *Quarterly Journal of the Geological Society, 82:* 94–100.
REYMANÓWNA, M. & WATSON, J., 1976. The genus *Frenelopsis* Schenk and the type species *Frenelopsis hoheneggeri* (Ettingshausen) Schenk. *Acta Palaeobotanica, Cracow, 17:* 17–26.
WATSON, J. 1977. Some Lower Cretaceous conifers of the Cheirolepidiaceae from the U.S.A. and England. *Palaeontology, 20:* 715–749.
WATSON, J. & ALVIN, K. L., 1976. Silcone rubber casts of silicified plants from the Cretaceous of Sudan. *Palaeontology, 19:* 641–650.

Botanical Journal of the Linnean Society (1983), *86:* 169–176. With 8 figures

Reconstruction of a Lower Cretaceous conifer

K. L. ALVIN

Department of Pure and Applied Biology, Imperial College, London SW7 2AZ.

Received April 1982, accepted for publication June 1982

Remains of *Pseudofrenelopsis parceramosa* (Fontaine) Watson (Cheirolepidiaceae) from the Wealden of the Isle of Wight have been examined from the point of view of the branching and mode of growth of the plant. Evidence is presented that the tree exhibited seasonal growth and that the young extension shoots bore numerous temporary ultimate branchets of limited growth. The tree was probably adapted to warm seasonally arid conditions.

KEY WORDS:—Conifer – Cretaceous – fossil reconstruction – *Pseudofrenelopsis*.

CONTENTS

INTRODUCTION

Pseudofrenelopsis parceramosa (Fontaine) Watson is a well-known and widespread Lower Cretaceous conifer attributed, on the basis of its *Classopollis*-containing male cone, to the Cheirolepidiaceae (Alvin, Spicer & Watson, 1978). Its shoot morphology and cuticle have been fully described by Watson (1977). Abundant remains, including material with internal structure, preserved in a silty bed interpreted as a river channel at Shippard's Chine, Isle of Wight, enabled Alvin, Fraser & Spicer (1981) to describe in some detail the stem and wood anatomy. In this paper, on the basis of material from the same bed, I attempt to give some account of the branching, mode of growth and habit of the tree.

MATERIAL AND METHODS

Portions of ultimate twigs and decorticated lower order axes are easily observed on blocks of matrix. They may also be obtained free of matrix by breaking down partially dried blocks in water and sieving; isolated fragments are then picked out and allowed to dry out slowly in a covered dish. The large compressed branch (Figs 1, 6) was removed from the bed in several pieces, allowed to dry out slowly while wrapped and then reassembled in the laboratory. Pyritized portions of this and other similar axes have been fractured and examined by SEM.

RESULTS

Ultimate branchlets: The great bulk of fragments of corticated twigs are believed to represent ultimate branchlets. They are usually more-or-less disarticulated even into single segments (internodes) but specimens with up to 14 articulations and some 120 mm in length are occasionally seen. The specimen in Fig. 2 is a typical example. In some parts of the bed there are coaly lenses consisting of an almost solid mass of more-or-less disarticulated shoots of this kind. Disarticulation is a common phenomenon in frenelopsid conifer shoots (Alvin, 1977; Watson 1977; Upchurch & Doyle, 1981); it probably occurred during transport due to the comparative thinness of the cuticle at the nodes. Commonly, portions consisting of only one or a few segments have matrix inside occupying the space where the cortical tissue would originally have been. If such specimens are split, the narrow woody cylinder, 1–2 mm in diameter, is often seen running up the centre (Fig. 3). Watson (1977) suggested that the shoots may have been succulent and Alvin *et al.* (1981) found evidence of an extensive thin-walled cortical tissue.

These presumed ultimate shoots are mostly about 4–6 mm in diameter but may occasionally be as narrow as 2 mm or as wide as 9 mm. They are parallel-sided or slightly constricted at the nodes with internodes typically somewhat longer than wide. Broken-down blocks rich in such remains usually yield a number of apices (Fig. 7I). These are narrower than typical shoots and show a gradual decrease in internode length towards the tip. Occasionally, at the base of an assembled portion of twig a few shorter internodes may occur as well. Thus, the ultimate branchlets were probably fleshy cylindrical shoots, unbranched, 120 mm or more in length and consisting, except at the extreme base and apex, of internodes typically longer than broad. I envisage that these branchlets were temporary 'dwarf-shoots' of limited growth and that they were shed as units at the end of their useful life, perhaps at the onset of an unfavourable season.

Extension shoots: These are represented by abundant, decorticated axes found throughout the bed (Figs 1, 6) and also a few corticated fragments which have been recovered from sievings of broken down blocks (Fig 7A–H). Most of the decorticated twigs and branches are probably attributable to *P. parceramosa* though a few, on the basis of differences in branching pattern, probably belong to different conifers. Those which I believe belong to *P. parceramosa* show a whorled arrangement of laterals and either a smooth surface (Fig. 6) or, as in some of the more slender specimens up to about 8 mm in diameter, with a number of scattered small scars about 1 mm in diameter (Fig. 6C, D). The large specimen (Figs 1, 6) has a total length of 925 mm. It is illustrated in two portions, but the longer, more slender portion was attached to the other at *x* (Fig. 6). The diameter at the base is 40 mm and at the distal extremity, 7.5 mm. Except for what appears to be an almost equal dichotomy at or a little above node III, branching is whorled, usually with three, occasionally only two, laterals of variable strength in the whorl. Whether the dichotomy was a true one in the sense that it involved an equal

Figures 1–5. Fig. 1. Decorticated branch attributed to *Pseudofrenelopsis parceramosa* (Fontaine) Watson. Scale bar equivalent to 30 mm. Fig. 2. Typical ultimate branchlet of *Pseudofrenelopsis parceramosa*. Scale bar equivalent to 5 mm. Fig. 3. Short portion of a similar branchlet split longitudinally to show the enclosed matrix and vascular cylinder. Scale bar equivalent to 4 mm. Fig. 4. Tracheid with compressed contiguous pitting from a pyritized portion of the branch in Fig. 1. Scale bar equivalent to 15 μm. Fig. 5. Tracheids and rays in RLS from the same specimen. Pitting is more spaced; biseriate-opposite pits occur at *b*: small cross-field pits may just be discernible at *c*. Scale bar equivalent to 25 μm.

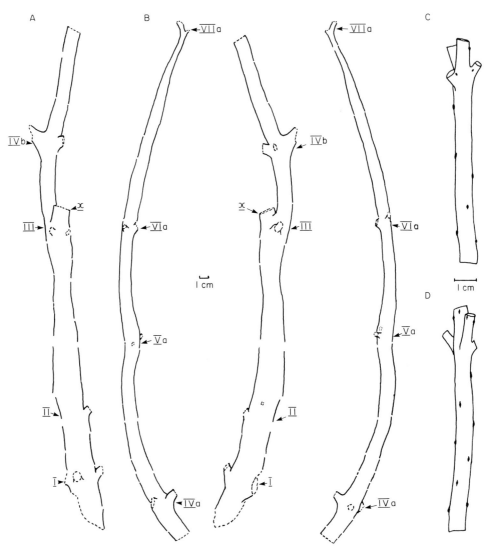

Figure 6. Decorticated branches attributed to *Pseudofrenelopsis parceramosa* (Fontaine) Watson. Drawn by tracing photographs. Fig. 6A & B. Drawings of the two sides of the branch shown in Fig. 1 showing the whorled lateral branching. (Successive branch nodes numbered in Roman numerals.) The slightly longer, more slender portion was attached to the other portion at *x*. C & D. Drawings of the two sides of another decorticated twig showing a single whorl of lateral branches and several small scars interpreted as the nearly occluded scars of ultimate branchlets.

division of the apical meristem must be regarded as doubtful; it could merely represent an unusually strong lateral. Whorls of lateral extension shoots were borne at rather irregular intervals along the main axis, intervals varying from 75 mm (branch nodes I–II) to 228 mm (nodes VIa–VIIa). This suggests an irregular pattern of growth perhaps depending on the relative favourability of successive growing seasons. This is also suggested by the variable width of the growth-rings in the secondary wood (Alvin *et al.*, 1981). Altogether the pattern seems comparable with that of many woody plants growing under seasonal conditions.

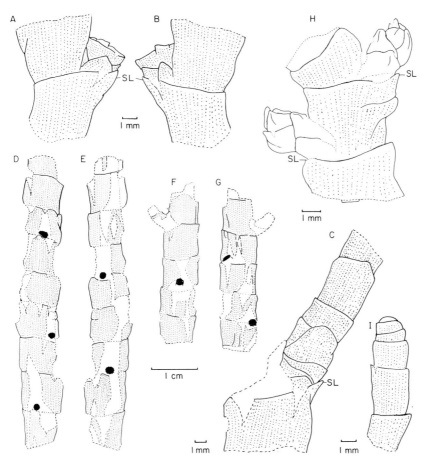

Figure 7. *Pseudofrenelopsis parceramosa* (Fontaine) Watson. All drawn by tracing photographs. A & B. Portion of young extension shoot with the base of an ultimate shoot attached. Two or perhaps three non-sheathing *Brachyphyllum*-like leaves occur at the base of the lateral. SL, subtending leaf. C. A similar specimen with more of the ultimate shoot present. D & E, F & G. Two rather older extension shoots with scars of ultimate branchlets. The outer cortex is perhaps being sloughed off by periderm development and thickening of the woody cylinder. H. Portion of young extension shoot with what appear to be buds possibly of lateral extension shoots with a number of *Brachyphyllum*-like leaves. I. Apex of an ultimate branchlet.

The major branch and other more fragmentary specimens showing similar features are attributed to *P. parceramosa* chiefly on the basis of association, but some confirmation of its identity is supplied by the tracheid pitting which has been observed by SEM of pyritized portions. Figs 4 and 5 show tracheid pitting in a pyritized portion of the large branch. The pitting is of the 'mixed' or 'protopinaceous' type like that of the mineralized or fusainized wood from the same bed and attributed to *P. parceramosa* (Alvin *et al.*, 1981). Such pitting is not characteristic of the other two (rarer) woods that these authors report from the same bed. Unfortunately, the cross-field pitting and other features are not well seen in these pyritized fragments, though it is possible that small cross-field pits are to be seen at *c* in Fig. 5.

If the bulk of the corticated shoot fragments in the deposit represent ultimate branchlets of limited growth and duration, the question remains as to how these were carried on the extension shoots. Some fragments recovered from sievings may supply the answer; a selection of these is shown in Fig. 7A–H). Those in Fig. 7A, B & C appear to represent basal portions of ultimate shoots attached to fragments of broader axes. In these and a few similar specimens the parent axis usually has segments which are either as broad or slightly broader than long in contrast to typical ultimate branchlets; they also contain a broader woody cylinder and are believed to represent fragments of young extension shoots. The rarity of such specimens in the deposit is presumably because extension shoots were not normally shed. The specimens in Fig. 7D, E, F & G probably represent portions of somewhat older extension shoots which have dropped their ultimate branchlets. They are about 8 mm in diameter with a woody cylinder of about 5–6 mm and are only partially corticated with the cuticle variously split and separating from the underlying coalified tissues. At most nodes there is a small round scar about 1–2 mm in diameter probably situated in the leaf axil, though the leaves themselves are obscure. It is suggested that these scars represent the scars of attachment of ultimate branchlets. The apparently persistent branch base at the top of the specimen in Fig. 7F & G may represent the base of an extension shoot branch.

The small scars seen on the surface of such decorticated specimens as that shown in Fig. 6C & D are also likely to represent scars of ultimate branchlets partially occluded by secondary wood development.

The specimen in Fig. 7H carried two apparent buds each of which bears a number (perhaps at least six) of more-or-less *Brachyphyllum*-like leaves which do not form complete sheaths around the axis. This contrasts with the bases of typical ultimate branchlets (Fig. 7A, B & C) which have only two or three such leaves below the typical, completely sheathing leaves. It has been suggested (Alvin, 1977; Watson, 1977) that the *Brachyphyllum*-type foliage may represent a juvenile form. Perhaps such foliage was more persistent at the base of extension shoots than ultimate 'dwarf-shoots'. The structures in Fig. 7H may represent buds of a pair of lateral extension shoots.

CONCLUSION

An attempted reconstruction of the tree is given in Fig. 8. Large logs in the same bed, some partially mineralized, reported by Alvin *et al.* (1981), if correctly assigned to *P. parceramosa*, provide evidence that it was a large tree. Decorticated branches and twigs, pyritized portions of which show the same kind of tracheid pitting as in the log wood and fusainized wood in the same bed, indicate a whorled pattern of branching of extension shoots though occasional dichotomies, equal or subequal, may have occurred. Young extension shoots still with cuticles, although only a few fragments have been found, were somewhat stouter than typical ultimate branchlets, had relatively short internodes and bore many axillary ultimate branchlets. The overwhelming abundance of these in the bed almost certainly indicates that they were dropped. They were probably shed as complete shoots, for although they are always preserved in a more-or-less disarticulated condition, their fragmentation probably occurred during transport. These cylindrical, unbranched and probably rather fleshy shoots must have represented the main photosynthetic organs. I do not think there is any evidence that the tree was deciduous; on the contrary, the thickly cutinized and indeed

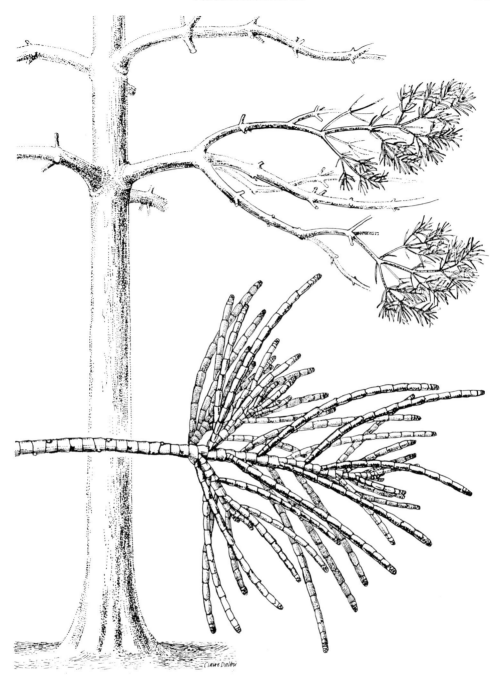

Figure 8. Suggested reconstruction of *Pseudofrenelopsis parceramosa* (Fontaine) Watson.

generally xeromorphic photosynthetic organs suggest an evergreen condition, though shedding may have been related to the occurrence of unfavourable, probably arid, periods. The extension shoots were also photosynthetic at first; how quickly they ceased to be so by periderm development and the sloughing off of the photosynthetic cortex cannot be determined.

The tree was probably well-adapted to seasonally arid conditions likely to have prevailed during the Lower Cretaceous in southern England and elsewhere (Batten, 1974; Allen, 1975, 1981; Harris, 1981). Although some frenelopsids are found in association with maritime conditions, (Daghlian & Person, 1977; Doludenko, 1978; Hluštík, 1978) and Upchurch & Doyle (1981) have recently presented evidence that *P. parceramosa* itself in America may have been associated with some maritime influence, there is yet no clear evidence of any such influence in the Isle of Wight deposit. Rather, I envisage the tree as having inhabited freshwater stream banks or lowlying areas dissected by braided streams which may have dried out seasonally or irregularly. The xeromorphic characters of the plant probably relate to true xerophytism rather than halophytism, bestowing physiological advantages during arid periods.

ACKNOWLEDGEMENTS

I am grateful to Mrs Claire Dalby for her interest, patience and artistic skill in drawing the reconstruction and to Mr Sinclair Stammers for photographic assistance. I also acknowledge stimulation and help with field work from Dr Robert Spicer.

REFERENCES

ALLEN, P., 1975. Wealden of the Weald: a new model. *Proceedings of the Geological Association, 86:* 389–437.

ALLEN, P., 1981. Pursuit of Wealden models. *Journal of the Geological Society, London, 138:* 375–405.

ALVIN, K. L., 1977. The conifers *Frenelopsis* and *Manica* in the Cretaceous of Portugal. *Palaeontology, 20:* 387–404.

ALVIN, K. L., SPICER, R. A. & WATSON, J., 1978. A *Classopollis*-containing male cone associated with *Pseudofrenelopsis. Palaeontology, 21:* 847–856.

ALVIN, K. L., FRASER, C. J. & SPICER, R. A., 1981. Anatomy and palaeoecology of *Pseudofrenelopsis* and associated conifers in the English Wealden. *Palaeontology, 24:* 759–778.

BATTEN, D. J., 1974. Wealden palaeoecology from the distribution of plant fossils. *Proceedings of the Geological Association, 85:* 433–458.

DAGHLIAN, C. P. & PERSON, C. P., 1977. The cuticular anatomy of *Frenelopsis varians* from the Lower Cretaceous of central Texas. *American Journal of Botany, 64:* 564–569.

DOLUDENKO, M. P., 1978. The genus *Frenelopsis* (Coniferales) and its occurrence in the Cretaceous of the U.S.S.R. *Palaeontological Journal, No. 3:* 107–112.

HARRIS, T. M., 1981. Burnt ferns from the English Wealden. *Proceedings of the Geological Association, 92:* 47–58.

HLUŠTÍK, A., 1978. *Frenelopsid plants (Pinopsida) from the Cretaceous of Czechoslovakia. Paleontologická Konference, 1977:* 129–141. Praha, Univerzita Karlova.

UPCHURCH, G. R. & DOYLE, J. A., 1981. Paleoecology of the conifers *Frenelopsis* and *Pseudofrenelopsis* (Cheirolepidiaceae) from the Cretaceous Potomac Group of Maryland and Virginia. In R. C. Romans (Ed.), *Geobotany II:* 167–202. New York: Plenum Press.

WATSON, J., 1977. Some Lower Cretaceous conifers of the Cheirolepidiaceae from the U.S.A. and England. *Palaeontology, 20:* 715–749.

Botanical Journal of the Linnean Society (1983) *86:* 177–225. With 68 figures

Palaeofloristic assemblages and palaeoecology of the Lower Oligocene Bembridge Marls, Hamstead Ledge, Isle of Wight

MARGARET E. COLLINSON

Department of Palaeontology, British Museum (Natural History), Cromwell Road, London SW7 5BD

Received December 1981, accepted for publication May 1982

A sequence of nine, plant-macrofossil-bearing horizons has been recognized in the Lower Oligocene from the Isle of Wight, England. The plant fossils are contained in finely laminated, olive-brown to black silty-clays and clays. Some horizons exhibit rhythmic alternation of sediment couplets; others show evidence of bioturbation and rootlet penetration. These horizons are included within a sequence of finely laminated greenish grey silty-clays and clays, largely devoid of plant macrofossil material. The plant-bearing horizons contain abundant fruits, seeds, fern sporangia, *Azolla* megaspores, and rare leaves. The nearest living relatives of the fossils all inhabit an aquatic or marginal aquatic environment. The associated fossil fauna includes lake bottom and lake margin species. The palynoflora of the plant-bearing and immediately underlying horizons includes large numbers of planktonic algae and limited representatives of a forest vegetation. It shows a strong relationship between the occurrences of *Sparganiaceaepollenites* pollen and of *Typha* macrofossils. Palaeoecological interpretations of this plant-fossil-bearing sequence are discussed with particular reference to modern environments.

One new genus and three new species are described. Emendations are presented for four other species. *Stratiotes* leaf margin teeth are described.

KEY WORDS:—Aquatic flora – fruit – lake deposits – Oligocene – palaeoecology – sedimentology – seed – Tertiary.

CONTENTS

0024–4074/83/010177 + 49 $03.00/0

INTRODUCTION

British Tertiary deposits were investigated (Collinson, 1978a) in an attempt to locate a succession suitable for sequential sampling of plant macrofossils. The sampling aimed to determine if data obtained could be used to interpret vegetational succession surrounding the depositional area.

Many interpretations of Quaternary deposits have been based on similar sampling (e.g. Watts & Wright, 1966; Watts & Winter, 1966; Birks, 1976). These authors interpret vegetational history from an analysis of the distribution of plant remains within successive horizons. Problems occur if an attempt is made to extrapolate this 'Quaternary approach' to Tertiary deposits. In Britain there are few sections which permit detailed sampling of plant material from successive strata. There is none of the floristically rich peat accumulations on which the studies of younger strata are based. Increased compaction results in there being less chance of determining absolute sedimentation rates. Borehole cores represent one possible approach to the sequential sampling problem, but they are unlikely to yield sufficient material for a sequential study of plant macrofossils (Collinson, 1978a).

Isolated fruit and seed concentrates (see references in Chandler, 1964) are unsuitable for ecological studies for two main reasons. First there is no reason to suppose that they have been deposited under similar conditions: two very different assemblages could represent the same vegetation affected by differential decay, transportation or sorting (Collinson, 1978a). Secondly the concentrates may vary internally such that two separate samples may yield different floras; furthermore successional change in vegetation cannot be revealed by isolated fossil occurrences. An additional problem is the presence of extinct plant genera and species in the Tertiary. Extrapolations from extant plant ecology are thus less reliable and require substantiation from other data wherever possible.

In this paper, which describes quantitative and qualitative analyses of macro- and microfloral remains, I follow Cooper (1976) and Curry et al. (1978) in placing the Eocene/Oligocene boundary at the base of the Bembridge Marls. Cavelier (1979), Châteauneuf (1980) and Ollivier-Pierre (1980) place the boundary at the base of the overlying Hamstead Beds. A discussion of the controversy may be found in Curry et al. (1978).

PREVIOUS WORK ON THE BEMBRIDGE MARLS

Flora: Plant fossils from the Bembridge Marls have been described by Reid & Chandler (1926) and Chandler (1963a). In both these works the majority of the records are from local concentrations within the Insect Limestone at Gurnard Bay. Machin (1971) discussed pollen and spores from this stratum. The depositional environment of this limestone is not fully understood and the plant remains occur in small patches, few of which have been located by recent workers. Jarzembowski (1976) relocated insect-bearing patches and these contained ?*Typha* leaves and fruits. Jarzembowski (1980) also recorded *Typha* in insect-bearing material from a new locality at St Helens (eastern Isle of Wight).

Palaeoenvironments: The palaeoenvironments of the Bembridge Marls have been investigated by Daley who based his conclusions on sedimentary data (Daley, 1973a, b) and macrofaunal remains (Daley, 1972). He recognized four depositional environments: estuarine, brackish lagoon, coastal lake and floodplain lake (Fig. 1).

The estuarine environment is represented by the Bembridge Oyster Bed (Forbes, 1856) and contains abundant *Ostrea*, generally as drifted, separated valves and shell debris. The sediments are fine, poorly-sorted and bioturbated sands. Studies of ostracods (Keen, 1966), Foraminifera (Bhatia, 1955) and other molluscs (Daley, 1972) support the existence of brackish or saline conditions. No plant macrofossils have been recorded from this facies which is represented only in the eastern Isle of Wight, at Whitecliff and Howgate Bays (National Grid Refs SZ 642863, SZ 648848) at the base of the cliff section.

The estuarine environment is superseded by a lagoonal environment extending over most of the outcrop area (see Daley, 1972). These sediments are predominantly blue-grey clays, homogeneous or with varve-like lamination containing a brackish fauna, dominated by the bivalve *Polymesoda* (Daley, 1972, 1973b). I have recovered plant remains from this facies at Thorness Bay (National Grid Ref. 44649322) but not at Hamstead Ledge. The flora from the Insect Limestone which occurs within this facies (Woodward, 1877, 1879) has been noted above.

The coastal lake environment characterized by tubes secreted by the polychaete worm 'Serpula', is thought to represent a period when freshwater conditions were established over a large area occasionally flooded by the sea. Rare plant remains have been recorded by Daley (1972: 22) in association with *Planorbina* and *Potamaclis*.

The flood plain lake environment occupies the middle and upper part of the Bembridge Marls and was interpreted as representing deposition in extensive, low-lying, flood plain lakes which were periodically incorporated into a river system. The environment is characterized by freshwater gastropods, small numbers of the bivalve *Unio* and olive-brown clays with some horizontal lamination. Daley (1972) subdivided this environment into lake margin and lake bottom. Both were observed to contain plant remains.

THE PALAEOECOLOGICAL STUDY AT HAMSTEAD LEDGE

The locality

Hamstead Ledge was chosen for a number of favourable features in spite of the tidally restricted access. The Bembridge Limestone forms a reef-like ledge out to sea (Figs 2–4) effectively protecting the overlying marls from erosion. Foreshore exposures of the marls extend up to 50 m seawards at equinoxial low tide (Fig. 4). A complete sequence is thus available, which is uncontaminated by slumping or by modern vegetation. The gentle dip means that a narrow horizon of about 40 mm in the cliff section may be exposed for up to 370 mm on the foreshore. Whilst it would have been difficult to excavate from such a thin horizon in the cliff face, a large volume may be collected from the foreshore. The foreshore exposure also facilitates lateral (along strike—within bed) as well as vertical (between bed) sampling.

The locality was reached via the private road past Hamstead Farm (National Grid Ref. SZ 39919126). A footpath leads down to the beach at the extreme eastern end of the ledge. The footpath via Bouldnor Cliff (Curry *et al.*, 1972) was

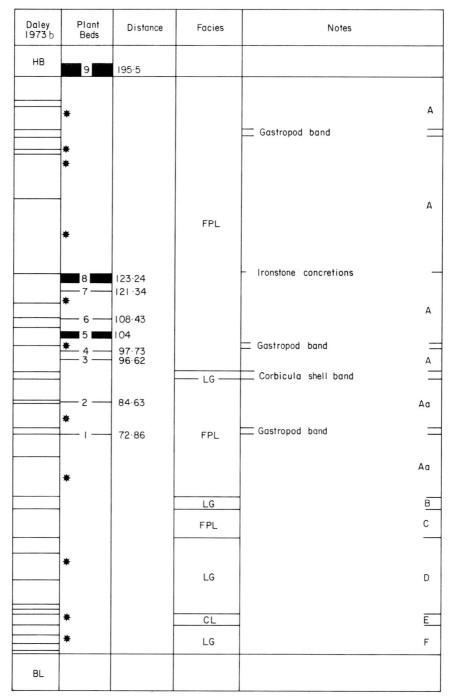

Figure 1. The position of the plant bearing horizons in the Bembridge Marls at Hamstead Ledge, Isle of Wight from the Hamstead Beds (HB); to the Bembridge Limestone (BL). The plant beds are numbered (see text) 1–9 and the distance column gives the distance (in m) along the foreshore of the base of each of the plant bearing horizons (from the top of the Bembridge Limestone). An asterisk indicates position of samples which were devoid of plant material. Facies indicates the depositional environment as interpreted by Daley (1973b): FPL—floodplain lake; LG—lagoon; CL—coastal lake. A—Olive to yellowish green clays with gastropods; Aa—as A but lacking gastropods; B—greenish

Figures 2–4. The foreshore exposure of the Bembridge Marls, at Hamstead Ledge, Isle of Wight. Fig. 2. View at mean low water showing the reef-like ledge created by the Bembridge Limestone. At extreme right centre the end of the ledge of the lowermost limestone horizon can be seen. The sea overlies marls between this and the uppermost limestone horizon which is well exposed. Above this, (to the left) are Bembridge Marls. Pole in foreground measures 1 m subdivided into 10 cm units. Fig. 3. View overlapping with Fig. 2 showing the full extent of the Bembridge Marls and a small amount of Hamstead Beds. Note the slumped and vegetated nature of the cliff section. Fig. 4. View at equinoxial low water taken from position of mean low water. Uppermost horizon of Bembridge Limestone is visible at top right. Dip and strike of the Bembridge Marls is clearly discernible. Dark coloured horizons are plant-bearing olive clays and silty-clays.

---

blue clay with *Polymesoda*; C—greenish clay with *Melanoides* and *Melanopsis*; D—bluish green to greenish blue clay with *Polymesoda*; E—blue-grey clay with *Serpula*; F—grey to greyish blue clay with *Polymesoda*.

Table 1. The distribution of plant macrofossils recovered from sieved residues of the plant bearing horizons from the Bembridge Marls, and lower Hamstead Beds Hamstead Ledge, Isle of Wight

Bed number	Total fruits and seeds	Typha latissima—seed	Potamogeton tenuicarpus	Limnocarpus forbesi	Stratiotes leaf teeth	Stratiotes neglectus	Alismaticarpum alatum	?Caricoidea obscura	C. cf. maxima	Cyperaceae genus indet.	Spirematospermum wetzleri
9t	202	A 175	VR		VR 5	VR	VR				R
9c	3	C 3	VR		VR	VR	VR				R
9b	24	A 20	VR	VR	VR 2	VR	VR 4			VR	R f
8t	1244	A 1161	R		C 9		C 34				
8b	334	A 255	R 1	VR 1	C 4		C 55				
7	7	R 6			VR		VR				
6	222	A 203			C 5	C 11				VR	
5	1512	A 1500	VR f		VR		VR 2				
4	1907	A 1790	C 21	C 112	C 8	VR 2	VR 1	VR 1			
3	1717	A 1601	C 17	A 18	R 12	VR 1	VR				
2	170	A 124	C 2	C 2	R 13		C 27	C 8			
1	6	C 2	VR 2			VR	VR 1				
Tot.	7346	6840	43	33	58	14	124	9			

Letters indicate abundance of species as recorded from qualitative studies. VR—less than 10 specimens recovered; R—10–20 specimens recovered; C—20–100 specimens recovered; A—more than 100 specimens recovered. Figures indicate the total numbers of specimens recorded from the quantitative studies, i.e. from 75 cc of sediment of each horizon. Figures in the 'cuticle' column indicate the volume of organic material (cc) within

impassable. Four visits were made, April 1975, September 1975, June 1976 and June 1977 and on none of these occasions was the cliff section exposed (Fig. 3) except at the very eastern end. These studies are therefore entirely based on the foreshore section (Figs 3, 4). The measurements of distance along the foreshore have been converted to the vertical scale on the assumption of a mean dip of 6°12′, the figure required to convert the total distance measured along the foreshore to the total recognized vertical extent (Bristow, 1889) of the Bembridge Marls at this site (Fig. 1). My own dip measurements varied from 4 to 8° (bearing 190°); Forbes (1856) quoted 5°. Because of the possible inaccuracies the distances along the foreshore (measured down dip), are quoted for each of the plant-bearing horizons (Fig. 1). Lithological and faunal studies (Daley, 1969, 1972, 1973b) have indicated relatively uniform depositional conditions through considerable parts of the Hamstead Ledge section (Fig. 1). These are most suitable for the proposed ecological studies.

Sampling, distribution and identification of floral remains

Plant macrofossils were not normally evident in the field but 250 cc test samples revealed that the darker brown clays contained plant material but the green, blue and some grey clays did not. Samples marked with an asterisk on Fig. 1 were devoid of plant material apart from one or two small cuticular fragments or fragments of opaque carbonaceous debris. Large volumes of sediment were

Table 1 contd.

Juncus vectensis	Carpolithes hamsteadensis	Sabrenia chandlerae	?Nymphaea liminis	Rhamnospermum bilobatum	Sambucus parvula	Carpolithes collumus	Acrostichum anglicum	Azolla prisca—massulae	Azolla prisca—megaspores	Harrisichara tuberculata	'Cuticle'
	R 22	R f			VR		A 71	A 79	A 99		8
		R			VR		A 2		VR 2		1
		R f	VR	VR	VR		A 8	C 12	C 32		26
		C 18			C 4	R 27	A 1415		VR 3		3
		A 9			A 3	C 10	A 451				3
	VR 1	VR					C 3		VR 4		½
	C 9	VR					C 15	VR 4	R 17		1
		VR			VR	C 10	A 3260				13
	C 15	A 5	VR		C	C 60	A 1405		VR 8	C	6
VR	C 32	C 5	VR	VR 2	R 1	C 40	A 1854		VR 2	R	'3
	C 6	VR 1					A 926				7
		VR				VR 1	A 120				15
	85	38		2	8	148	9530	95	167		

each 75 cc sample measured after 30 min settling). Apart from the fruits and seeds this is mainly small cuticular debris, much of which is monocotyledonous or small, narrow cylindrical structures which are probably rootlets. Letters b, c and t indicate base, centre and top of bed respectively. For details of the species see explanations to Figs 17–44 and Appendix 1.

subsequently collected from all horizons which contained recognizeable plant macrofossils.

Reid & Chandler (1926) noted that different information may be obtained from bedding planes and disaggregated sediments. All sediment samples were split along bedding planes prior to disaggregation. *Dicotylophyllum* and *?Acrostichum lanzaeanum* were recognized only on bedding planes. Disaggregation and subsequent treatment follows that described in Collinson (1978a, b, 1980a). Sediment samples were disaggregated in H_2O_2 solution and plant fossils cleaned in HF prior to SEM study and identification. Initial studies were qualitative (Table 1—letters). Subsequently, approximately 150 gm (75 ml measured by displacement) of each horizon were disaggregated and all plant remains above 125 μm were recorded. These quantitative results are given in Table 1—numbers. Two studies of intra-horizon variation (Tables 2 and 3) demonstrate the uniformity of the horizons.

Samples for palynological study were taken from each plant-bearing horizon and from the immediately underlying horizon (Fig. 1). Preparations were made following standard techniques, removing unwanted mineral matter with HCl and HF and Streel sieving, retaining material between 10 and 105 μm. Samples were mounted using 'Ockol' on a coverglass, which, when dry, was inverted on to a pool of a standard mounting medium on a slide. No attempt was made to remove 'unwanted' organic material which may be informative in an ecological study (Manum, 1976, Scott & Collinson, 1978).

Although the size fraction between 105 and 125 μm appears to have been

13

Table 2. Lateral sampling of bed 5, Bembridge Marls, Hamstead Ledge, Isle of Wight

Taxon	Sample number and distance down strike (m)					
	5a	5b	5c	5d	5e	5f
	High water mark	3.4	4.0	31.4	40	45
Typha latissima—seed	A	A	A	A	A	A
Potamogeton tenuicarpus	VR	VR	VR	VR	VR	VR
Stratiotes—leaf teeth		VR				VR
Sabrenia chandlerae	VR	VR			VR	VR
Alismaticarpum alatum	VR		VR		VR	VR
Sambucus parvula	VR				VR	VR
Carpolithes collumus	VR	VR	R	VR	R	VR
Acrostichum anglicum	A	A	A	A	A	A
Acrostichum lanzaeanum		VR			VR	VR
Dicotylophyllum pinnatifidum		VR			VR	
Gastropods (at the base of bed)	G.		G.V.		G.Pl.	G.V.Pl.
Freshwater ostracods	A	A	A	A	A	A

Abbreviations as for Table 1; G—*Galba*, V—*Viviparus* and Pl.—*Planorbina*.

Table 3. Lateral sampling of bed 8 top; Bembridge Marls, Hamstead Ledge, Isle of Wight

Taxon	Sample number and distance down strike (m)		
	8a	8b	8c
	12	15	24
Typha latissima—seed	889.A	1161.A	1028.A
Potamogeton tenuicarpus	1.VR	1.R	2.VR
Alismaticarpum alatum	29.C	34.C	57.C
Sabrenia chandlerae	8.C	18.C	6.C
Sambucus parvula	1.C	4.C	2.C
Carpolithes collumus	14.R	27.C	32.C
Acrostichum anglicum	1312.A	1415.A	1582.A
Azolla prisca—megaspores	1	3	1
Stratiotes—leaf teeth	2.C	9.C	25.C
Potamaclis and *Planorbina*	VR	VR	VR
Freshwater ostracods	A	A	A

Abbreviations as for Table 1. Actual numbers counted in quantitative studies are given alongside estimates from qualitative studies.

ignored it was, in fact, collected during palynological preparation. It contained only plant material that had already been recognized in the larger size fraction.

Counts were made of 200 grains of identifiable pollen and spores from three transects, on each of two slides. Results are plotted on Fig. 45. In some cases pollen and spores were rare and the entire slides had to be scanned. For bed 9 three slides had to be studied whilst for beds 3, 4 and 8 more than 200 grains were encountered within the six transects (Fig. 45). Numbers of algae, fungal spores and unidentifiable pollen and spores encountered during the transects are given in Fig. 45.

Macrofossils have been identified by the 'nearest living relative' method. This involves comparing the fossil organ with equivalent organs of living plants, searching for the closest morphological similarity. This is considered to indicate systematic affinity at either modern generic or family rank. No living species have been recognized. For detailed examples and discussion of this method see Dolph & Dilcher (1979) and Collinson (1978a, b, 1980a). Systematic notes and authors for the fossil species may be found on the figure explanations (Figs 17–44) and in the Appendix. New species are described in the Appendix where emendations of existing species are also given.

Plant microfossils have been identified using Jansonius & Hills (1976) and works cited in the figure explanations (Figs 45–67). Molluscs and ostracods have been identified using the works of Daley (1972) and Keen (1971, 1977) respectively.

Descriptions of plant-bearing horizons

Wet sediment colours cited below are taken from the *Rock-Color Chart* (Geological Society of America, 1970).

Bed 1, Figs 5 & 15: Plant-bearing horizon laterally discontinuous, olive-grey (5Y3/2) finely parallel-laminated, colour banded, silty-clays and clays. The major colour change at the base of this horizon discordant with the bedding (Fig. 5). In places junction sharply defined but elsewhere bands of the underlying greenish grey clays (5GY6/1) interspersed with the olive clays. Small-scale cross-laminations (1–2 mm) occur. Bedding planes at the basal junction and within the horizon (Fig. 15) with shell debris and intact gastropods of the genera *Melanoides* and *Melanopsis*.

Fruits and seeds rare in this horizon but cuticular debris common. Large pieces of monocotyledon (*?Typha*) leaves observed on bedding planes in the field.

Bed 2, Fig. 16: Plant-bearing horizon olive-black (5Y2/1), finely parallel-laminated, colour banded silty-clays and clays. At the junction with underlying greenish grey (5GY6/1) clays, gastropods of the genus *Potamaclis* (Fig. 16) abundant with occasional *Melanoides*, mostly intact and showing some degree of current orientation in places but locally variable (Fig. 16). The olive-black clays conforming with the topography of the underlying gastropods. Occasional specimens of *Viviparus*, *Melanopsis* and *Melanoides* present within the plant-bearing horizon but shell debris absent. Organic foraminiferid tests found in disaggregated sediments.

Fruits, seeds and cuticular debris quite common in this horizon. All organic material partially pyritized.

Bed 3, Figs 6–8: UNDERLYING HORIZON—Greenish to light olive-grey (5Y5/2), finely laminated, colour banded with darker greenish grey (5GY4/1), silty-clays containing sparse cuticular debris. JUNCTION—variable: either a sharp colour change parallel with the lamination; or a sharp colour change but with a silty horizon intercalated (Fig. 8); or a disturbed junction with a shell concentrate intercalated (Fig. 7). Shell concentrate containing the bivalve *Unio* along with gastropods of the genera *Viviparus*, *Galba* and *Potamaclis* intact but often crushed. Many shell fragments and vertebrate debris present along with a few thick walled seeds. Some of the shells orientated vertical to the bedding where still preserved. PLANT BEARING HORIZON—Laminated silty-clays and clays with silty laminae up to 300 μm thick. Silty layers, pinkish (5YR8/1) to yellowish (5Y8/1) grey; clay layers

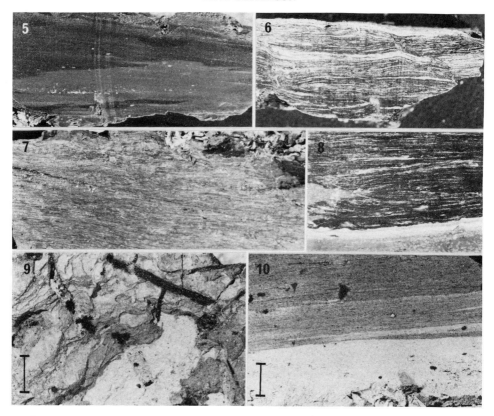

Figures 5–10. Sedimentological features of the Bembridge Marls, Hamstead Ledge, Isle of Wight. Vertical lines on sections are due to cutting technique. These are razor-sliced faces, not thin sections. Scale bars equivalent to 3 mm, Figs 5–8, 10; to 5 mm, Fig. 9. Fig. 5. Vertical section through junction at the base of bed 1. Note the colour change discordant with the bedding and the visible fine lamination in the overlying olive (dark) clays. Figs 6–8. Vertical sections of bed 3. Fig. 6. Part of the plant-bearing horizon showing organic rich (dark) and inorganic (light) laminae with small-scale cross-lamination, lenticular lamination, microfaulting and evidence of rootlet penetration. Note variable thickness of laminae. Fig. 7. Colour banded underlying green clays showing small scale lenticular lamination and disturbed shell concentrate in the uppermost part at the junction. Fig. 8. Detail of junction in an area where shell concentrate is absent and a thick silty-clay (white) layer is present. Note zone of green (light) clay at left within upper plant-bearing olive (dark) clays. Fig. 9. A small portion of the junction at the base of bed 5 degaged to show rootlets descending from the olive (dark) into the green (light) clays. Fig. 10. Vertical section through the junction between underlying green (light) and plant-bearing olive (dark) clays in bed 5. Note lack of obvious lamination in green clays, sharp junction with overlying clays, and gradual decrease in green laminae in olive clays. Larger black structures are rootlet sections.

dark olive-black (5Y2/1) to brownish black (5YR2/1). Laminae variable in thickness, some lenticular or wavy with small-scale cross-lamination and faulting (Figs 6, 8). Lens-shaped units of the underlying lithology occurring rarely near the junction (Fig. 8). An abundant and diverse fruit and seed flora present. Cuticular debris rare. Positive evidence of rootlet penetration (Fig. 6).

Bed 4: Plant-bearing horizon laminated silty-clays and clays as in bed 3 above but with a variable silty content, in places almost absent. Underlying clays finely laminated, grading into homogeneous, pale olive (10Y6/2) to light olive-grey (5Y5/2). Junction sharp with the overlying olive-black clays; no shell concentrate.

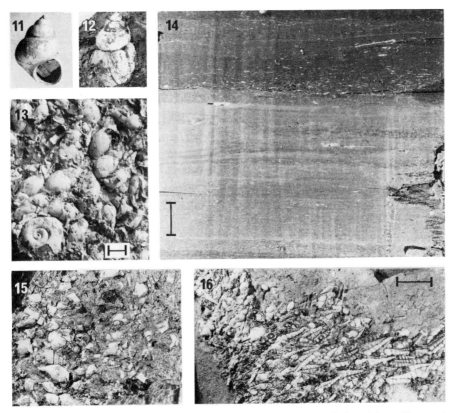

Figures 11–16. Faunal and sedimentological features of the Bembridge Marls and lower Hamstead Beds, Hamstead Ledge, Isle of Wight. Vertical lines on Fig. 14 are due to cutting technique. This is a razor sliced face, not a thin section. Scale bars equivalent to 10 mm, Figs 11, 12, 15 & 16; to 3 mm, Fig. 14; and 10 mm, Fig. 13. Figs 11 & 12. *Viviparus lentus* Solander from the plant-bearing horizon of bed 9. 12 *in situ*, upstanding from a bedding plane. Fig. 13. Bedding plane at the junction at the base of bed 5 showing shell concentrate, *Planorbina* (bottom left) and *Galba*. The latter show down dip orientation (to top of figure). Fig. 14. Vertical section through junction at the base of bed 9. Near the base are apparently non-laminated green clays. These are overlain by a zone of laminated clays, predominantly green, which are overlain by the plant-bearing horizon of predominantly olive clays (dark). In these olive clays very fine, scarcely visible, horizontal lamination is present. White flecks are calcareous fossils. Fig, 15. Portion of bedding plane within the plant-bearing horizon of bed 1 showing shell debris, *Melanoides* (ornamented) and *Melanopsis* (smooth). Fig. 16. Bedding plane at the junction at the base of bed 2 showing *Potamaclis* with some current orientation to the right of the figure.

Potamaclis the only intact gastropod present but shell and vertebrate debris common. Ostracods particularly abundant including *Hemicyprideis*.

Fruits and seeds abundant and diverse, cuticular debris quite common.

Bed 5; Figs 9, 10 & 13: Plant-bearing horizon finely parallel-laminated colour banded olive-grey (5Y3/2) silty-clays and clays. Basal junction initially sharp overlying greenish grey (5GY6/2) silty-clays and clays. A decreasing proportion of the green clays continuing into the plant-bearing horizon (Fig. 10). Rootlet (Fig. 9) penetration of the clays, extending from the olive clay through into the green clay. Shell remains absent within deposit but in places at junction a shell parting comprising *Galba*, *Planorbina* (Fig. 13) and *Viviparus* present. Ostracods (*Candona* and *Virgatocypris*) present.

Cuticular debris common and leaf fragments of *?Acrostichum lanzaeanum* and *Dicotylophyllum pinnatifidum* found. Fruits and seeds common.

Bed 6: One small lens, 3.42 m in lateral extent of olive-black (5Y2/1) silty-clays and clays within greenish grey (5GY4/1) silty-clays and clays. In places with a sharp basal junction, in others with units of the underlying colour within the plant-bearing lens. Shell and vertebrate debris common at the junction and crushed material of *Viviparus* occurring within the horizon.

Fruits and seeds quite common but cuticular debris rare. Large monocotyledon leaves (*?Typha*) up to 150 mm in length observed in the field.

Bed 7: Finely laminated, light olive-grey (5Y5/2) silty-clays and clays, overlying dark greenish grey (5GY4/1) silty-clays and clays with a gradual transition. No shells, shell debris or ostracods observed. Fruits, seeds and cuticular debris rare.

Bed 8: Bioturbated olive-grey (5Y5/2–5Y3/2) silty-clays and clays intermixed with greenish grey (5GY6/1–5GY4/1) silty-clays and clays at the base and top of this bed. Bluish clays are in the intermediate part (5Y6/1–5GY4/1). These containing few fruits and seeds (a sparse equivalent of the base and top of the bed). Also containing *Potamaclis*, ostracods, carbonaceous plant fragments and large monocotyledon leaves (*?Typha*).

The upper and lower bioturbated plant-bearing strata containing *Planorbina* and *Viviparus*, some a little crushed, intact *Potamaclis* and abundant ostracods. Abundant fruits and seeds occurring but cuticular debris rare.

Bed 9, Figs 11, 12 & 14: Plant-bearing horizon finely laminated, olive-black (5Y2/1) silty-clays and clays with a small amount of small scale lenticular lamination at the base; overlying dark greenish grey (5G6/1) silty-clays and clays. Transition gradual, with a decreasing proportion of the green clays in the overlying brown clays (Fig. 14). *Planorbina* and *Unio* present near the base along with *Viviparus* (Figs 11 & 12), the latter lying at various angles to the bedding (Fig. 12). Ostracods present. Bedding planes showing evident comminuted plant debris.

Abundance of plant material variable through this thick bed. *Azolla* common at the base, abundant at the top, but rare in the middle. Fruits and seeds rare at the base and middle but common at the top. Cuticular debris abundant at the base.

DISCUSSION

Sedimentary features

Lamination and couplets: In some horizons, it has been possible to count the numbers of silty-clay and clay laminae e.g. 104 for bed 3 plant-bearing horizon. This horizon is 1 cm in vertical extent. The 21 m of the Bembridge Marls at Hamstead Ledge were probably deposited over 1.5 My (from data in Berggren, 1971; Cavelier, 1979). Assuming regular influx, bed 3 may therefore represent up to 750 years of deposition, 7 years for each recognizable lamina. Alternatively the silty-clay/clay couplets might represent annual varves deposited over only 52 years.

Such alternating layers of different texture, colour or composition with parallel lamination, form in areas of slack water in channels (Coleman, 1969); shallow marine seas; lagoons and freshwater lakes (Fairbridge & Bourgeois, 1978; Picard

& High, 1979). The differences between the layers may reflect a change in biological productivity or in chemical or physical parameters such as temperature; precipitation; evaporation; pH/EH relationship, and horizontal or vertical water motions. (Fairbridge & Bourgeois, 1978).

Interpretations of the causes of modern and ancient varving include increased erosion resulting from fire (Cwynar, 1978); chemical changes resulting in higher calcite precipitation in spring and early summer (Ludlam, 1969, 1981; Ryder, Fouch & Elison, 1976); chemical changes resulting in the precipitation of iron oxides in spring and early summer (Anthony, 1977); increased biological productivity resulting in precipitation of diatom and other skeletal elements of algae in spring and summer (McLeroy & Anderson, 1966); the periodic development of sediment surface algal mats (Eugster & Hardie, 1975); increased erosion due to precipitation resulting in periodic increase in the influx of silt (Picard & High, 1979) or simply increased organic productivity (Bergström & Levi-Setti, 1978). The dark layers are considered to be rich in organic matter and are often referred to as sapropel or oil shale. The dark organic and light inorganic couplets are often considered to be annual, the dark deposited in winter and the light in spring and summer (Ludlam 1969, 1981; Tolonen, 1978). However they are not necessarily seasonally produced (Fraser, 1929; Eugster & Hardie, 1975; Lambert & Hsü, 1979). A range of periodicities from 12 to 500,000 years for so-called non-glacial varves is documented by Picard & High (1979).

For these alternating couplets to have been preserved in the Bembridge Marls it is likely that they were deposited in the absence of bioturbating organisms in very tranquil conditions probably in an anaerobic environment (Fairbridge & Bourgeois, 1978). Such is typical of a stratified (meromictic) water body where there is no overturn, stratification resulting from temperature (cold bottom water) or water chemistry (heavy minerals concentrated in bottom water, e.g. saline waters in a lagoon), (Fairbridge & Bourgeois, 1978). Davis (1973) recorded that from 6 to 12 mm of sediment in the littoral zone and 1 mm in deeper lake waters were disturbed by the annual overturn. Zangerl & Richardson (1963) reported that even slight wave action disrupted bottom sediments in Lake Pontchartrain. Similarly, bioturbation, if present, will destroy evidence of couplet deposition (Twenhofel & Mckelvey, 1941).

Some beds, such as 3, 4 and 9, have wavy and lenticular lamination in association with a rhythmic alternation of silty-clay and clay couplets. These sedimentary features are most commonly associated with tidal influences (Reineck & Singh, 1973), but can also occur on lake margins particularly where the wind has a large fetch and where prevailing winds can build up a seiche motion (Hutchinson, 1957). Coleman (1966) recorded the occurrence of small scale lenticular bedding such as is seen in parts of beds 3, 4 and 9, on lake bottom sediments in front of small deltas.

The green to olive colour change: The majority of the sediments in the sequence studied are green clays (Fig. 1) but the plant-bearing horizons themselves are olive-brown-black clays. Such a striking colour difference in sediments is probably due to the chemistry of the overlying waters. Absence of plant material in sediments has been considered to be due to overlying clear, oxygenated waters (Lambert, 1976).

In eutrophic lakes, organic matter is rapidly broken down, and, under slight reducing conditions, a green coloration is imparted to the sediments because of the high content of ferrous ion. In Oligotrophic lakes organic matter accumulates and chocolate-brown or black sulphurous sediments are produced (Mortimer, 1941, 1942; Jessen, 1949; Reeves, 1968; McBride, 1974; Peterson, 1976). With thermal stratification, the hypolimnion rapidly becomes anaerobic and breakdown of organic matter is retarded or prevented. Reeves (1968) suggested that the change from green to brown sediments indicated a change from an open (mixed) to a stagnant (stratified) lake. According to Strøm (1955) green muds themselves indicate moderately stagnant conditions and black muds indicate deoxygenated conditions. Stratification is usually associated with deep waters but, in shallow water bodies cessation of through flow (cut off) (Peterson 1976) or encroaching aquatic vegetation could cause development of stagnant, anaerobic conditions.

The macroflora

For authorship, taxonomic notes, affinities etc. see figure explanations and Appendix 1. For distribution see Tables 1–3.

The Hamstead Ledge, Bembridge Marls: The macrofloral association in the Bembridge Marls consists of one dicotyledon leaf form, *Dicotylophyllum*; one monocotyledon leaf form *?Typha*; a number of thick-walled fruits and seeds, *Spirematospermum* (Fig. 25), *Stratiotes* (Fig. 40), *Sabrenia* (Figs 29, 30), *Caricoidea* cf. *maxima*; smaller, thinner-walled fruits and seeds, *Limnocarpus* (Fig. 38), *Potamogeton* (Fig. 44), *?Caricoidea obscura* (Fig. 35), Cyperaceae, genus indeterminable, *Sambucus* (Fig. 42), *Alismaticarpum* (Figs 33, 34, 36, 37, 39 & 41); large, thin-walled *Rhamnospermum*; small thin-walled *?Nymphaea liminis* (Fig. 28); very thin-walled, cuticular seeds, *Typha* (Fig. 24), cf. *Juncus vectensis* (Figs 17, 32); leaf teeth of *Stratiotes* (Fig. 23) and sporangia of *Acrostichum* (Fig. 19). Megaspores and massulae of *Azolla* (Fig. 31) and gyrogonites of *Harrisichara* (Fig. 43) also occur. Two species of *Carpolithes* (Figs 18, 21, 22 & 20, 26) are also recorded.

Of these 22 taxa, extant *Sambucus* is a shrub which frequently grows around lakes and streams although not exclusively associated with this environment. *Dicotylophyllum pinnatifidum*, if related to Myricaceae, may have occupied a similar habitat. Members of the Cyperaceae and Juncaceae are also common associates of marginal aquatic floras. Possible habitats for *Rhamnospermum bilobatum*, *Spirematospermum wetzleri* and the two *Carpolithes* taxa are more or less unknown. A near living relative of *Spirematospermum*, *Cenolophon* is found along streams and water courses in Thailand (Friedrich & Koch, 1970).

Of the remaining species, extant members of the Alismataceae (*Alismaticarpum*), Characeae (*Harrisichara*), Hydrocharitaceae (*Stratiotes*), Nymphaeaceae (*Sabrenia* & *?Nymphaea*), Potamogetonaceae and Ruppiaceae (*Potamogeton* and *Limnocarpus*), Salviniaceae (*Azolla*), and Typhaceae (*Typha*), are exclusively aquatic or marginal aquatic plants. The genus *Acrostichum* also grows as a marginal aquatic.

The flora includes free-floating plants (*Azolla* and *Stratiotes*), rooted plants with floating leaves (Nymphaeaceae and Potamogetonaceae) anchored, submerged *Harrisichara* and emergent plants (*Typha*, *Acrostichum*, Alismataceae, Cyperaceae and Juncaceae).

Many of the *Typha* seeds had probably germinated as they lack the operculum which is pushed off the seed at germination. Other seeds in an equally flattened and compressed state still possess this operculum (Fig. 24). *?Nymphaea* (Fig. 28) and some *Sabrenia* seeds (Fig. 29) lack the cap; some *Potamogeton* and *Limnocarpus* fruits

Figures 17–25. All light micrographs, Fig. 25 by reflected light, others by transmitted light. Scale bars equivalent to 0.05 mm, Figs 17 & 19; to 0.25 mm, Figs 18, 21 & 22; to 0.10 mm, Figs 20 & 23; to 0.10 mm Fig. 24; and to 5 mm, Fig. 25. Macroflora from the Bembridge Marls, Hamstead Ledge, Isle of Wight. For detailed notes see Appendix 1. (See Collinson, 1978a, for detailed descriptions and additional illustrations). Fig. 17. cf. *Juncus vectensis*. Seed; affinity Juncaceae—possibly *Juncus* (rushes). Holotype V.60909. Seed showing operculum (at apex) and hilum (at base). Figs 18, 21 & 22. *Carpolithes hamsteadensis*. Affinity unknown monocotyledon. Fig. 18. V.60913. Urceolate specimen with intact outer layer. Fig. 21. Holotype V.60912. Specimen with intact outer layer. Fig. 22. V.60911. Specimen with outer layer splitting longitudinally revealing the inner layer. Fig. 19. *Acrostichum anglicum* Collinson 1978b. Fern sporangium containing spore. Affinity Adiantaceae, *Acrostichum* (leather fern). V.59102. Fig. 20. *Carpolithes collumus*. Affinity unknown. Holotype V.60915. Specimen with central body. Fig. 23. *Stratiotes* sp. Leaf margin tooth; affinity Hydrocharitaceae—*Stratiotes*. V.60905. Fig. 24. *Typha latissima* (Al.Br.) Reid & Chandler, 1926. Seed; affinity Typhaceae. Operculum still present at apex. Only fragments of outer cell layer are present. V.60901. Fig. 25. *Spirematospermum wetzleri* (Heer) Chandler emend. Kock & Friedrich 1971. Fruit, basal part; affinity Zingiberaceae. V.60903.

lack the germination valve; some *Stratiotes* and *Sambucus* seeds are split into two valves and some *Alismaticarpum* achenes show splitting (Fig. 37). All of these seeds could therefore have germinated and contributed to the contemporaneous vegetation. Many of the *Acrostichum* sporangia have shed their spores which may have germinated around the depositional site.

Fruit and seed numbers recorded from the quantitative studies (Table 1) vary

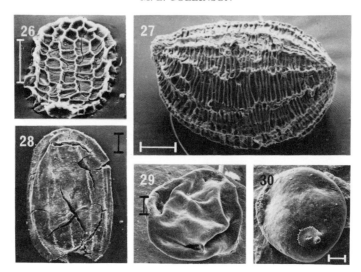

Figures 26–30. All SEM. Scale bars equivalent to 0.1 mm, Fig. 26; to 0.05 mm, Fig. 27; to 0.25 mm, Figs 28 & 30; to 0.5 mm, Fig. 29. Macroflora from the Bembridge Marls, Hamstead Ledge, Isle of Wight. For detailed notes see Appendix 1 (see Collinson 1978a for detailed descriptions and additional illustrations). Fig. 26. *Carpolithes collumus*. Affinity unknown. V.60900(2). Specimen showing surface cells and basal neck. Fig. 27. *Juncus compressus* Jacq. A modern seed (outer testa removed) for comparison with fossil in Figs 32 & 17. Fig. 28. *?Nymphaea liminis* Collinson 1980a. Seed lacking germination cap; affinity Nymphaeaceae. V.60249. Figs 29 & 30. *Sabrenia chandlerae* Collinson 1980a. Seeds; affinity Nymphaeaceae. Fig. 29. Specimen lacking cap which may therefore have germinated. V.60228. Fig. 30. Specimen with cap in position. V.60227.

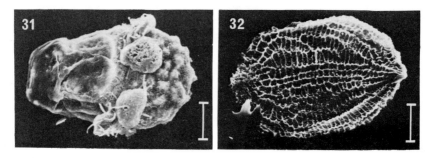

Figures 31 & 32. Both SEM. Scale bars equivalent to 0.10 mm, Fig. 31; to 0.05 mm, Fig. 32. Macroflora from the Bembridge Marls, Hamstead Ledge, Isle of Wight. For detailed notes see Appendix 1 (see Collinson 1978a for detailed descriptions and additional illustrations). Fig. 31. *Azolla prisca* Reid & Chandler emend. Fowler, 1975. (See also Collinson, 1980b). Megaspore with microspore massulae attached, apex to left; affinity Salviniaceae. Specimen now lost. Comparable material on stub V.60900. Fig. 32. cf. *Juncus vectensis*. Seed; affinity Juncaceae—possibly *Juncus* (rushes). V.60900(1). Seed with operculum to left showing central rows of square cells indicating raphe position and subdivision of cells by subsidiary thickenings.

Potamogetonaceae/Ruppiaceae. Large germination valve still present at right. V.60898. Fig. 40. *Stratiotes neglectus* Chandler, 1923. Seed; affinity Hydrocharitaceae—*Stratiotes*. V.60904. Fig. 42. *Sambucus parvula* Chandler, 1926 herein emend. Seed, ventral face; affinity Caprifoliaceae—*Sambucus*. V.60907. Fig. 43. *Harrisichara tuberculata* (Lyell) Grambast, 1957 (see also Feist-Castel, 1977). Gyrogonite; affinity Characeae. Apex to right. V.60906. Fig. 44. *Potamogeton tenuicarpus* Reid & Reid herein emend. fruit; affinity Potamogetonaceae. Large germination valve still present to left. V.60899.

Figures 33–44. All SEM except Fig. 33 which is transmitted light micrograph. Scale bars equivalent to 0.2 mm, Figs 33, 34, 37, 41 & 42; to 0.1 mm, Fig. 35; to 0.5 mm, Fig. 36; to 0.3 mm, Figs 38 & 46; to 0.25 mm, Fig. 39; to 1.0 mm, Fig. 40; to 0.2 mm, Fig. 43. Macroflora from the Bembridge Marls, Hamstead Ledge, Isle of Wight. For detailed notes see Appendix 1 (see Collinson 1978a for detailed descriptions and additional illustrations). Figs 33, 34, 36, 37, 39 & 41. *Alismaticarpum alatum*. Fruit; affinity Alismataceae. Fig. 33. Syntype V.60917. Fruit showing thin wing and basal notch. Fig. 34. Syntype V.60919. Fruit with outer cell layer intact showing lateral wing. Fig. 36. Syntype V.60920(2). Group of eight fruits. Fig. 37. Syntype V.60918. Fruit revealing central cell layer and showing possible germination split and lateral wing (apex to right). Fig. 39. Syntype V. 60920(3). Group of four fruits. Fig. 41. V.60920(1). Fruit showing central cell layer and remnants of outer layer in position of wing at right. Fig. 35. *Caricoidea obscura* Chandler, 1963a. Fruit; affinity Cyperaceae. Apex to right. V.60902. Fig. 38. *Limnocarpus forbesi* (Heer) Chandler emend. Collinson, in press. Fruit; affinity

from 3 in the middle of bed 9 to 1907 in bed 4. Of the total count of 7346 fruits and seeds 6840 are *Typha*. Although *Typha* seeds are rare in beds 1, 2, 6, 7 and 9 compared with other horizons they still comprise at least 50% and generally more than 70% of the macroflora. Such an abundance would be expected from the large (e.g. 600 000, Howard-Williams, 1975) numbers of fruits produced by one fruiting spike. H. H. Birks (1973) demonstrated that large numbers of *Typha* seeds only occur within the growth zone of the plant with very few transported into the open waters of lakes.

The abundance of a potential fossil seed does not always reflect the abundance of the parent plant in the vegetation. For example, H. H. Birks (1973) noted that three seeds $(100 \text{ cm}^3)^{-1}$ indicated a moderate cover for *Brasenia* but a very low cover for *Najas*. *Potamogeton* fruits were characteristically rare whatever their role in the vegetation. Annual species whose survival is by seed produce large numbers of seeds compared with perennials. Vegetative propagation is also common in aquatic plants (Sculthorpe, 1967) and reduces the need for high seed production. Thus it is to be expected that many aquatic plant seeds will only occur in low numbers in sediments.

Other factors including flotation properties (H. H. Birks, 1973) and animal dispersal may affect the distribution of seeds within lakes (Ridley 1930). Potential bird dispersal-agents existed in Britain in the Tertiary (Walker, 1980).

It is very difficult to interpret the Hamstead Ledge macrofossil quantitative distribution except in a general manner. High numbers of *Typha* seeds indicate close proximity to growing plants and occur along with high numbers of *Acrostichum* sporangia (beds 3, 4, 5 and 8 top) representing another emergent, shoreline-rooted plant. In contrast, in beds 6 and 9, low numbers of *Typha* and *Acrostichum* are associated with high numbers of either *Stratiotes* or *Stratiotes* and *Azolla*, the free-floating plants.

Beds 3, 4 and 9 contain the most diverse assemblages and show sedimentological features indicating a nearshore site, or possible influence of an inflow stream. Some of the floral elements which only occur in these beds may have arrived via an inflow, e.g. *?Nymphaea*, *Rhamnospermum*, *Caricoidea* cf. *maxima*, cf. *Juncus* and *Spirematospermum*, (Table 1). Equally this diversity of vegetation could be due to proximity of inorganic influx. Provided such influx is not excessive, these sites are most favourable for the extensive growth of marginal aquatic vegetation (Misra, 1938).

Bed 6 is the most distinct from beds 3, 4 and 9. *Potamogeton*, *Limnocarpus*, *Alismaticarpum*, *Sambucus* and *Sabrenia* are all absent from bed 6. This bed contains the only specimens of Cyperaceae, genus indeterminable, and very high numbers of *Stratiotes* seeds. It also contains *Azolla* but very low numbers of the shoreline plants *Typha* and *Acrostichum*. This bed was represented only by one isolated lens of plant-bearing sediment. The presence of free-floating plants and rarity of shoreline plants suggest that this may be a more open water deposit but the presence of large mono-cotyledon (*?Typha*) leaves is not easily compatible with this interpretation. It seems possible that bed 6 may represent deposition of parts of a floating mat of vegetation, or deposition beneath a small floating mat within otherwise open water. Such a mat can incorporate material torn away from the edge of marginal aquatic vegetation as well as material of free-floating species (Meriläinen & Toivonen, 1979). Similar floating mats in modern floodplain environments have been recorded by Yount (1963).

The overall rarity of *Stratiotes* seeds is unusual, as the presence of plants is demonstrated by leaf margin teeth (Table 1). West & Gay-Wilson (1968) record *Stratiotes* leaf teeth in the absence of seeds within an aquatic, Quaternary flora. *Stratiotes* today is a dioecious genus and fruits well only under certain conditions, spreading frequently by vegetative propagation. Perhaps similar reproductive vagaries could account for the absence of seeds in many of the Bembridge Marls horizons.

Stratiotes seeds and leaf teeth are both absent only from bed 5 which contains abundant shoreline plants, with evidence of *in situ* growth in the form of rootlets but no evidence of even the slight disturbance in sedimentation indicated by wavy lamination in beds 3 and 4 or bioturbation in bed 8. Bed 5 therefore is probably the most near-shore deposit representing the most stable vegetation and is least likely to include free-floating plants.

Notes on other Bembridge Marls localities and floras from the immediately overlying Hamstead Beds

The significance of the Hamstead Ledge associations becomes apparent from comparisons with the floras from other Bembridge Marls sites. Samples from the basal part of the section at Burnt Wood (western end of Thorness Bay) within blue-grey clays (Daley, 1973b: bed xiii) containing *Melanoides*, *Melanopsis*, *Potamaclis*, and vertebrate debris, yielded isolated specimens of *Sambucus*, *Stratiotes*, *Sabrenia*, *Potamogeton* and *Limnocarpus*. Although these plants may have been growing locally in the lagoonal environment (as suggested by Daley, 1973b) this, and other comparable assemblages, have probably been subjected to transportation and sorting with retention in an aerobic environment conducive to decay of less durable plant remains.

At Gurnard Ledge (national Grid Ref. SZ 46299467) a plant-bearing horizon was sampled from the cliff some 6 m below the top of the marls. This sample contained *Dicotylophyllum*, *Typha*, *Potamogeton*, *Limnocarpus Alismaticarpum*, *Sabrenia*, *Acrostichum*, *Harrisichara*, *Carpolithes hamsteadensis*, *Rhamnospermum* and *Stratiotes* (seeds and leaf teeth). A new seed form (*Carpolithes*, sp. C, Collinson, 1978a), probably a member of the Limnocharitaceae, was also recorded. *Azolla* and *Carpolithes collumus* were absent. No gastropods were recorded but the ostracods *Candona daleyi* and *Virgatocypris edwardsi* were abundant along with vertebrate debris. Rootlets penetrated the olive-brown plant-bearing silty-clays into the underlying green silty-clays. This horizon is bed xxviii Daley (1973b) which he considered to be the lateral equivalent of his bed xxiii at Hamstead Ledge. My Hamstead bed 5 maps (Fig. 1) just below Daley's bed xxiii but I feel sure that these are all the same horizon. The flora at Gurnard is very similar to that at Hamstead with one additional species which may be another emergent aquatic plant.

The Hamstead Beds in Bouldnor cliff (National Grid Ref. SZ390910) I found to be largely obscured. Approximately 80 m were exposed in the foreshore immediately overlying the Bembridge Marls at Hamstead Ledge, including two plant-bearing horizons. The first, 59.94 m along the shore from the base of the Hamstead beds comprises irregularly laminated, partly bioturbated, colour-banded olive-black (5Y2/1–5Y4/1) silty-clays. A basal shell concentrate of *Potamides* and *Nystia* is present with ostracods in the shell apertures. The ostracod

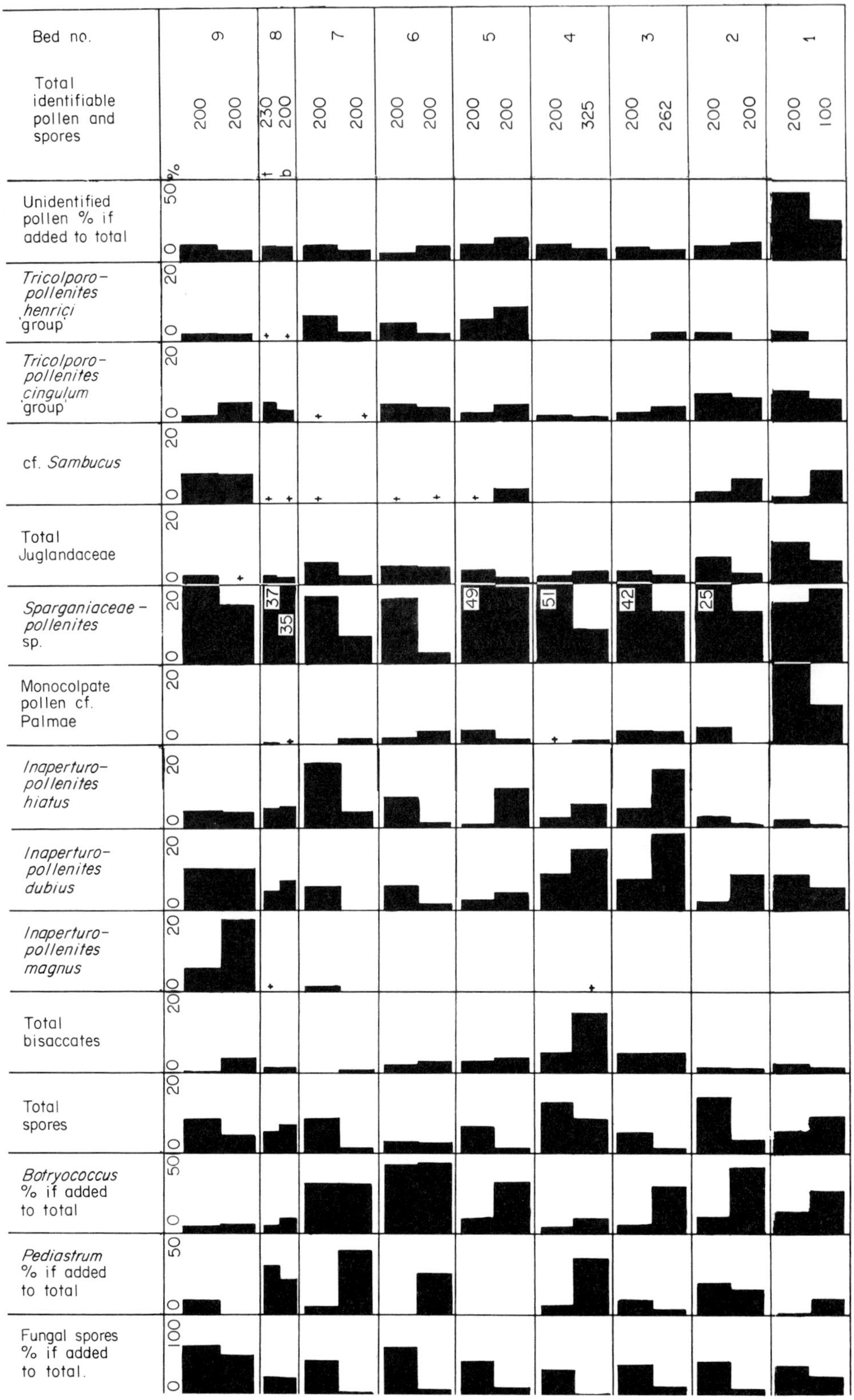

Bed no.
9
8
7
6
5
4
3
2
1
Total identifiable pollen and spores
200
200
230
200
200
200
200
200
200
200
200
325
200
262
200
200
200
100
%
t
b
Unidentified pollen % if added to total
Tricolporo-pollenites 'henrici group'
Tricolporo-pollenites 'cingulum group'
cf. Sambucus
Total Juglandaceae
Sparganiaceae-pollenites sp.
37
35
49
51
42
25
Monocolpate pollen cf. Palmae
Inaperturo-pollenites hiatus
Inaperturo-pollenites dubius
Inaperturo-pollenites magnus
Total bisaccates
Total spores
Botryococcus % if added to total
Pediastrum % if added to total
Fungal spores % if added to total.
0
20
50
100

assemblage, typical of the Bembridge Marls, is present through the horizon along with the gastropods *Galba* and *Viviparus*. Plant macrofossils include *Harrisichara*, *Sabrenia*, *Rhamnospermum*, *Sambucus* and *Typha*.

A second horizon at 63.3 m contains comminuted plant debris in finely parallel, laminated clays with *Typha*, *Alismaticarpum*, *Stratiotes* leaf teeth, *Carpolithes collumus* and a few fragmentary remains of *Potamogeton* and *Sabrenia*. The ostracod assemblage is present but gastropods are absent.

These higher Hamstead Beds assemblages show no significant differences from those of the underlying Hamstead Beds (bed 9) and Bembridge Marls (beds 1–8). The macrofloras and their palaeoecology support the views of Bristow (1889) and Chandler (1963a) that no real distinction exists between the Bembridge Marls and the Hamstead Beds.

The microflora

For authorship, reference to taxonomic studies and notes on affinities see explanations to the figures (Figs 45–67).

One significant feature is the apparent lack of pollen which can be securely tied to any of the macrofossil taxa. Pollen of *Aglaoreidia* (Fig. 65) may possibly be from the *Limnocarpus* plant (Collinson, in press) and *Sparganiaceaepollenites* (Figs 66, 67) may be from the *Typha* plant (though it might belong to a member of the Sparganiaceae). Pollen referred to cf. *Sambucus* (Figs 55, 57, 58) could be *Sambucus* but might belong to a number of other families. Pollen of Cyperaceae (three species of macrofossil); Nymphaeaceae (two species of macrofossil); Alismataceae

---

Figure 45. The distribution and percentage occurrence of abundant pollen, spores and algae in the Bembridge Marls and lower Hamstead Beds at Hamstead Ledge, Isle of Wight. For each bed the upper column represents the plant-bearing horizon the lower column the underlying horizon; except in bed 8 where no precise lower junction could be observed. In bed 8 both the top and base of the plant bearing horizon were studied indicated 8t and 8b. + indicates present but less than 1%. Pollen, spores and algae listed are figured in Figures 46–67. Details of the 'total' columns are as follows. Total spores—Fig. 53 and *Cicatricosisporites* sp. (Pot. & Gell.) R.Pot. 1956; (Schizaeaceae): *Verrucosisporites quintus* (Thomson & Pflug) Krutsch, 1959; *Baculatisporites* (Osmundaceae): *Polypodiaceoisporites* aff. *potonei* (R. Pot. & Gell.) Kedves, 1961; (Pteridaceae, Cyatheaceae): *Verrucatosporites* sp., *Laevigatosporites haardti* (R.Pot. & Ven.) Thomson & Pflug, 1953; (Polypodiaceae): *Azolla prisca* Reid & Chandler emend. Fowler, 1975; massulae (Salviniaceae). Total bisaccates—*Pityosporites* Seward emend. Manum, 1960; *Abietineaepollenites* Pot. 1958; (Pinaceae) see Manum, 1962. Total Juglandaceae—Fig. 61 and *Triatriopollenites engelhardtioides* Roche, 1973 and other species of the genus; *Plicatopollis* species and *Caryapollenites* cf. *simplex* (R.Pot. & Ven.) Krutsch, 1960. Total Monocolpate pollen cf. Palmae—Figs 63 & 64 and *Arecipites pseudotranquillus sensu* Nichols, Ames & Traverse 1973. cf. *Sambucus* is the total of Figs 55, 57 & 58. A total of 44 pollen taxa, eight spore taxa, four algae and four acritarchs were recorded. The acritarchs were one specimen each of four different kinds. The algae were Figs 46, 48 & 50. Spores and pollen are figured (Figs 49. 51–67) and listed above. Additional pollen taxa are as follows—(they represent together less than 6% of the total in any one horizon, usually less than 2%) *Pandaniidites* Elsik, 1968; *Cycadopites* Wodehouse, 1933; *Monocolpopollenites oviformis* Kedves, 1965; *M. ligneolus* (Pot. 1931) Weyland, Pflug & Mueller, 1960; *Liliacidites* Couper, 1953; *Plicapollis pseudoexcelsus* Krutsch, 1960, subsp. *minor* Pflug, 1953, (affinity Myricaceae); *Chenopodopollis* Krutsch, 1966 (Chenopodiaceae); *Tricolporopollenites pseudolaesus* R. Pot. 1931; *T. iliacus* subsp. *medius* Pflug & Thomson, 1953 (Aquifoliaceae); *T. megaexactus* (Pot.) Thomson & Pflug, 1953; *T.* other species; *Tricolpopollenites anguloluminosus* (Anderson) Elsik, 1968; *T. geranioides* (Couper) Elsik, 1968, (?Oleaceae); *Tetracolporopollenites* spp. (Sapotaceae); *Ericipites* Wodehouse, 1933 (Ericaceae). These pollen are figured in Collinson 1978a. Dispersed organic matter including cuticular debris is more common in the plant-bearing horizons than in the underlying horizons.

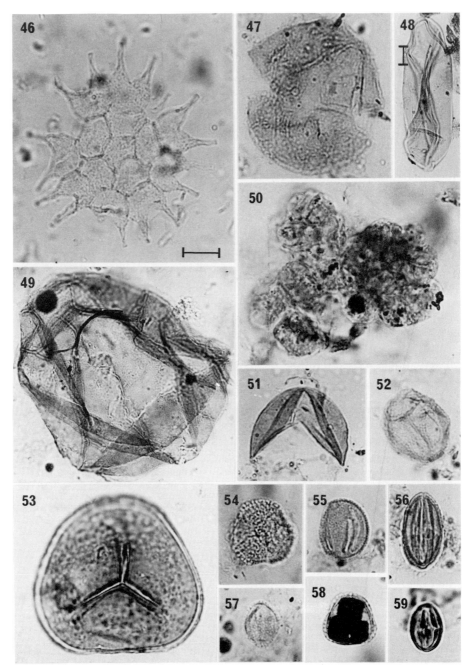

Figures 46–59. All transmitted light micrographs. Distributions of 2% or above have been noted here for species not given on Fig. 45. Scale bars equivalent to 0.01 mm for all Figs. Scale bar for Fig. 46 applies to all other Figs except Fig. 48. Some major components of the microfloral assemblage from the Bembridge Marls and lower Hamstead Beds, Hamstead Ledge, Isle of Wight (for further notes, distribution and records of other species, see Fig. 45 and Figs 60–67). Fig. 46. *Pediastrum* sp. Affinity Algae-Chlorococcales. References Borge & Erdtman, 1954, Millington & Gawlik, 1975. Fig. 47. Dinoflagellate cyst, cf. Peridiniaceae, cf. *Vectidinium*. References Davey *et al.*, 1966, Liengjarern, Costa & Downie, 1980. (Restricted to bed 9, 6% in the plant bearing horizon, 3% in the underlying horizon.) Fig. 48. *Schizosporis* cf. *parvus* Cookson & Dettman, 1959. Affinity Algae-Zygnemetales. References Norvik & Burger, 1976; Van Geel, 1976; Van Geel, Bohncke & Dee, 1981. (Reaches a

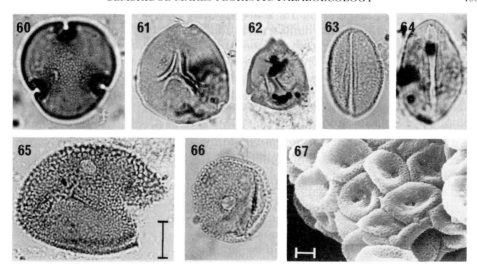

Figures 60–67. All transmitted light micrographs except Fig. 67 which is SEM. Distributions of 2% or above have been noted here for species not given on Fig. 45. Scale bars equivalent to 0.01 mm. Scale bar on Fig. 65 applies to all other Figs except Fig. 67. Some major components of the microfloral assemblage from the Bembridge Marls, Hamstead Ledge, Isle of Wight (for further notes, distribution and records of other species see Fig. 45 and Figs 46–59). Fig. 60. *Intratriporopollenites pseudoinstructus* Mai, 1961. Affinity Tiliaceae. References Gruas-Cavagnetto, 1976, 1977. (Reaches a maximum of 2% in plant bearing horizon of bed 9.) Fig. 61. *Plicatopollis plicatus* (R.Pot.) Krutsch, 1962. Affinity Juglandaceae. Reference Gruas-Cavagnetto, 1977. Fig. 62. *Triatriopollenites bituitus* (Pot.) Thomson & Pflug, 1953. Affinity Myricaceae. Reference Gruas-Cavagnetto, 1974. (Reaches a maximum of 2.5% in bed 9.) Fig. 63. *Sabalpollenites aereolatus* (Pot.) R. Pot. 1958. Affinity Palmae. Reference Durand & Ollivier-Pierre, 1972. Fig. 64. *Monocolpopollenites tranquillus* ((R.Pot.) Thomson & Pflug, 1953) Nichols *et al.*, 1973. Affinity Palmae. Reference Kedves & Bohony, 1966. Fig. 65. *Aglaoreidia cyclops* Erdtman emend. Fowler, 1971. Affinity ?Potamogetonaceae/Ruppiaceae. References Machin, 1971; Collinson, in press; Sittler *et al.*, 1975. (Only occurs in the underlying horizon of bed 4 and the base of bed 8.) Figs 66 & 67. *Sparganiaceaepollenites* sp. Thiergart, 1938. Affinity Typhaceae or Sparganiaceae. References McAndrews *et al.*, 1973; Machin, 1971; Krutsch, 1970; Ollivier-Pierre, 1970. Fig. 66. Individual pollen grain, Fig. 67. Detail from part of a pollen sac.

maximum of 3% in underlying horizon of bed 2. Occurs also in beds 3, 8 top and 9 plant-bearing horizons and beds 3, 7 and 9 underlying horizons.) Fig. 49. *Inaperturopollenites magnus* (R.Pot.) Thomson & Pflug, 1953. Affinity Araucariaceae, *Pseudotsuga* or *Larix*. Fig. 50. *Botryococcus* cf. *braunii* Kützing. Affinity Algae-Chlorococcales. References Pacltova, 1966; Gray, 1960; Travers, 1955. Fig. 51. *Inaperturopollenites hiatus* (R.Pot.) Thomson & Pflug, 1953. Affinity Taxodiaceae. Fig. 52. *Inaperturopollenites dubius* (R.Pot. & Ven.) Thomson & Pflug, 1953. Affinity Taxodiaceae or Cupressaceae. Fig. 53. *Deltoidospora* sp. Affinity *Acrostichum anglicum* Collinson, 1978b; or other Filicales. Fig. 54. cf. Salicaceae/Platanaceae. References Martin & Rouse, 1966; Srivastava, 1972. (Reaches 2–3% in bed 8 top, bed 5 and bed 2 underlying horizon.) Figs 55 & 57. cf. *Tricolpopollenites hians* (Stanley) Elsik, 1968. Affinity Salicaceae or *Sambucus*. References Schumaker, Lambry & Roche, 1973; McAndrews, Berti & Norris, 1973. Fig. 56. *Tricolporopollenites* cf *microhenrici* (Pot.) Thomson & Pflug, 1953 subsp. *intrabaculatus* Pflug, 1953. Affinity Fagaceae. References Elsik, 1968; Kedves, 1969. Fig. 58. cf. *Sambucus*. References Machin, 1971; Gruas-Cavagnetto, 1976. Note pyrite crystals within grain, a feature common to much of the plant fossil material studied. Fig. 59. *Tricolporopollenites cingulum* subsp. *oviformis* (R.Pot.) Thomson & Pflug, 1953. Affinity Fagaceae – *Castanea*. References Kedves, 1965; McAndrews *et al.*, 1973.

(one species of macrofossil); Hydrocharitaceae (one species of macrofossil); Juncaceae (one species of macrofossil); and Zingiberaceae (one species of macrofossil) have not been found.

Of the spores, some of the smooth deltoid forms and some smooth-scabrate forms (Fig. 53) most probably represent dispersed spores of *Acrostichum anglicum*. Others may belong to other ferns. The Schizaeaceae, Pteridaceae, Cyatheaceae and Osmundaceae from the spore flora have not been recorded from the macroflora. *Azolla* megaspores and massulae have been recorded from the macroflora and massulae have also been encountered in the microfloral preparations. No isolated microspores have been seen.

The leaf form *Dicotylophyllum pinnatifidum* has been tentatively referred to the Myricaceae, and pollen of *Triatriopollenites* species (Fig. 62) has been recovered which may also be referable to the Myricaceae.

In Quaternary strata, Watts & Winter (1966) demonstrated that pollen and seed curves for the aquatic genera *Sagittaria*, *Typha*, *Potamogeton*, *Myriophyllum* and *Nuphar* corresponded closely. Watts & Wright (1966) also noted a similar correspondence for *Typha*. These were invoked in support of a local origin for the aquatic species. Dodson (1977) found little relationship between abundance of *Typha* pollen and the position of source plants. Cohen (1975) showed that certain macrofossils such as *Nymphaea* were abundant with their pollen in peat deposits.

Interpretation of pollen data is complicated in the Tertiary by the poor preservation of pollen with a thin exine and by the problems of recognition of extinct species. The lack of correspondence between the macro- and microfloras is not unexpected since a number of the aquatic families represented as macrofossils possess pollen with a thin exine which might not have been preserved. Pollination is not usually hydrophilous so that pollen is not directly introduced into the water (Sculthorpe, 1967).

Of particular note is the possible relationship between *Typha* macrofossils and *Sparganiaceaepollenites* pollen. The pollen diagram (Fig. 45) shows an increase in the proportions of *Sparganiaceaepollenites* in the plant-bearing horizons (except in bed 1 where *Typha* macrofossils are very low in number). Also the pollen, and macrofossil, maxima coincide, in beds 3, 4, 5 and, to a lesser extent, bed 8. Overall abundance and occurrence of clusters and pollen sacs (Fig. 67) strongly suggests that the plant bearing *Sparganiaceaepollenites* pollen was locally abundant in the depositional environment. A local abundance has also been strongly indicated for *Typha* further supporting the affinity of the two organs. Frazier & Osanik (1969) demonstrated that abundant pollen of *Typha* was characteristic of the interdistributary floodplain lakes environment.

Abundance of the algae *Pediastrum* and *Botryococcus* fluctuates considerably but shows a general decrease in the plant-bearing horizons, *Pediastrum* is absent from bed 5, which is considered to be a very nearshore deposit. Chlorophycean aquatic algae are also represented by *Schizosporis*, a probable resting stage of members of the Zygnemetales (Fig. 48).

Bed 9 contains the only recorded dinoflagellate (Fig. 47) and large numbers of *Inaperturopollenites magnus* (Fig. 49). *Pediastrum* is absent from the underlying horizon. Apart from this bed the microflora shows a high degree of uniformity throughout the sequence. Small variations are likely to be the result of sampling error in that the number of grains counted was too small for detailed statistical analyses.

Interpretation of the depositional environment

Origin and interpretation of the plant bearing horizons: Daley (1973b) considered that the plant-bearing horizons were the result of deposition of allochthonous debris brought in by storms and floods. If this were correct it seems unlikely that there should be no change in lithology within the plant-bearing horizons, merely the pronounced change in colour. Furthermore, the change is often gradual with an increasing olive component in the plant-bearing horizon. Undoubtedly, as Daley (1973b) suggested, some of the gastropod partings represent thanatocoenoses, but the species found concentrated in the partings often occur in small numbers within the plant-bearing horizon. Warme (1971) emphasized that the most abundant 'dead' shells in concentrates in modern lagoons generally represented the most abundant live species in the lagoon. The thanatocoenoses could have been brought about by winnowing (where current oriented) or by natural accumulation of 'dead' shells on the lake bottom. In the present examples, the shells are embedded equally in the underlying and overlying sediments, the overlying laminations conforming with the shell topography. This indicates that they settled out in very tranquil conditions. Any change in the local environment (e.g. the development of anaerobic conditions) could have caused the death of a number of gastropods. It is unnecessary to invoke storm input to explain their presence.

There is no evidence of an increased component of allochthonous pollen in the olive clays as compared with the green clays. The systematic composition of the microflora is uniform. The only features which distinguish the plant-bearing horizons are an increase in the proportion of fungal spores, a slight increase in the spore content and a decrease in the proportions of the algae.

The macroflora consists of an association of plants whose nearest living relatives include a range of aquatic and marginal aquatic plants. No members of the flora occur exclusively outside this habitat. The assemblage includes free floating as well as marginal, emergent plants.

The fauna (see Table 4) is composed of planktonic and benthic, aquatic, lake-margin-dwelling ostracods (Keen, 1975) and includes lake-mud-browsing gastropods along with pulmonate gastropods which live amongst vegetation in shallow water, with depths up to 3 m (Baker, 1945, Daley, 1972). In all the ostracod assemblages various moult stages have been found indicating that the assemblages are biocoenoses (Keen, 1971, 1977).

The association of such a community of aquatic and subaquatic fauna and flora can, in my opinion, only be derived from original association in life. Despite the information given (Ridley, 1930) concerning storm and flood transportation of aquatic taxa from quiet water environments, it seems unlikely that this association would have survived such transportation. Some degree of sorting and separation would have occurred, particularly of the smaller plant parts and the ostracods. (Scholl (1963) showed that freshwater gastropod shells were often transported into coastal bays but noted no recognizable plant debris transported along with them.) More fragmentation would be expected and retention in an aerobic environment would have caused decay of less resistant plant material. Some direct evidence of other flora and fauna would also be expected. Darnell (1961) documented considerable influx of woody plant debris from *Taxodium* and *Nyssa* swamp after flooding of the Mississippi river into Lake Pontchartrain.

One further feature supports the above interpretation. Many of the seeds had

Table 4. Distribution of faunal remains in the plant-bearing horizons, Bembridge Marls and lower Hamstead Beds, Hamstead Ledge, Isle of Wight

Bed number	Planorbina	Viviparus	Potamaclis	Melanoides	Melanopsis	Galba	Unio	Candona daleyi	Virgatocypris edwardsi	Hemicyprideis montosa	Fish & other vertebrate debris	Foraminiferid tests	Shell debris
9t								x	x				
9c		x						x	x				
9b	x	x					x	x	x		C		R
8t	x	x	x					x	x				R
8b	x	x	x					x	x				R
7													
6		x									R		C(J)
5	x(J)	x(J)				x(J)		x	x		R		R(J)
4		x						x	x	x	C		C
3		x(J)	x(J)			x(J)	x(J)				C		A(J)
2		x	x(J)	x	x							C	R(J)
1				x	x						C		C

(J)—Indicates present only at the basal junction; A—abundant; C—common; R—rare. Determinations follow Daley (1972) and Keen (1971, 1977).

probably germinated (see pp. 190–191) and these would have been attached to the seedling at least for a short period of time during which their ungerminated equivalents could have been separated from them. An empty testa, once free, would have different hydrodynamic properties. Obviously, the germinated seeds are not in their growth position and the empty testas must have shifted within the water body. Redistribution of plant material within the lake system is well-documented (Darnell 1961; Jessen, 1949; Watts & Bright, 1968) and does not affect my interpretation.

All the features of the plant-bearing horizons indicate autochthonous accumulation with gradual increase in the component of preservable organic material in the depositional environment. Encroachment of local floating aquatic and marginal aquatic vegetation would create anaerobic conditions in the underlying sediments and in turn permit preservation of plant material in olive-black, organic-rich deposits. Floating vegetation would also give the tranquil conditions necessary to produce fine lamination and preserve sediment couplets (Zangerl & Richardson, 1963).

Evidence for regional terrestrial vegetation: At lake margins, as sedimentation proceeds, rooted vegetation becomes established. This entraps further sediment and a succession of floristic zones develops around the lake according to water depth. I confirm the statement (Daley, 1973a) that there are no large *in situ* fossil roots to provide evidence of local swamp vegetation (*sensu* Coleman, 1966). Large logs are known (Insole, personal communication, and see Forbes, 1856) from the marls west of Hamstead Ledge, and palm leaves have been recorded (Gardner, 1888).

These may indicate local woody vegetation. One must assume that during most Bembridge Marls deposition the succession did not proceed to swamp vegetation within the area of present day exposures. Possible woody plants are represented by the seeds of *Sambucus*, fruits of *Spirematospermum* and leaves of *Dicotylophyllum*. The latter was thought by Reid & Chandler (1926) to be a marginal aquatic shrub because of its constant association with an aquatic flora. This is not inconsistent with its proposed affinities with the Myricaceae. The habitat of *Spirematospermum* is not known but a near living relative is a shrub, inhabiting evergreen forest at altitudes of 800–1000 m along water courses (Friedrich & Koch 1970). Extant *Sambucus* is a shrub with a variety of habitats.

The microflora includes pollen referable to a number of families with woody members such as Juglandaceae, Myricaceae, Fagaceae, Salicaceae, Platanaceae, Palmae, Tiliaceae (Figs 54–57, 59–64). Inaperturate pollen (suggested affinity with Taxodiaceae, Cupressaceae and Araucariaceae, Figs 49, 51 & 52) occurs throughout the deposits though *I. magnus* appears in sudden abundance in bed 9. Bisaccate pollen grains are rare except in the horizon underlying bed 4. Palmae are abundant in bed 1. These fluctuations may indicate the occasional proximity of certain kinds of woodland or swamp vegetation. The pollen assemblage (Figs 45, 49, 51–67) is composed of grains, typical of wind transportation; (Proctor & Yeo, 1973) unornamented, with a thin exine and generally small which may have been transported some distance. Apart from *Tricolpopollenites hians*, cf. *Sambucus*, *Sparganiaceaepollenites* and *Aglaoreidia* even reticulate grains are absent. The last two probably represent marginal aquatic species. *Sambucus* is present in the macroflora, that is, probably of local origin and *T. hians*, if referable to *Sambucus* or Salicaceae could be a marginal aquatic shrub. All the ornamented grains may therefore be of local origin.

Numbers of fern spores found have been low. The total spore count is misleading (Fig. 45) as this includes those (Fig. 53) which probably belong to *Acrostichum* (Fig. 19) from the local vegetation. Other spores are scarce. As Colinvaux & Schofield (1976) have shown, where extensive non-aquatic vegetation is developed around a lake, fern spores form a large component of the palynoflora. If a lake is surrounded by forest or woodland, this vegetation is represented in both modern (Drake & Burrows, 1980; Spicer, 1980) and fossil (Bradley, 1963; Wilson, 1980) sediments.

Combined evidence indicates the absence of local woodland, swamp or extensive terrestrial vegetation during deposition of the Bembridge Marls. A distant forest vegetation is indicated by the microflora. Vegetation containing one or two tree or shrub species, may have occasionally grown close to the depositional site, for example Araucariaceae in bed 9 and Palmae in bed 1. Shrubs such as *Sambucus*, *Spirematospermum* and *Dicotylophyllum* probably grew locally in limited areas with sufficiently elevated surface to remove them from constant inundation. This need only be a fraction of a metre.

Fresh or brackish water?: There is very little evidence of brackish influence. Organic foraminiferid tests have been found in bed 2 and the association of the gastropods *Melanoides* and *Melanopsis* in beds 1 and 2 may suggest some brackish influence (Daley, 1972). One dinoflagellate has been found in bed 9 but this is not a typical marine or brackish form and may be a freshwater species of Peridineaceae. Dodson (1974a) recorded that *Pediastrum*, an alga abundant throughout this sequence, was restricted to 3.5°/$_{oo}$ maximum salinity. The ostracods belong to assemblage I of Keen (1977) with a salinity range of less than 3°/$_{oo}$. In bed 4 an additional ostracod *Hemicyprideis montosa* occurs. This tolerated a wide range of salinities and probably inhabited organic-rich lake muds in areas with aquatic vegetation (Keen, 1971). The ostracod assemblage is absent from beds 1 and 2 where possible brackish influences have been noted (foraminiferids and gastropods). Exclusively brackish ostracods (Keen, 1971) and molluscs such as *Polymesoda* (Parker, 1960, Daley, 1972) are absent from the plant-bearing horizons. As these occur at other horizons (Fig. 1) at Hamstead Ledge this argues in favour of freshwater conditions during the deposition of the plant-bearing horizons.

Many of the species of the macroflora do have living relatives which tolerate brackish water e.g. *Acrostichum* (Collinson 1978b); *Chara* (Wood, 1952a, b; Cole, 1963: 417; Burne, Bauld & de Deckker, 1980); *Potamogeton* (Martin & Uhler, 1951) and *Typha* (Carl, 1937; Penfound & Hathaway, 1938). However, Howard-Williams (1975) and Moss (1980) record that *Typha* seeds did not germinate under saline conditions (caused by evaporation) in freshwater lakes. *Limnocarpus* has been considered a brackish water plant from occurrences in marine sediments (Chandler, 1961a, b, 1963b), but is associated with freshwater biota (this study and Chandler, 1963a). Near living relatives (Potamogetonaceae *sensu lato* occupy freshwater, brackish and marine environments. Carl (1937) noted that Nymphaeaceae did not survive in a brackish marsh. None of the macrofloral components has a relative that is confined to brackish habitats.

The aquatic flora could have occurred inland. Only *Acrostichum* (and possibly *Limnocarpus*) have associations with a coastal habitat and neither are restricted to that environment (Collinson, 1978b and in press). That the water body was coastal is deduced from the sequence of underlying strata including estuarine and lagoonal deposits and from faunal remains (Daley, 1972, 1973b; Keen, 1977). It is unlikely that salinities exceeded 3°/$_{oo}$ in the depositional environment of the plant-bearing horizons. The lowest beds, 1 and 2, show the greatest tendency to brackish influence.

The nature of the water body: According to Picard & High (1972) a lake is enclosed entirely by fluviatile settings and fluviatile rocks must intervene between a possible lacustrine unit and known marine rocks. If the unknown unit is in close proximity with marine rocks then a lagoonal setting which opened to the sea is most likely.

There is no evidence for salinity above 3°/$_{oo}$ during deposition of the plant-bearing horizons. The upper reaches of lagoons frequently have such low salinities especially if there is a narrow or constricted opening to the sea. Finely laminated sediments including couplets are common to both lakes and lagoons as are the minor tidal effects which distinguish these two sedimentary environments from those of shallow seas, which show major tidal effects. Faunal and floral elements are unlikely to offer distinguishing features except where salinity is a barrier. Baer (1969), for example, considered palaeontological criteria as the best indicators of environments in interpreting a lake transgressing over a lagoon.

Three-dimensional geology (e.g. Peterson, 1976; Fisher & McGowan, 1969; Van Veen, 1971; Mullens, 1977) revealing coastal barriers, sand bars etc. would be necessary for a deeper understanding of the depositional environment. This is not available in the restricted outcrop area of the Bembridge Marls. From the distribution of bed 9 (=Black Band, see Bristow, 1889), the water body may have extended at least once across the entire outcrop area. Nevertheless it remained shallow enough to support extensive aquatic vegetation. No distinct change in depositional environment is indicated by contemporaneous deposits of the Paris Basin (Cavelier, 1979; Châteauneuf, 1980) or Provence (Sittler *et al.*, 1975).

The flood plain lake environment (Daley, 1973b) was said to contain evidence of the incorporation of lakes into a sluggish river system, and fluvio-lacustrine cyclothems indicating river influence were recorded (Daley, 1973a). The basal part of the marls have an estuarine environment but this is overlain by a lagoonal phase which includes a coastal lake. The coastal lake presumably represents a lagoon finally cut off from the sea. The lagoonal environment passes gradually into the floodplain lake environment at Hamstead Ledge. However, at Whitecliff Bay the two are separated by an erosional boundary (Daley, 1973b). There is no clear evidence that parts of Daley's flood plain lake environment are not merely the upper reaches of a lagoon where salinities had dropped below 3°/$_{oo}$. In fact gradual transition and reversions to lagoonal conditions in places tend to support this latter hypothesis.

The water body may have been either a floodplain lake or the upper reaches of a lagoon with a distant, restricted connection to the sea. It probably extended to cover at least the northern half of the Isle of Wight on some occasions, while remaining shallow enough to support extensive aquatic and marginal aquatic vegetation.

The depth of the water: Rooted aquatic macrophytic vegetation is restricted to relatively shallow water depths, from 0.5 to 2.3 m (Dodson, 1977). According to Moss (1980) emergent plants extend to depths of 1 m, floating leaved plants to 3 m and totally submerged vascular plants to 6.5 m. According to Sculthorpe (1967) these figures are 1.5 m, 3.5 m and 11 m respectively. Charophytes may extend to 17 m (Langangen, 1974) and exceptionally to 164 m (Moss, 1980).

Lake fringing vegetation generally extends to depths of 1–1.5 m (Dodson 1974b, Sculthorpe 1967). *Typha* extends to depths of 0.5–1.0 m (Dodson, 1974b, 1977; Howard-Williams & Liptrot, 1980).

Russell (1942) gave a graphic description of aquatic plant colonization of the Louisiana marshes. In this area plants invade standing waters up to depths of 1.5 m but usually only to about 0.6 m. This critical depth of 0.6 m allows "almost instantaneous plant invasion".

The nearest living relatives of the molluscs associated with the aquatic plants are restricted to the same depth ranges as the plants on which they are dependent (Baker, 1945). Watts & Bright (1968) recorded no live molluscs at depths of more than 4 m, and noted a limit of 3 m for most species.

The plant-bearing horizons were probably deposited in waters up to 3 m in depth. The horizons (e.g. beds 3, 4, 5, 8 top and 9 base) with indication of nearshore conditions, probably deposited in water 0.3–1.0 m in depth, bed 5 being the shallowest.

The absence of the peat-forming environment by comparison with the Everglades: Many investigations of depositional environments and of accruing plant remains in areas of modern marsh vegetation have been based on peats. Those of the Everglades (Florida) have been most extensively studied (Spackman, Dolsen & Riegel, 1966; Spackman, Riegel & Dolsen, 1969; Cohen, 1970, 1973, 1974, 1975; Cohen & Spackman, 1972, 1977, 1980; Cohen & Wiedemann, 1973).

It is difficult to compare these results with those of the present study and to understand why, in an area apparently supporting extensive aquatic and marginal aquatic vegetation, no peat deposits have been formed. Some of the fossiliferous horizons do have a high organic content (Table 1–cuticle) but this never approaches that of a peat.

Howard-Williams (1975) suggested that drying out of lakes periodically was one of the principal causes preventing high organic matter accumulation. However, there is no evidence of drying out, such as mudcracks, red/green couplets or multicoloured sediments (Braunagel & Stanley, 1977; Gorter, 1978) in the area of the Bembridge Marls under consideration.

Lind & Uiser (1962) considered that, under tropical conditions, even if ground was waterlogged, decomposition would be too rapid to allow peat to accumulate. Nevertheless peat does accumulate in some areas of subtropical conditions. It seems unlikely that conditions were tropical during the Bembridge Marls deposition (Collinson, Fowler & Boulter, 1981).

Accumulated peat may be destroyed by fire even if the water table is only rarely lowered and this might happen without sedimentological evidence of drying out. Evidence of the fire in the form of fusain (i.e. charcoal, Scott & Collinson, 1978) would be expected but none has been found in this study.

Davis (1946) observed that peats did not develop beneath swamps along many rivers and estuaries because of high alluvial input. Only very low organic matter peats were formed along other rivers and in lagoons. He also recorded that calcareous water swamps produced little or no peat and that marly clays were the only sediments in some lakes. In other areas 'soil muck' (less than 50% organic matter) was found to border areas of higher organic matter peats, marking a change in the vegetation. Soil muck for example, accumulated beneath the custard apple swamp vegetation which once occupied a zone around Lake Okeechobee.

Davis (1946) also recorded many examples of peat interbedded with marl (calcium carbonate rich clay) both in the freshwater Everglades and in the south-eastern coastal areas of mangrove swamp and salt marsh. He considered that the general occurrence of marl beneath saw grass peat indicated conditions which were formerly shallow, alkaline, hard water containing numerous shells and calcareous plants. He noted that such conditions prevailed today in areas where marl is being deposited.

However, in places the southern marls were so intimately associated with peat that some of them must have developed at the time of peat formation. The conditions which caused the alternation of peat and marl forming environments were said to be "not well known". The dilemma is further expressed in his description (Davis, 1946: 177–178) of the sediments of Mud Lake. Here a thick sapropel overlay fibrous, saw grass peat which contained an interbedded clay layer about 3 m from the base of the lake basin. Davis states that this clay layer was formed "due to some as yet undetermined cause". In some areas, such as the Tamiami canal, marls and peats alternated each being only a few inches thick. Some of the marls in various areas were noted to contain organic matter mixed with the calcareous material.

The Everglades Basin has a gradient of about 0.3 m in every 7 or 8 miles. Water draining from the hinterland passes very slowly across the area and carries only a fine-grained sediment load (Davis, 1943). Lake Okeechobee itself extends over 725 square miles and is only 4 m above sea level. Very few parts of it extend to 5 m in depth. The Everglades cover a further 3600 square miles with various types of marsh (Davis, 1943).

The deposition of marl over the Fort Thompson Limestone (during the Pleistocene) was probably crucial to the development of the Everglades. The limestones provided effective underground drainage, even if some distance below surface sediments, provided that these remained relatively pervious. The marls formed a layer infilling surface irregularities retarding water seepage and water loss as a result of gradient drainage. The major change from marl to peat deposition over the northern and central parts of the Everglades was probably due to the gradual sea level rise (see also Spackman *et al.*, 1966) coupled with sediment redistribution blocking the previous drainage systems (Davis, 1943).

Since the last Wisconsinan glaciation (*c.* 35 000 years) many varied peat-forming environments have developed in the Everglades. Each has particular conditions of formation, depending on the balance of the water table throughout the year and often requiring precise seasonal variations to maintain the equilibrium (Davis, 1943, 1946). The major controlling features for the different soil and vegetation types were considered to be: seasonal character of the rainfall, differences in soil water content and fluctuations in soil and surface water levels, the more organic soils forming in areas of higher water table. Spackman *et al.* (1969) drew attention to the complicated geological and biological interactions involved in the Everglades. Geological factors such as differential sediment compaction, differential bedrock solution and effects of marine transgression all influence the type of sediment deposited.

Natural plant succession is also a major controlling factor. The Saw Grass (*Mariscus*-dominated) environment sediment is typically a black peat. In adjacent Spike Rush (*Eleocharis*-dominated) environment the sediment is a 'fresh water marl' or lime mud. The latter environment prevails in areas of slightly deeper water, where a minor difference of elevation reduces the competitive advantage of the saw grass. The freshwater marl appears to be genetically related to the formation of an algal mat, which has not been observed in the Bembridge Marls. The Bembridge Marls environment is not considered to be open like that of the spike rush. However, the marl (spike rush) and peat (saw grass) are interconvertible, the peat transgressing over the marl due to marginal invasion during plant succession. Such succession is typical of that envisaged for the Bembridge Marls plant bearing horizons.

Such systems may explain interbedding of marls and peats in other areas, but in each there will be a unique complex of controlling factors. The sediments produced may depend ultimately on plant species and natural plant succession, while initial colonization may be governed by the water table and its fluctuations.

Neither of the two main emergent aquatic plants in the Bembridge Marls flora (*Typha* and *Acrostichum*) is dominant in the Everglades region today. However, one of the sub-environments of the 'Everglades and freshwater mash' vegetation is dominated by *Typha*. This sub-environment often occurs immediately adjacent to brackish water making the boundary (between *Typha*-dominated fresh marsh and *Typha*-dominated brackish marsh) difficult to draw (Davis, 1943). Kenoyer (1929) recorded an association of *Typha* and *Acrostichum* in a Panama shoreline marsh developed following deliberate flooding in 1914 and not then having reached the swamp stage of the succession. Details of these sediments would be of value in interpretation of the Bembridge Marls environments.

It seems necessary to assume that the precise water table conditions required by the *Typha/Acrostichum* association were not conducive to peat formation. An additional contributary factor preventing peat formation was probably inorganic influx. Presumably neither the *Typha/Acrostichum* association nor the other emergent vegetation was able to filter the sediment influx, restricting it to the margins of the vegetation stands.

Interpretation of individual horizons

The interpretation, from sedimentological evidence, of the deposition of beds 3, 4 and 9-base in shoreline situations and/or near an influx stream, is supported by the presence of a large and diverse flora. Beds 3, 4, 5 and 8 top have high numbers of *Typha* and *Acrostichum*, the shoreline rooted plants. In contrast, beds 6 and 9 have low numbers of *Typha* and *Acrostichum* and high numbers of *Stratiotes*, or *Azolla* and *Stratiotes*, the free-floating aquatic plants.

The ostracod fauna of beds 4, 5, 8 and 9 is characteristic of the lake edge assemblage of Keen (1975) and various moult stages have been found, indicating that the assemblages are biocoenoses (Keen, 1971, 1977). According to Daley (1972) the gastropods *Galba*, *Planorbina* and *Potamaclis* in association occur in the lake margin assemblage. These together with *Unio*, *Viviparus* and occasional specimens of *Melanoides* and *Melanopsis*, occur in the lake bottom assemblage. According to Picard & High (1972) presence of unionids and viviparids indicates fluviatile influence.

Bed 3 therefore is probably a lake bottom deposit near an influx stream, while beds 4 and 9-base may be transitional lake bottom to lake edge situations, possibly also near influx streams.

Bed 8, also a lake edge situation, is the only horizon which shows bioturbation, so characteristic of modern freshwater marsh sediments (Coleman, 1966). The diversity of fruits and seeds in this horizon is relatively low and comparable with that recorded by Coleman (1966—13 spp.) for similar sediments in a modern freshwater marsh. Bottom conditions during deposition of this horizon must have been sufficiently aerated to permit activity of infauna but the reasons for this are not evident from the present study.

In bed 9 the sedimentary features and floral diversity vary. The base is like beds 3 and 4 but the higher parts are more like bed 6. The disappearance of the features indicating shoreline or stream influence higher in the horizon may indicate the expansion of the water body, cut off from the stream, or the onset of more tranquil conditions at the shore. Different floral associations occur through the horizon and perhaps these reflect successional changes in the vegetation. The upper part of the horizon indicates the extensive development of a cover of floating *Azolla* and so may represent the greatest extent of the water body with deeper water over the depositional site at Hamstead Ledge.

Bed 6 is an isolated lens of plant-bearing sediment. An abundance of free-floating plants is associated with large monocotyledonous leaf fragments (?*Typha*). This horizon probably represents deposition of a floating mat of vegetation which was once anchored to the shoreline vegetation. At this time it was invaded by shoreline plants (most likely to be *Typha*) the vegetative parts of which were carried away when it subsequently floated into open water (Russell, 1942; Meriläinen & Toivonen, 1979).

Bed 5, which occurs above a parting of *Galba* and *Planorbina*, contains the ostracod assemblage but lacks the planktonic alga *Pediastrum*. It lacks the free-floating *Azolla* and *Stratiotes* but contains the only leaves of *Dicotylophyllum*, fragments of *Acrostichum lanzeanum*, large numbers of *Acrostichum anglicum* and *Typha* and is penetrated by rootlets. This horizon probably represents a nearshore deposit, well within the zone of rooted vegetation. It is probably the most stable and well-established of these nearshore deposits as it lacks even the slight sedimentary disturbance of beds 3, 4 and 9 base. It also extends at least as far as Gurnard Ledge with little change in floristic composition.

CONCLUSIONS

Palaeoenvironments during deposition of nine plant-macrofossil-bearing horizons in the Bembridge Marls and lowest Hamstead Beds (Lower Oligocene) of Hamstead Ledge, Isle of Wight have been analysed using plant macro- and microfossils, macro- and microfaunas and sedimentological features.

Plant macrofossils are represented mainly by fruits and seeds but also by fern sporangia, water fern megaspores, leaf margin teeth and rare leaves. These plant organs range from large to small; sclerotic, thick-walled to cuticular, thin-walled; and include both germinated and non-germinated seeds. There is no evidence of significant pre- or postdepositional transport and sorting or of prolonged retention in an environment favouring decay. Of the nearest living relatives of the plant fossils, all are known to inhabit aquatic or marginal aquatic environments and most are restricted to these habitats. Representatives of free-floating, submerged and emergent plants and plants substrate-rooted but with surface floating leaves are included in the nearest living relatives of the plant fossils.

The microflora is dominated by *Sparganiaceaepollenites* pollen and includes large numbers of planktonic algae. The plant-bearing horizons have a proportionate increase in spores and fungal spores and decrease in algae compared with the underlying horizons. The species composition is more or less uniform throughout the sequence. There is occasional local abundance of tree species and some indication of distant forest vegetation.

The fauna from the plant-bearing horizons consists mainly of molluscs and ostracods. The molluscs have previously been interpreted as inhabiting lake margins or shallow water amongst vegetation and as bare lake mud browsers (Daley, 1972). The ostracod biocoenoses include lake-margin-dwelling, planktonic and benthic species (Keen, 1975).

Sedimentological features of the sequence include fine lamination, rhythmic alternation of sediment couplets, some small-scale lenticular lamination and microfaulting and a distinct colour change from green to olive-brown or olive-black coinciding with the occurrence of plant macrofossils. Evidence of rootlet penetration, from the plant-bearing into the underlying horizons, and of bioturbation is sometimes present.

It is concluded that the plant-bearing horizons represent deposition during a period when invasion of aquatic vegetation over the surface of a water body has resulted in the onset of anaerobic bottom conditions. Such conditions have, in turn, permitted the preservation of plant remains in olive-brown clays.

The macroflora contained in these sediments reflects the local aquatic vegetation, affected only by some slight winnowing and redistribution within the water body. The faunas represent the natural associates of these aquatic plants. Abundant planktonic algae occurred in the water body but the microflora of the plant-bearing horizons reflects the presence of macrophytic vegetation.

The vegetation is interpreted as an extensive herbaceous marsh with occasional scattered 'tree islands' and/or areas where slight topographic variation allowed invasion of more 'terrestrial' species. The marsh itself was dominated by a *Typha/Acrostichum* association (Bullrush or Cattail/Leather Fern). The association grew under conditions unfavourable for peat formation. (Comparable marshland vegetation today, e.g. in the Everglades area, is usually dominated by different genera and frequently forms peat.) Subsidiary associates of the *Typha/Acrostichum* marsh were emergent Cyperaceae and Juncaceae. Marginal aquatic plants were represented by abundant Nymphaeaceae, Potamogetonaceae and Alismataceae. Free-floating aquatics included abundant *Azolla* and *Stratiotes*. The locally developed more 'terrestrial' vegetation included *Sambucus*, *Spirematospermum* and probably Myricaceae.

Each plant-bearing horizon represents a slightly different sub-environment, for example, nearshore, more open water, deposition of a floating mat, but it has not been possible to confirm the lake marginal succession in any single horizon.

Water salinity during deposition of the plant-bearing horizons did not exceed $3.5\%_{oo}$; water depth probably varied from 0.3 to 3 m but the precise nature of the water bodies remains enigmatic. They may have been floodplain lakes or, perhaps more likely, the innermost reaches of coastal lagoons with a distant, restricted connection to the sea. At maximum extent they covered at least the present outcrop area of the Bembridge Marls.

ACKNOWLEDGEMENTS

I thank the Keeper of the Palaeontology Department, British Museum (Natural History), London for permission to study and sample material; the Director of the Royal Botanic Gardens, Kew for material from the herbarium and seed collections; Drs B. Daley, N. Edwards, E.-M. Friis, K. Fowler, J. Hooker, A. Insole, M. Keen, A. Scott, R. Spicer; Mr T. Windle and Ms M. Schaffer for helpful discussions; the Royal Society for a grant to Professor Chaloner which made possible the purchase of the SEM used during this study, and Mr A. Davis of Bedford College for printing the photographs. The advice and continued encouragement from Professor W. G. Chaloner were of the greatest value during this work, and my deepest thanks are to Professor W. S. Lacey, who introduced me to palaeobotany.

Much of this work was undertaken during the tenure of an N.E.R.C. research studentship.

APPENDIX

Systematic notes, descriptions and diagnoses of new material from the Bembridge Marls and lower Hamstead Beds, Hamstead Ledge, Isle of Wight— Lower Oligocene.

Stratiotes sp. Leaf margin teeth (Fig. 23), (Collinson, 1978a).

Triangular leaf margin fragments, length 660–2250 µm, basal breadth 375–2600 µm. Basal part always torn, very variable in extent, composed of elongate-oblong or hexagonal cells, with particularly thickened anticlinal walls. Upper and lower cuticles fused at base of single, large, thick-walled, triangular, terminal cell which is 100–425 μm long and 100–375 μm in basal breadth. Occasionally paired terminal cells occur.

Gay-Wilson (1973) and Katz, Katz & Kipiani (1965) have described leaf margin teeth assigned to *S. aloides* L. and *S. intermedius* (Hartz) Chandler from Quaternary deposits. The Bembridge leaf teeth probably belong to the plant which bore the *S. neglectus* Chandler seeds. Gay-Wilson (1973) noted the occurrence of nine seeds to 150 leaf teeth. A similar discrepancy has been recorded here (Table 1).

Typha latissima (Al. Braun) Reid & Chandler, (Reid & Chandler, 1926: 60, pl. III figs 4–11).

OTHER REFERENCES: Chandler, 1963a: 369, pl. 34 figs 166–170; Nikitin, 1965: 52, pl. iv figs 1 & 2.

This species was originally based on leaves, with an emended diagnosis by Reid & Chandler (1926) to include fruit data. Chandler (1963a) described seeds from the Hamstead Beds which she assigned to this species. I have been unable to reconcile that description either with the original or with new material. The conspecificity of the leaves with either the fruits or the seeds may be questioned but there is no sound reason for rejecting it. What follows is a diagnosis of the seeds of the species.

Typha latissima (Al. Braun) Reid & Chandler. Seed.

Seed narrowly elliptic; translucent, cuticular seed coat composed of two layers. Outer layer (?testa) of longitudinally elongate, oblong cells, inner layer (?tegmen) of transversely elongate hexagonal cells. One end of seed rounded or narrowing gradually with a broad mucro. Opposite end truncate with a circular aperture which may be closed by an operculum with a central, stalk-like mucro. Seed pendulous anatropous. Raphe seen as a narrow longitudinal fold. Length of seed 487–1162 μm, breadth 235–467 μm.

ILLUSTRATION: Figure 24.

Potamogeton tenuicarpus C. Reid & E. M. Reid, *emend.* M. E. Collinson *loc. cit.* (Reid & Reid, 1910: 173, figs 53 & 54) (Fig. 44).

OTHER REFERENCES: Chandler, 1957: 85, Chandler, 1963a: 371.

Endocarp obovoid, ventral margin convex with a deep indentation above which a broad stout spine occurs. Endocarp curved around a central oval or comma-shaped aperture with the two limbs meeting at the ventral indentation. The limbs not necessarily abutting so that the aperture may be open at the ventral margin. Style base terminal or subterminal on the ventral margin. Dorsal margin occupied by a germination valve extending from the base of the endocarp at the ventral margin almost to the style base. The germination valve bearing a broad central ridge with a groove on either side, the ridge bearing small spine bases. Endocarp concave with central aperture and germination valve bounded by upstanding ridges. Cells of the endocarp rectangular to cuboid, aligned in concentric rows. Seed curved around the central aperture, testa cells elongate along this curvature. Hilum terminal on the upper limb. Length of endocarp 775–150 μm, breadth 625–1225 μm.

ILLUSTRATION: Figure 44.

Potamogeton tenuicarpus may be distinguished from *P. pygmaeus* Chandler (Chandler, 1925–1926) as follows.

P. tenuicarpus	*P. pygmaeus*
Aperture surrounded by ridge.	Aperture situated in declevity.
Endocarp concave.	Endocarp convex.
Germination valve with a groove on either side of a broad central ridge which bears small spines.	Germination valve convex with narrow central crest on which are situated long spines or their bases.

Potamogeton pygmaeus seems to be replaced in the Tertiary floras of southern Britain by *P. tenuicarpus* during Bembridge Marls deposition. The former is recorded from the Insect Limestone by Reid & Chandler (1926). (See also Chandler, 1961a, 1963a, b for records of *P. pygmaeus*.)

The above emended diagnosis is based on re-examination of the type specimens and all British Eocene and Oligocene *Potamogeton* material. No attempt has been made to make the diagnosis differential to the numerous other records of the genus the fossil record of which is clearly in need of major revision.

?*Caricoidea obscura* Chandler (Chandler, 1963a: 343, pl. 28 figs 29, 30). [Excluding other material listed there in synonymy.]

This material is represented only by endocarps, lacking the exocarp. Its definite affinity with *C. obscura* Chandler (Chandler, 1960) cannot therefore be confirmed. Collinson (1978a: 311–327) suggested emendation of *Caricoidea* to include only those fruits with a thick parenchymatous exocarp. *Caricoidea* as potentially emended was considered to represent a clearly defined, extinct cyperacean genus. The illustration (Collinson, 1978a: fig. 8) is essentially the same as that of Mai & Walther (1978: taf. 1 fig. 7a) although the two were arrived at independently. The revision of Mai & Walther (1978) of this and related genera is not in total accordance with that of Collinson (1978a). The Bembridge material, and others lacking the exocarp should perhaps be placed in another form-genus. There is some confusion as to whether the entire group should be included in the modern genus *Cladium* (see Chandler, 1963a). Clearly more work is needed, especially modern comparative studies, to clarify generic and specific relationships in this group.

Endocarp oburceolate, one seeded, narrowing to a mucronate or pointed apex and contracting abruptly to a narrow, well-defined, basal neck. Two longitudinal ridges occurring on the endocarp surface, opposite one another. Surface obscurely pitted, composed of irregularly arranged, small parenchymatous cells. Endocarp wall uniformly thick *c.* 100 μm. Locule lining obscure. Length of endocarp 855–1625 μm, breadth 685–1140 μm.

ILLUSTRATION: Figure 35.

Caricoidea cf. *maxima* Chandler *emend.* Chandler (Chandler, 1963a: 341).

Only two half specimens of this species were recorded in the present study (Collinson, 1978a: 320). They possess a thick parenchymatous exocarp but are too poorly preserved for precise specific determination.

Cyperaceae *genus indet.* (Collinson, 1978a: 327) V.60908.

Two specimens of an elongate ovoid endocarp with an elongate mucronate apex; truncate base pierced by a large aperture, sealed by a plug; ill-defined basal neck and surface with rows of conspicuous rounded pits up to 40 μm in diameter. They resemble *Cladiocarya* Reid & Chandler (1926) but lack the longitudinal ridges of this genus.

Sambucus parvula Chandler (Chandler, 1926).

Collinson (1978a) observed that the *Sambucus* seeds from the Bembridge Marls transgressed a supposedly distinct (Chandler, 1963a: 353) size difference betweeen the two species *S. parvula* and *S. colwellensis* Chandler (Chandler, 1963a). (*Sambucus mudensis* Chandler (Chandler, 1963b) is distinctly broader.) In addition the Bembridge material revealed detail of the sclerotesta which had not previously been clearly seen (Collinson, 1978a; Scott & Collinson, 1978). Subsequent detailed statistical analyses of fossil and modern *Sambucus* assemblages indicate that species cannot be separated on their seeds. Size ranges overlap between species; within species inter-assemblage variation can be as great as that between species. Other morphological features are generically distinctive but show no useful interspecific variation. It is not possible to distinguish between *S. parvula* and *S. colwellensis.*

Sambucus parvula Chandler (Chandler, 1926: 43, pl. 7, fig. 9a–c).

SYNONYM: *S. colwellensis* Chandler (Chandler, 1963a: 353, pl. 31 figs 108–113). *S. parvulus.*

OTHER REFERENCES: Chandler, 1961a: 150; Chandler, 1962: pl. 23, figs 3–10. The single seed (Chandler, 1962: 144, pl. 23, figs 1 & 2) is here tentatively excluded from this species.

TYPE: Holotype. V. 20096; Chandler, 1962: pl. 23 fig. 6.

TYPE LOCALITY: Lower Headon Beds, Hordle, Hampshire.

Other localities. ?Dorset Pipe Clays, Lake. Upper Headon Beds, Colwell Bay, Isle of Wight; Bembridge Marls and Hamstead Beds, Hamstead Ledge, Isle of Wight.

GEOLOGIC RANGE: ?Early Eocene. Late Eocene–early Oligocene.

Seed subovoid, subtriangular or elongate ovoid, bisymmetrical and flattened concavo-convex with the hilum a small slit-like or circular scar, terminal or subterminal at the pointed end of the concave ventral face. Seed surface ornamented with 7–12 sinuous, interrupted, nodular corrugations or ridges. Ridges becoming obscure to absent at one or both margins of the ventral face leaving a flat margin. Testa two-layered, external layer columnar in section, one cell deep, composed of thick-walled cells, polygonal in surface view with an angular lumina and papillate anticlinal walls. Internal layer 2–4 cells deep composed of longitudinally and transversely oriented fibres. Length of seed 980–1750 μm, breadth 600–1250 μm.

ILLUSTRATION: Figure 42.

Rhamnospermum bilobatum Chandler (Chandler, 1925).

OTHER REFERENCES: Chandler, 1963a: 355, 379; Chandler, 1963b: 134; Chandler 1962: 146.

This species has previously been recorded from the London Clay to the Hamstead Beds and has also recently been found in the Reading Beds at Newbury and Felpham. It ranges therefore almost throughout the British Tertiary but its modern affinity is still unknown. The Bembridge material shows an outer coat, with a carbonaceous layer upon a cuticular layer. Within this is an inner, cuticular layer, fused to the outer coat around the circular scar in the centre of the groove between the two lobes. I interpret this outer coat as the testa, the inner coat as the tegmen and the circular scar as the chalaza of a seed.

Acrostichum lanzaeanum (Visiani) Chandler (Chandler, 1925).

OTHER REFERENCES: Chandler, 1963a, b; Reid & Chandler, 1926: 33 (detailed description).

The specimens recorded here are fragments of pinnules with a reticulate nervation characteristic of the modern fern genus *Acrostichum*.

Dicotylophyllum pinnatifidum Reid & Chandler (Reid & Chandler, 1926).

OTHER REFERENCE: Chandler, 1963a.

Narrow, coriaceous, pinnatifid leaves with opposite or alternate segments. No cuticular details obtainable. Possibly referable to the Myricaceae or Proteaceae.

Juncaceae cf.

Juncus vectensis M. E. Collinson **sp. nova**

DERIVATION: Specific epithet after the type occurrence on the Isle of Wight.

Seed broadly ovoid–subovoid, narrowing at one end, rounded at the other. Rounded end bearing a small operculum, delimited from the seed body by a groove. Seed surface composed of straight sided cells hexagonal–quadrangular in surface view arranged in irregular longitudinal rows, 7–14 on each flattened face of the seed. Anticlinal walls thickened and upstanding, longitudinally orientated walls forming undulating ribs on the seed surface. Cell rows merging on to the narrow end of the seed but truncated at the operculum. Each cell subdivided by 4–6 subsidiary longitudinal ribs. Cellular detail of operculum obscure. In one position 1–4 rows of regularly aligned cells, square in surface view, indicate the passage of the raphe from a basal hilum (narrow end of seed) to an apical chalaza (rounded end). Length of seed 295–337 μm, breadth 163–237 μm.

TYPE: Holotype V.60909, Fig. 17; Paratypes V.60910; V.60900(1), Fig. 32; and other specimens on V.60900. Specimen numbers British Museum (Nat. Hist.) eight seeds.

ILLUSTRATIONS: Figures 17 & 32.

LOCALITY: Hamstead Ledge, National Grid Refs SZ 400918–404920. Isle of Wight, England.

STRATUM: Bembridge Marls, Lower Oligocene.

GEOLOGIC RANGE: Type locality only.

Seeds are flattened as a result of compression, and folded and distorted at the margins indicating original inflation. Darkening of the seed at either end indicates the hilum and chalaza. The surface cells are arranged in longitudinal rows, sometimes regular, sometimes bifurcating from and/or merging with, one another. Each cell is subdivided into 4 or 6 smaller units (Fig. 32), the significance of which is not clear. They may be periclinal wall thickenings or an inner layer of cells fused to the outer layer. Dissection has not revealed a separate inner layer. The narrowed hilar end is distinctly pointed in some specimens (V.60910.). No seeds lack the operculum but it is presumed that this is shed during germination as in modern *Juncus* (Welch, 1966). All features of these seeds are consistent with assignment to the family Juncaceae and to the extant genus *Juncus* (Fig. 27). Dickson (1970), Watts (1959) and Körber-Grohne (1964) have provided keys to identify certain species of *Juncus* seeds. There are, however, over 300 species in this genus and, in view of the limited features available, it is very unlikely that this Oligocene seed could be clearly excluded from all, or included in any one, modern species. (The problem is similar to that for *Rhododendron* seeds as discussed by Collinson & Crane, 1978.) Some genera in the family such as *Luzula* have distinct seeds but others may be similar to those of *Juncus*. For this reason the determination is only provisional. As far as I am aware this is the earliest fossil record for the Juncaceae.

Alismataceae

Alismaticarpum M. E. Collinson **gen. novum**

DERIVATION: Fruit with relationship to Alismataceae.

Fruit subtriangular or elongate ovoid in lateral view, strongly flattened in the bilateral plane of symmetry. One long margin rounded, the other more-or-less straight and with a thin wing. Wing extending from near the narrow, notched base of the fruit, around the apex and on to the opposite margin. Basal notch separating two limbs of an almost equally lobed, single seed. The slightly shorter limb on the straight, winged side of the fruit. Seed recurved at about one-sixth of length of fruit from apex. Fruit wall of three distinct layers: outer layer, one cell deep, of narrow, longitudinally elongate cells, oblong or hexagonal in surface view with convex, folded, outer, periclinal walls. The outer thin-walled layer and one cell depth of the central layer form the wing. Central layer 1–2 cells deep, of large, thick-walled polygonal cells with sharply-angled lumina. Inner layer of narrow, longitudinally and transversely oriented, thick-walled fibres. Length of fruit 615–1010 μm, breadth 312–695 μm. Fruits may be aggregated in groups of up to 8.

TYPE SPECIES: *Alismaticarpum alatum* M. E. Collinson

Alismaticarpum alatum M. E. Collinson **sp. nova**

DERIVATION: the epithet refers to the lateral wing of the seed.

Diagnosis as for genus.

TYPE: Syntypes V.60917, Fig. 33; V.60918, Fig. 37; V.60919, Fig. 34; V.60920(2), Fig. 36; V.60920(3), Fig. 39; Paratypes V.60921 and V.60920(1), Fig. 41. Specimen numbers British Museum (Nat. Hist.). Over 100 individual fruits and 36 aggregates of two or more fruits.

ILLUSTRATIONS: Figures 33, 34, 36, 37, 39, 41 & 68A–E.

15

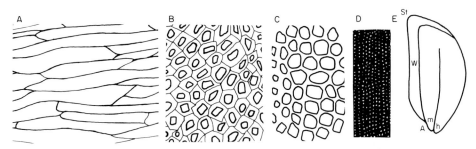

Figure 68. A–D; Cellular detail of the fruit wall of *Alismaticarpum alatum*. A—Outer layer, thin-walled, convex elongate cells; B—least abraded view of central cell layer; C—most abraded view of central cell layer; D—inner layer. E—Interpretation of *Alismaticarpum alatum*. Lateral view of fruit (achene). Ventral side to left. A—Attachment to receptacle; St—style; W—lateral wing; m—micropylar limb of seed; h—hilar limb of seed.

LOCALITY: Hamstead Ledge, National Grid Refs SZ 400918–404920, Isle of Wight, England.

STRATUM: Bembridge Marls, Lower Oligocene.

GEOLOGIC RANGE: Bembridge Marls and lower Hamstead Beds; Lower Oligocene; Hamstead Ledge and Gurnard Ledge; Isle of Wight, England.

The fruits occur in a vast range of compression states (V.60921) many of which obscure their true nature. Only cell structure and dissection has revealed that all represent the same species. About 20% of specimens are compressed dorsoventrally or obliquely (part of Figs 36, 39). About 30% are compressed laterally but lack the wing (Fig. 41) and the remaining specimens are compressed laterally but possess the wing (Figs 33, 34, 37). The fruiting heads (Fig. 36) provide some explanation for the range of compression states. In a group each fruit will experience slightly different compression forces. Individually dispersed fruits, being so thin, would probably always compress laterally.

The lateral wing (ventral margin) does not reach the base but broadens obliquely from near the base (Figs 33, 34). It extends around the apex, broadest at or near the apex of the ventral margin. A very narrow, short style may have been situated here (Fig. 68E) but it has not been seen on any specimen. The wing is often partially (Fig. 37) or almost totally (Fig. 41) lost or its presence may be obscured by compression (Figs 36, 39).

The narrowed base of the fruit is notched and the ventral limb is very slightly shorter than the dorsal. The fruit contains a single seed, recurved at about one-sixth of the fruit length from the apex. Several specimens (e.g. Fig. 37) are splitting along the line of this junction. This may be the result of germination. In living Alismataceae achenes the seedling emerges from the weakest point of the fruit, the tip of the micropylar limb of the seed at the base of the ventral margin (Kaul, 1976, 1978; Björquist, 1967). Compression has obscured detail of the seed. The fruit attachment to the receptacle is presumed to have been at the oblique base of the wing (Fig. 68E).

The outer layer of the fruit wall is composed of narrow longitudinally elongate cells (Figs 34, 68A) from 5 to 20 μm broad and up to 10 times longer than broad. The cell walls are thin and this layer is frequently partially lacking, presumably because of biodegradation. The central layer is, in contrast, composed of very thick-walled cells (Figs 37, 41, 68B, C). These are pentagonal or hexagonal from 70 to 120 µm in diameter with a wall thickness often equalling the breadth of the lumen. The thickening is unevenly distributed so that a pentagonal cell may have a quadrangular lumen. The lumen is sharply angled. This layer exhibits variable preservation (Figs 68B, C). The innermost layer as revealed by loss of outer layers is a regular criss-cross network of fibres with foveolae from 1 to 5 µm in diameter (Fig. 68D).

Of the specimens aggregated in fruiting heads, 16 paired specimens, four groups of three, 12 groups of four (Fig. 39), one group of five and three groups of eight (Fig. 36) have been found.

The structure of these fossil fruits along with their aggregation into fruiting heads supports an affinity with the Alismataceae. A brief survey of 11 genera of extant Alismataceae (Collinson, 1978a) revealed that no modern genus was closely comparable with the fossil. *Echinodorus* Rich. *Ranalisma* Stapf. and *Sagittaria* L. have thin achenes with lateral wings. The wings usually pass all around the achene, although they may be much narrower at one side than the other. According to Kaul (1976) and personal observations, species of *Damasonium* Mill, *Limnophyton* Mig., *Luronium* Rafin and *Weisneria* M. Mich. have aggregates of as few as 6 or 8 fruits in fruiting heads. *Weisneria* species have thick-walled cells in their fruit wall and most genera have a thin outer layer. A basal notch indicating the separation of the seed's limbs can be seen on *Alisma, Echinodorus, Ranalisma* and *Sagittaria*.

Previous fossil records of Alismataceae are largely based on recognition of the modern genera in late Oligocene, Miocene and younger floras (Dorofeev, 1958, 1977; Katz *et al.* 1965). Nikitin (1965) described *Sagisma* for fossil unwinged fruits with small oblong surface cells and a shallow thick-walled layer in the fruit wall. The only other record from the British Tertiary *?Alisma* sp. (Chandler, 1963b) is represented by a single specimen about which no further comments can be made.

Carpolithes hamsteadensis M. E. Collinson **sp. nova**

DERIVATION: specific epithet from the type locality.

Elongate ovoid fruits or seeds, narrowing to a rounded apex, truncate, torn and pierced by a circular aperture at the base. Outer cell layer, frequently also split longitudinally and torn at the apex, composed of longitudinally elongate cells with upraised anticlinal walls arranged in longitudinal rows. Inner cell layer narrowing suddenly to a rounded apex with a thin, hair-like prolongation reaching to tip of outer layer. Inner cell layer with a collar-like darkened zone around the distinct basal neck which terminates within the outer layer. Inner cell layer composed of longitudinally elongate cells with beaded, upraised anticlinal walls arranged in longitudinal rows. Length 1050–1975 µm, breadth 475–930 µm.

TYPE: Holotype V.60912, Fig. 21; Paratypes V.60911, Fig. 22; V.60913, Fig. 18; V.60914. Specimen numbers British Museum (Nat. Hist.), 69 fruits.

ILLUSTRATIONS: Figures 18, 21 & 22.

LOCALITY: Hamstead Ledge, National Grid Ref. SZ 400918–404920, Isle of Wight, England.

STRATUM: Bembridge Marls, Lower Oligocene.

GEOLOGIC RANGE: Bembridge Marls and lower Hamstead beds; Lower Oligocene; Hamstead Ledge and Gurnard Ledge; Isle of Wight, England.

These specimens could be interpreted as one seeded fruits, or as bitegmic seeds. The former interpretation seems more likely, the outer layer being a fruit wall and the inner layer a testa with the apical prolongation indicating a micropyle and the basal collar being the hilum (? and chalaza). Attachment of the fruit would be at the torn base, with the narrowing apex being the style. Folding of the cuticular layers indicates original inflation of the fruits. The morphology strongly suggests that these are monocotyledonous fruits but no similar extant material has yet been seen.

Carpolithes collumus M. E. Collinson **sp. nova**

DERIVATION: specific epithet from the basal neck.

Specimen oblong, subovoid, rarely reniform, flattened, narrowing abruptly to a distinct hollow neck at one end, rounded at the opposite end. Thin cuticular wall composed of quadrangular square or oblong cells with thickened, upstanding anticlinal walls arranged in longitudinal rows, 5–8 rows across each flattened face, 9–13 cells along each row. Cells 20–40 µm in breadth. Length 288–457 µm, breadth 125–262 µm.

TYPE: Holotype V.60915, Fig. 20; Paratypes V.60900(2) Fig. 26 and other specimens on V.60900, V.60916. Specimen numbers British Museum (Nat. Hist.), 66 specimens.

ILLUSTRATIONS: Figures 20 & 26.

LOCALITY: Hamstead Ledge National Grid Ref. SZ 400918–404920, Isle of Wight, England.

STRATUM: Bembridge Marls, Lower Oligocene.

GEOLOGIC RANGE: Bembridge Marls and lower Hamstead beds, Lower Oligocene, Hamstead Ledge, Isle of Wight, England.

In spite of considerable searching (Collinson, 1978a) no modern seeds comparable with *C. collumus* have been seen. It is assumed that the basal neck represents an attachment but there is no indication of other structures. The central body (Fig. 20) has only been seen on two specimens. *Carpolithes collumus* may not be a seed as it shows some resemblance to Salviniaceous sporocarps (Kräusel, 1920; Katz *et al.*, 1965). However, it is difficult to understand the absence of preserved spores within such a structure.

REFERENCES

ANTHONY, R. S., 1977. Iron rich rhythmically laminated sediments in Lake of Clouds, north eastern Minnesota. *Limnology and Oceanography, 22:* 45–54.

BAER, J. L., 1969. Palaeoecology of cyclic sediments of the Lower Green River Formation, Central Utah. *Brigham Young University Geology Studies, 16:* 3–95.

BAKER, F. C., 1945. *The Molluscan Family Planorbidae.* Urbana: University of Illinois Press.

BERGGREN, W. A., 1971. Tertiary boundaries and correlations. In B.F. Funnell & W.R. Riedel (Eds), *Micropaleontology of the Oceans:* 693–809. Woods Hole Oceanographic Institution Contribution 2016.

BERGSTRÖM, J. & LEVI-SETTI, R., 1978. Phenotypic variation in the Middle Cambrian trilobite *Paradoxides davidis* Salter at Manuels, S.E. Newfoundland. *Geological et Palaeontologica, 12:* 1–40.

BHATIA, S. B., 1955. The foraminiferal fauna of the late Palaeogene sediments of the Isle of Wight, England. *Journal of Paleontology, 29:* 665–693.

BIRKS, H. H., 1973. Modern macrofossil assemblages in lake sediments in Minnesota. In H.J.B. Birks & R.G. West (Eds), *Quaternary Plant Ecology:* 172–189. Oxford: Blackwells.

BIRKS, H. J. B., 1976. Late-Wisconsinan vegetational history at Wolf Creek, Central Minnesota. *Ecological Monographs, 46:* 395–429.

BJÖRKQUIST, I., 1967. Studies in *Alisma* L. I. Distribution, variation and germination. *Opera Botanica a Societate Botanica, 17:* 1–128.

BORGE, O. & ERDTMAN, G., 1954. On the occurrence of *Pediastrum* in the Tertiary strata of the Isle of Wight. *Botaniska Notiser, 2:* 112–113.

BRADLEY, W. H., 1963. Palaeolimnology. In D. G. Frey (Ed.), *Limnology in North America:* 621–652. Madison: University of Wisconsin Press.

BRAUNAGEL, L. C. & STANLEY, K. O., 1977. Origin of variegated red beds in the Cathedral Bluffs Tongue of the Wasatch Formation (Eocene), Wyoming. *Journal of Sedimentary Petrology, 47:* 1201–1219.

BRISTOW, H. W., 1889. *The Geology of the Isle of Wight,* 2nd Ed., revised by C. Reid and A, Strahan. *Memoirs of the Geological Survey of Great Britain.*

BURNE, R. V., BAULD, J. & DE DECKKER, P., 1980. Saline lake charophytes and their geological significance. *Journal of Sedimentary Petrology, 50:* 281–293.

CARL, G. C., 1937. Flora and Fauna of brackish water. *Ecology, 18:* 446–453.

CAVELIER, C., 1979. La Limité Éocène—Oligocène en Europe occidentale. *Sciences Géologiques (Strasbourg), Mémoire 54:* 1–280.

CHANDLER, M. E. J., 1923. The geological history of the genus *Stratiotes. Quarterly Journal of the Geological Society of London, 79:* 117–138.

CHANDLER, M. E. J., 1925–1926. *The Upper Eocene Flora of Hordle, Hants.* London: Palaeontographical Society.

CHANDLER, M. E. J., 1957. The Oligocene flora of the Bovey Tracey lake basin, Devonshire. *Bulletin of the British Museum (Natural History). Geology, London, 3:* 71–123.

CHANDLER, M. E. J., 1960. Plant remains of the Hengistbury and Barton Beds. *Bulletin of the British Museum (Natural History), Geology, London, 4:* 191–238.

CHANDLER, M. E. J., 1961a. Flora of the Lower Headon Beds of Hampshire and the Isle of Wight. *Bulletin of the British Museum (Natural History), Geology, London, 5:* 91–158.

CHANDLER, M. E. J., 1961b. Post-Ypresian plant remains from the Isle of Wight and the Selsey Peninsula, Sussex. *Bulletin of the British Museum (Natural History), Geology, London, 5:* 13–41.

CHANDLER, M. E. J., 1962. *The Lower Tertiary Floras of Southern England, II, Flora of the Pipe Clay series of Dorset (Lower Bagshot).* London: British Museum (Natural History).

CHANDLER, M. E. J., 1963a. Revision of the Oligocene Floras of the Isle of Wight. *Bulletin of the British Museum (Natural History). Geology, London, 6:* 321–384.

CHANDLER, M. E. J., 1963b. *The Lower Tertiary Floras of Southern England, III, Flora of the Bournemouth Beds; the Boscombe and the Highcliff Sands.* London: British Museum (Natural History).

CHANDLER, M. E. J., 1964. *The Lower Tertiary Floras of Southern England, IV, A Summary and Survey of Findings in the Light of Recent Botanical Observations.* London: British Museum (Natural History).

CHÂTEAUNEUF, J-J., 1980. Palynostratigraphie et Paléoclimatologie de l'Éocène supérieur et de l'Oligocène du Bassin de Paris. *Mémoires du Bureau de recherches géologiques et minières, 116:* 1–360.

COHEN, A. D., 1970. An allochthonous peat deposit from southern Florida. *Bulletin of the Geological Society of America, 81:* 2477–2482.

COHEN, A. D., 1973. Petrology of some Holocene peat sediments from the Okefenokee swamp-marsh complex of southern Georgia. *Bulletin of the Geological Society of America, 84:* 3867–3878.

COHEN, A. D., 1974. Petrography and paleoecology of Holocene peats from the Okefenokee swamp-marsh complex of Georgia. *Journal of Sedimentary Petrology, 44:* 716–726.

COHEN, A. D., 1975. Peats from the Okefenokee swamp-marsh complex. *Geoscience and Man, 11:* 123–131.

COHEN, A. D. & SPACKMAN, W., 1972. Methods in peat petrology and their application to reconstruction of paleoenvironments. *Bulletin of the Geological Society of America, 83:* 129–142.

COHEN, A. D. & SPACKMAN, W., 1977. Phytogenic organic sediments and sedimentary environments in the Everglades-mangrove complex Part II. The origin, description and classification of the peats of southern Florida. *Palaeontographica, B, 162:* 71–114.

COHEN, A. D. & SPACKMAN, W., 1980. Phytogenic organic sediments and sedimentary environments in the Everglades-mangrove complex of Florida Part III. The alteration of plant material in peats and the origin of coal macerals. *Palaeontographica, B, 172:* 125–149.

COHEN, A. D. & WIEDEMANN, H. V., 1973. Distribution and depositional history of some pre-lagoonal Holocene sediments in the Ciénaga Grande de Santa Marta, Columbia. *Mitteilungen aus dem Instituto Colombo-Alemán de Investigaciones Cientificas "Punta de Betín", 7:* 139–154.

COLE, G. A., 1963. The American Southwest and Middle America. In D. G. Frey (Ed.), *Limnology in North America:* 393–434. Madison: University of Wisconsin Press.

COLEMAN, J. M., 1966. Ecological changes in a massive fresh water clay sequence. *Transactions. Gulf Coast Association of Geological Sciences, 16:* 159–174.

COLEMAN, J. M., 1969. Brahmaputra River: Channel process and sedimentation. *Sedimentary Geology, 3:* 131–239.

COLINVAUX, P. A., 1976. Historical ecology in the Galapagos Islands, i) A Holocene pollen record, from El Junco Lake, Isla San Cristobal. *Journal of Ecology, 64:* 989–1012.

COLLINSON, M. E., 1978a. *Palaeocarpology and related Palaeobotanical Studies of Palaeogene Sediments from Southern Britain.* Unpublished Ph.D. Thesis, University of London.

COLLINSON, M. E., 1978b. Dispersed fern sporangia from the British Tertiary. *Annals of Botany, 42:* 233–250.

COLLINSON, M. E., 1980a. Recent and Tertiary seeds of the Nymphaeaceae *sensu lato* with a revision of *Brasenia ovula* (Brong.) Reid and Chandler. *Annals of Botany, 46:* 603–632.

COLLINSON, M. E., 1980b. A new multiple floated *Azolla* from the Eocene of Britain with a brief review of the genus. *Palaeontology, 23:* 213–229.

COLLINSON, M. E., in press. A reassessment of fossil Potamogetoneae fruits with description of new material from Saudi Arabia. *Tertiary Research, 4.*

COLLINSON, M. E. & CRANE, P. R., 1978. *Rhododendron* seeds from the Palaeocene of southern England. *Botanical Journal of the Linnean Society, 76:* 195–205.

COLLINSON, M. E., FOWLER, K. & BOULTER, M. C., 1981. Floristic changes indicate a cooling climate in the Eocene of southern England. *Nature, London, 291:* 315–317.

COOKSON, I. C. & DETTMANN, M. E., 1959. On *Schizosporis,* a new form genus from Australian Cretaceous deposits. *Micropaleontology, 5:* 213–216.

COOPER, J., 1976. British Tertiary Stratigraphical and rock terms formal and informal, additional to Curry, 1958. Lexique Stratigraphique International. *Tertiary Research Special Papers, 1:* 1–37.

COUPER, R. A., 1953. Upper Mesozoic and Cainozoic spores and pollen grains from New Zealand. *Bulletin. Geological Survey of New Zealand. Palaeontology, 22:* 221–277.

CURRY, D., DALEY, B., EDWARDS, N., MIDDLEMISS, F. A., STINTON, F. C. & WRIGHT, C. W., 1972. *The Isle of Wight, Geologists' Association Guide 25.* Colchester: Geologists' Association.

CURRY, D., ADAMS, C. G., BOULTER, M. C., DILLEY, F. C., EAMES, F. E., FUNNELL, B. M. & WELLS, M. K., 1978. A correlation of Tertiary rocks in the British Isles. *Geological Society of London Special Report, 12:* 1–72.

CWYNAR, L. C., 1978. Recent history of fire and vegetation from laminated sediment of Greenleaf Lake, Algonquin Park, Ontario. *Canadian Journal of Botany, 56:* 10–21.

DALEY, B., 1969. *A Palaeoenvironmental Study of the Bembridge Marls (Oligocene) of the Isle of Wight, Hampshire.* Unpublished Ph.D. Thesis, University of Reading.

DALEY, B., 1972. Macroinvertebrate assemblages from the Bembridge Marls (Oligocene) of the Isle of Wight, England and their environmental significance. *Palaeogeography, Palaeoclimatology and Palaeoecology, 11:* 11–32.

DALEY, B., 1973a. Fluvio-lacustrine cyclothems from the Oligocene of Hampshire. *Geological Magazine, 110:* 235–242.

DALEY, B., 1973b. The palaeoenvironments of the Bembridge Marls (Oligocene) of the Isle of Wight, Hampshire. *Proceedings of the Geologists' Association, 84:* 83–93.

DARNELL, R. M., 1961. Trophic spectrum of an estuarine community, based on studies of Lake Pontchartrain, Louisiana. *Ecology, 42:* 553–568.

DAVEY, R. J., DOWNIE, C., SARJEANT, W. A. S. & WILLIAMS, G. L., 1966. Studies on Mesozoic and Cainozoic dinoflagellate cysts. *Bulletin of the British Museum (Natural History), Geology, London, Supplement 3:* 1–248.

DAVIS, J. H., 1943. The natural features of southern Florida, especially the vegetation and the Everglades. *Bulletin of the Florida Geological Survey, 25:* 5–311.

DAVIS, J. H., 1946. The peat deposits of Florida; their occurrence, development and uses. *Bulletin of the Florida Geological Survey, 30:* 1–247.

DAVIS, M. B., 1973. Redeposition of pollen grains in lake sediments. *Limnology and Oceanography, 18:* 44–52.

DICKSON, C. A., 1970. The study of plant macrofossils in British Quaternary deposits. In D. Walker & R. G. West (Eds), *Studies in the Vegetational History of the British Isles:* 233–254. Cambridge: Cambridge University Press.

DODSON, J. R., 1974a. Vegetation and climatic history near Lake Keilambete, Western Victoria. *Australian Journal of Botany, 22:* 709–717.

DODSON, J. R., 1974b. Vegetation history and water fluctuations at Lake Leake, south eastern South Australia. I. 10 000 BP to present. *Australian Journal of Botany, 22:* 719–741.

DODSON, J. R., 1977. Pollen deposition in a small closed drainage basin lake. *Review of Palaeobotany and Palynology, 24:* 179–193.

DOLPH, G. E. & DILCHER, D. L., 1979. Foliar physiognomy as an aid in determining paleoclimate. *Palaeontographica, B, 170:* 151–172.

DOROFEEV, P. I., 1958. New data on the Oligocene flora near the village Rezhenka in West Siberia. *Doklady Akademii Nauk SSSR, 123:* 171–174. (In Russian.)

DOROFEEV, P. I., 1977. On Tertiary *Caldesia* in Byelorussia. *Doklady Akademii Nauk belorusskoĭ SSR, 21:* 176–178. (In Russian.)

DRAKE, H. & BURROWS, C. J., 1980. The influx of potential macrofossils into Lady Lake, North Westland, New Zealand. *New Zealand Journal of Botany, 18:* 257–274.

DURAND, S. & OLLIVIER-PIERRE, M. F., 1972. Analyse palynologique. *Bulletin d'information des Geólogues du Bassin de Paris, 32:* 41–52.

ELSIK, W. C., 1968. Palynology of a Paleocene Rockdale Lignite, Milam County Texas. I. Morphology and Taxonomy. II. Morphology and Taxonomy end. *Pollen et Spores, 10:* 263–314, 599–664.

EUGSTER, H. P. & HARDIE, L. A., 1975. Sedimentation in an ancient Playa-Lake complex: The Wilkins member of the Green River Formation of Wyoming. *Bulletin of the Geological Society of America, 86:* 319–334.

FAIRBRIDGE, R. W. & BOURGEOIS, J., (Eds) 1978. *The Encyclopedia of Sedimentology. The Encyclopedia of Earth Sciences Vol. VI.* Stroudburg Pennsylvania: Dowden, Hutchinson & Ross.

FEIST-CASTEL, M., 1977. Evolution of the charophyte floras in the Upper Eocene and Lower Oligocene of the Isle of Wight. *Palaeontology, 20:* 143–157.

FISHER, W. L., & McGOWEN, J. H., 1969. Depositional systems in the Wilcox Group (Eocene) of Texas and their relationship to occurrence of oil and gas. *Bulletin of the American Association of Petroleum Geologists, 53:* 30–54.

FORBES, E., 1856. On the Tertiary fluvio-marine formation of the Isle of Wight. *Memoirs of the Geological Survey of Great Britain,* 162 pp.

FOWLER, K., 1971. A new species of *Aglaoreidia* Erdtm. from the Eocene of southern England. *Pollen et Spores, 13:* 135–147.

FOWLER, K., 1975. Megaspores and massulae of *Azolla prisca* from the Oligocene of the Isle of Wight. *Palaeontology, 18:* 483–507.

FRASER, H. J., 1929. An experimental study of varve deposition. *Transactions of the Royal Society of Canada, 23.*(Section IV): 49–60.

FRAZIER, D. E. & OSANIK, A., 1969. Recent peat deposits—Louisiana coastal plain. *Special papers Geological Society of America, 114:* 63–85.

FRIEDRICH, W. L. & KOCH, B. E., 1970. Comparison of fruits and seeds of fossil *Spirematospermum* (Zingiberaceae) with those of living *Cenolophon. Bulletin of the Geological Society of Denmark, 20:* 192–195.

GARDNER, J. S., 1888. Third report of the committee for the purpose of reporting on the fossil plants of the Tertiary and Secondary Beds of the United Kingdom. *Report of the British Association for the Advancement of Science, 1887:* 229.

GAY-WILSON, D., 1973. Notable plant records from the Cromer Forest Bed. *New Phytologist, 72:* 1207–1234.

GEOLOGICAL SOCIETY OF AMERICA, 1970. *Rock—color chart.* Boulder: Colorado.

GORTER, J. D., 1978. Triassic environments in the Canning Basin, Western Australia. *Bureau of mineral resources. Journal of Australian Geology and Geophysics, 3:* 25–33.

GRAMBAST, L., 1957. Ornamentation de la gyrogonite et systématique chez les charophytes fossiles. *Revue Générale de Botanique. 64:* 339–362.

GRAY, J., 1960. Fossil chlorophycean algae from the Miocene of Oregon. *Journal of Paleontology, 34:* 453–463.

GRUAS-CAVAGNETTO, C., 1974. Associations sporopolléniques et microplanctoniques de l'Éocène et de l'Oligocène inférieur du Bassin de Paris. *Paléobiologie continentale, 5:* 1–20.

GRUAS-CAVAGNETTO, C., 1976. Étude palynologique du paléogène du sud de l'Angleterre. *Cahiers de Micropaléontologie, 1:* 5–49.

GRUAS-CAVAGNETTO, C., 1977. *Étude palynologique de l'Éocène du Bassin anglo-parisien.* Unpublished thesis, Université Pierre et Marie Curie, France.

HOWARD-WILLIAMS, C., 1975. Vegetation changes in a shallow African lake: response of the vegetation to a recent dry period. *Hydrobiologia, 47:* 381–398.

HOWARD-WILLIAMS, C. & LIPTROT, M. R. M., 1980. Submerged macrophyte communities in a brackish South African estuarine lake system. *Aquatic Botany, 9:* 101–116.

HUTCHINSON, G. E., 1957. *A Treatise on Limnology. 1 Geology, Physics & Chemistry.* New York: John Wiley.

JANSONIUS, J. & HILLS, L. V., 1976. *Genera File of Fossil Spores.* Calgary: University of Calgary Geology Department.

JARZEMBOWSKI, E. A., 1976. Report of Easter field meeting: the Lower Tertiaries of the Isle of Wight. *Tertiary Research, 1:* 11–16.

JARZEMBOWSKI, E. A., 1980. Fossil insects from the Bembridge Marls, Palaeogene of the Isle of Wight, southern England. *Bulletin of the British Museum (Natural History), Geology, London, 33:* 237–293.

JESSEN, K., 1949. Studies in the late Quaternary deposits and flora-history of Ireland. *Proceedings of the Royal Irish Academy B, 52:* 85–290.

KATZ, N. J., KATZ, S. V. & KIPIANI, M. G., 1965. *Atlas and Keys of Fruits and Seeds Occurring in the Quaternary Deposits of the U.S.S.R.* Moscow: Nauka. (In Russian.)

KAUL, R. B., 1976. Conduplicate and specialised carpels in the Alismatales. *American Journal of Botany, 63:* 175–182.

KAUL, R. B., 1978. Morphology of germination and establishment in seedlings in Alismataceae and Hydrocharitaceae. *Aquatic Botany, 5:* 139–147.

KEDVES, M., 1961. Étude palynologiques dans le bassin de Dorog II. *Pollen et Spores, 3:* 101–153.

KEDVES, M., 1965. Palynological investigations on the Lower Eocene layers in the surrounding country of Iszkaszentgyorgy III. *Acta Biologica, 11:* 33–50.

KEDVES, M., 1969. *Palynological Studies on Hungarian Early Tertiary Deposits.* Budapest: Akadémiai Kiado.

KEDVES, M. & BOHONY, E., 1966. Observations sur quelques pollen de palmiers provenant des couches Tertiaires de Hongrie. *Pollen et Spores, 8:* 141–147.

KEEN, M. C., 1966. *Ostracoda and the Eocene Oligocene boundary in North West Europe.* Unpublished Ph.D. Thesis, University of Leicester.

KEEN, M. C., 1971. A palaeoecological study of the ostracod *Hemicyprideis montosa* (Jones and Sherborn) from the Sannoisian of North West Europe. *Bulletin du Centre de Recherches de Pau Société nationale des Pétroles d'Aquitaine, Supplement 5:* 523–543.

KEEN, M. C., 1975. The palaeobiology of some Upper palaeogene fresh-water ostracods. *Bulletin of American Paleontology, 65:* 272–283.

KEEN, M. C., 1977. Ostracod assemblages and the depositional environments of the Headon, Osborne and Bembridge Beds (Upper Eocene) of the Hampshire Basin. *Palaeontology, 20:* 405–445.

KENOYER, L. A., 1929. Genera and successional ecology of the lower tropical rain-forest at Barro Colorado Island, Panama. *Ecology, 10:* 201–222.

KOCH, B. E. & FRIEDRICH, W. L., 1971. Früchte und Samen von *Spirematospermum* aus der Miozanen Fasterholt-Flora in Danemark. *Palaeontographica, B, 136:* 1–46.

KÖRBER-GROHNE, V., 1964. Bestimmungsschlüssel für subfossile *Juncus*-Samen und Gramineen-Früchte. *Probleme der kustenforschung im Nordseegebiet; Schriftenreihe Niedersächs Landesanst Marschen-u Wurtenforschrifte, 7:* 1–47.

KRÄUSEL, R., 1920. Ein Beitrage zur Kenntnis der Diluvialflora von Ingramsdorf in Schlesien. *Neues Jahrbuch für Mineralogie, 1:* 104–110.

KRUTSCH, W., 1959. Mikropaläontologische (Sporenpaläontologische) untersuchungen in der Braunkohle des Geiseltales. I Die Sporen und die Sporenartigen sowie ehemals im Geiseltal zu Sporites gestellten Formeinheiten der Sporae dispersae der Mitteleozänen Braunkohle des mittloren Geiseltales (Tagebau Neumark-west i.w. S.) unter Berücksichtigung und Revision weiterer Sporenformen aus der bisherigen Literatur. *Beihefte zur Zeitschrift Geologie, 21/22:* 1–425.

KRUTSCH, W., 1960. Beitrage zur Sporenpaläontologie der präoberoligozäne kontinentalen und marinen Tertiarablarungen Brandenburgs. *Bericht der Geologischen Gesselsschaft in der Deutschen Demokratischen Republik, 5:* 290–343.

KRUTSCH, W., 1962. Stratigraphisch bzw. botanisch wichtige neue Sporen und Pollenformen aus dem deutschen Tertiär. *Geologie. Zeitschrift für das Gesamtgebiet der Geologischen Wissenschaften, 11;* 261–392.

KRUTSCH, W., 1966. Zur kenntnis der praquartaren periporaten pollenformen. *Geologie Beihefte, 55:* 16–71.

KRUTSCH, W., 1970. *Atlas der mittel-und-jungtertiären dispersen Sporen und Pollen sowie der Mikroplankton-formen des nördlichen Mitteleuropas VII Monporate, monocolpate, longicolpate, dicolpate et ephedroide (polyplicate) Pollenformen,* Berlin: Zentrale Geolgische Institut.

LAMBERT, A., & HSÜ, K. J., 1979. Non annual cycles of varve-like sedimentation in Watensee, Switzerland. *Sedimentology, 26:* 453–461.

LAMBERT, D. J., 1976. A detailed study of initial deposition of Tertiary lacustrine sediments near Mills, Utah. *Brigham Young University Geology Studies, 23:* 9–35.

LANGANGEN, A., 1974. Ecology and distribution of Norwegian charophytes. *Norwegian Journal of Botany, 21:* 31–52.

LIENGJARERN, M., COSTA, L. & DOWNIE, C., 1980. Dinoflagellate cysts from the Upper Eocene–Lower Oligocene of the Isle of Wight. *Palaeontology, 23:* 475–499.

LIND, E. M. & UISER, S. A., 1962. A study of a swamp at the North end of Lake Victoria. *Journal of Ecology, 50:* 599–613.

LUDLAM, S. D., 1969. Fayettville Green Lake, New York III The laminated sediments. *Limnology and Oceanography, 14:* 848–857.

LUDLAM, S. D., 1981. Sedimentation rates in Fayettville Green Lake, New York, U.S.A. *Sedimentology, 28:* 85–96.

MACHIN, J., 1971. Plant microfossils from the Tertiary deposits of the Isle of Wight. *New Phytologist, 70:* 851–872.

MAI, D. H., 1961. Über eine fossile Tiliaceen-Blüte und Pollen aus dem deutschen Tertiär. *Geologie, 1:* 54–93.

MAI, D. H. & WALTHER, H., 1978. Die Floren der Haselbacher Serie im Weisselster—Becken (Bezirk Liepzig, DDR). *Abhandlungen des Staatlichen Museums für Mineralogie und Geologie zu Dresden, 28:* 1–101.

MANUM, S., 1960. On the genus *Pityosporites* Seward 1914 with a new description of *Pityosporites antarcticus* Seward. *Nytt Magasin for Botanikk, 8:* 11–15.

MANUM, S., 1962. Studies in the Tertiary flora of Spitsbergen, with notes on Tertiary floras of Ellesmere Island, Greenland, and Iceland: A palynological investigation. *Norsk Polarinstitut, 125:* 1–127.

MANUM, S., 1976. Palynomorphs and palynodebris in relation to environment in the Tertiary Norwegian sea (D.S.D.P. leg 38 material). *Courier Forschungsinstitut Senckenburg, 17:* 86.

MARTIN, A. C. & UHLER, F. M., 1951. Food of game ducks in the United States and Canada. *United States Fish and Wildlife Service Research Report, 30.*

MARTIN, H. A. & ROUSE, G. F., 1966. Palynology of Late Tertiary sediments from Queen Charlotte Islands, British Columbia. *Canadian Journal of Botany, 44:* 171–208.

McANDREWS, J. H., BERTI, A. A. & NORRIS, G., 1973. Key to the Quaternary pollen and spores of the great lakes region. *Life Sciences Miscellaneous Publications, Royal Ontario Museum.*

McBRIDE, E. F., 1974. Significance of color in red, green, purple, olive brown and grey beds of Difunta Group, northeastern Mexico. *Journal of sedimentary Petrology, 44:* 760–773.

McLEROY, C. A. & ANDERSON, R. Y., 1966. Laminations of the Oligocene Florissant Lake deposits, Colorado. *Bulletin of the Geological Society of America, 77:* 605–618.

MERILÄINEN, J. & TOIVONEN, H., 1979. Lake Keshimmainen, dynamics of vegetation in a small, shallow lake. *Annales Botanici Fennici, 16:* 123–139.

MILLINGTON, W. F. & GAWLIK, S. R., 1975. Cell shape and wall pattern in relation to cytoplasmic organization in *Pediastrum simplex*. *American Journal of Botany, 62:* 824–832.

MISRA, R. D., 1938. Edaphic factors in the distribution of aquatic plants in English lakes. *Journal of Ecology, 26:* 411–451.

MORTIMER, C. H., 1941. The exchange of dissolved substances between mud and water in lakes. I, II. *Journal of Ecology, 29:* 280–329.

MORTIMER, C. H., 1942. The exchange of dissolved substances between mud and water in lakes. III and IV. *Journal of Ecology, 30:* 147–201.

MOSS, B., 1980. *Ecology of Freshwaters*. Oxford: Blackwell.

MULLENS, M. C., 1977. Bibliography of the geology of the Green River Formation, Colorado, Utah and Wyoming to March 1 1977. *United States Geological Survey Circular, 754.*

NICHOLS, D. J., AMES, H. T. & TRAVERSE, A., 1973. On *Arecipites* Wodehouse, *Monocolpopollenites* Thomson and Pflug and the species "*Monocolpopollenites tranquillus*". *Taxon, 22:* 241–256.

NIKITIN, P. A., 1965. *An Aquitanian seed flora from Iagernogo Sad, Tomsk*. Tomsk: Tomsk University. (In Russian.)

NORVICK, M. S. & BURGER, D., 1976. Palynology of the Cenomanian of Bathurst Island, Northern Territory, Australia. *Bulletin of Mineral Resources Geology and Geophysics, Australian Government Publishing Service, 151:* 1–169.

OLLIVIER-PIERRE, M-F., 1970. *Contribution à l'étude palynologique de niveau sapropelien de la sonnetierre en la Bernerie (Loire Atlantique).* Unpublished Ph.D. Thesis, University of Rennes, France.

OLLIVIER-PIERRE, M-F., 1980. Étude palynologique (spores et pollens) de gisements paléogènes du Massif Armoricain. Stratigraphie et paléogéographie. *Mémoires de la Société géologiques et minéralogique de Bretagne, 25:* 1–239.

PACLTOVA, B., 1966. The results of micropalaeobotanical studies of the Chattian-Aquitainian complex in Slovakia. *Rozpravy Ceskoslovenské Akademie Ved, 76:* 1–78.

PARKER, R. H., 1960. Ecology and distributional patterns of marine macro-invertebrates, Northern Gulf of Mexico. In F. P. Shepard, F. B. Phleger & T. H. Van Andel (Eds), *Recent Sediments, Northwest Gulf of Mexico:* 302–337. Tulsa: American Association of Petroleum Geology.

PENFOUND, W. & HATHAWAY, E. S., 1938. Plant communities in the marshlands of south eastern Louisiana. *Ecological Monographs, 8:* 1–56.

PETERSON, A. R., 1976. Paleoenvironments of the Colton Formation, Colton, Utah. *Brigham Young University Geology Studies, 23:* 3–35.

PFLUG, H. D., 1953. Zur Entstehung und Entwicklung des angiospermiden Pollens in der Erdgeschichte. *Palaeontographica, B, 95:* 60–171.

PICARD, M. D. & HIGH, L. R., 1972. Criteria for recognizing lacustrine rocks. In J. K. Rigby & W. K. Hamblin (Eds), *Recognition of Ancient Sedimentary Environments:* 108–145. *Society of Economic Palaeontologists and Mineralogists Special Publication 16.*

PICARD, M. D. & HIGH, L. R., 1979. Lacustrine Stratigraphic Relations. In A. M. Anderson & W. J. Biljon (Eds), *Some Sedimentary Basins and Associated Ore Deposits of South Africa:* 1–21. *Geological Society of South Africa Special Publication 6.*

POTONIÉ, R., 1931. Pollen formen der Miozänen Braunkohle. *Sitzungsberichte der Gesellschaft Naturforschender Freunde zu Berlin, 1–3:* 24.

POTONIÉ, R., 1956. Synopsis der Gattungen der Sporae dispersae I Teil: Sporites. *Beihefte zum Geologischen Jahrbuch, 23:* 1–103.

POTONIÉ, R., 1958. Synopsis der Gattungen der Sporae dispersae II Teil. *Beihefte zum Geologischen Jahrbuch, 31:* 1–114.

PROCTOR, M. & YEO, P., 1973. *The Pollination of Flowers*. London: Collins.

REEVES, C. C., 1968. *Introduction to Paleolimnology*. Amsterdam: Elsevier.

REID, C. & REID, E. M., 1910. The lignite of Bovey Tracey. *Philosophical Transactions of the Royal Society of London B, 201:* 161–178.

REID, E. M. & CHANDLER, M. E. J., 1926. *Catalogue of Cainozoic Plants in the Department of Geology I The Bembridge Flora*. London: British Museum (Natural History).

REINECK, H.-E. & SINGH, I. B., 1973. *Depositional Sedimentary Environments*. Heidelberg: Springer-Verlag.

RIDLEY, H. N., 1930. *The Dispersal of Plants Throughout the World*. Ashford, Kent: Reeve & Co.

ROCHE, E., 1973. Étude des sporomorphes du Landénien de Belgique et de quelques gisements du Sparnacian Français. *Mémoires pour Servir à Explication des Cartes Géologiques et Minières de la Belgique, 13:* 13–138.

RUSSELL, R. J., 1942. Flotant. *Geographical Review, 32:* 74–98.

RYDER, R. T., FOUCH, T. P. & ELISON, J. H., 1976. Early Tertiary sedimentation in the Western Uinta Basin, Utah. *Bulletin of the Geological Society of America, 87:* 496–512.

SCHOFIELD, E. K., 1976. Historical ecology in the Galapagos Islands, (ii) A Holocene spore record, from El Juneo Lake, Isla San Cristobal. *Journal of Ecology, 64:* 1013–1028.

SCHOLL, D. W., 1963. Sedimentation in modern coastal swamps, southwestern Florida. *Bulletin of the American Association of Petroleum Geologists, 47:* 1581–1603.

SCHUMACKER-LAMBRY, J. & ROCHE, E., 1973. Étude palynologique (pollens et spores) des marnes à emprientes de Gelinden (Palaeocene, Belgique). *Annales de la Société Géologique de Belgique, 96:* 413–433.

SCOTT, A. C. & COLLINSON, M. E., 1978. Organic sedimentary particles: Results from scanning electron microscope studies of fragmentary plant material. In W. B. Whalley (Ed.), *Scanning Electron Microscopy in the Study of Sediments:* 137–167. Norwich: Geo Abstracts.

SCULTHORPE, C. D., 1967. *The Biology of Aquatic Vascular Plants.* London: Edward Arnold.

SITTLER, C., SCHULER, M., CARATINI, C., CHATEAUNEUF, J. J., GRUAS-CAVAGNETTO, C., JARDINE, S., OLLIVIER-PIERRE, M-F., ROCHE, E. & TISSOT, C., 1975. Extension stratigraphique repartition geographique et ecologie de deux genres polliniques paleogenes observes en Europe occidentale.: *Aglaoreidia & Boehlensipollis. Bulletin. Société Botanique de France Colloquium Palynologie, 122:* 231–245.

SPACKMAN, W., DOLSEN, C. P. & RIEGEL, W. L., 1966. Phytogenic organic sediments and sedimentary environments in the Everglades-mangrove complex. Part I: Evidence of a transgressing sea and its effects on environments of the Shark River area of Southwestern Florida. *Palaeontographica, B, 117:* 135–152.

SPACKMAN, W., RIEGEL, W. L. & DOLSEN, C. P., 1969. Geological and biological interactions in the swamp-marsh complex of Southern Florida. *Special papers of the Geological Society of America, 114:* 1–35.

SPICER, R. A., 1980. The importance of depositional sorting to the biostratigraphy of plant megafossils. In D. L. Dilcher & T. N. Taylor (Eds), *Biostratigraphy of Fossils Plants. Successional and Paleoecological Analyses:* 171–183. Stroudburg, Pennsylvania: Dowden, Hutchinson & Ross.

SRIVASTAVA, S. K., 1972. Some spores and pollen from the Paleocene Oak Hill Member of the Naheala Formation Alabama (U.S.A.). *Review of Palaeobotany and Palynology, 14:* 217–285.

STRØM, K. M., 1955. Land locked waters and the deposition of black muds. In P. D. Trask (Ed.), *Recent Marine Sediments, A Symposium:* 357–372. Society of Economic Palaeontologists and Mineralogists Special Publication, 4.

THIERGART, F., 1938. Die pollen flora der Niederlansitzer Braunkohle, besonders im Profil der Grube Marga bei Senftenberg. *Jahrbuch der (königlich) Preussischen Geologischen Landesanstalt und (Bergakademie), 58:* 282–351.

THOMSON, P. W. & PFLUG, H., 1953. Pollen und Sporen des mitteleuropaischen Tertiärs. *Palaeontographica, B, 94:* 1–138.

TOLONOEN, M., 1978. Palaeoecology of annually laminated sediments in lake Ahuenainen, South Finland, I, II, III. *Annales Botanici Fennici, 15:* 177–208, 209–222, 223–240.

TRAVERSE, A., 1955. Pollen analyses of the Brandon lignite of Vermont. *United States Department of the Interior Bureau of Mines Report, 5151:* 1–107.

TWENHOFEL, W. H. & McKELVEY, V. E., 1941. Sediments of fresh-water lakes. *Bulletin of the American Association of Petroleum Geologists, 25:* 826–849.

VAN GEEL, B., 1976. Fossil spores of Zygnemataceae in ditches of pre historic settlements in Hoogkarpel (The Netherlands). *Review of Palaeobotany and Palynology, 22:* 337–344.

VAN GEEL, B., BOHNCKE, S. J. P. & DEE, H., 1981. A palaeoecological study of an upper late glacial and Holocene sequence from 'De Borchert' the Netherlands. *Review of Palaeobotany and Palynology, 31:* 367–448.

VAN VEEN, F. R., 1971. Depositional environments of the Eocene Mirador and Misera Formations, Maracaibo Basin, Venezuela. *Geologie en Mijnbouw, 50:* 527–546.

WALKER, C. A., 1980. The distribution of birds in the English Palaeogene. *Tertiary Research 3:* 25–30.

WARME, J. E., 1971. Palaeoecological aspects of a modern coastal lagoon. *University of California Publications in Geological Sciences.*

WATTS, W. A., 1959. Interglacial deposits at Kilbeg and Newtown Co. Waterford. *Proceedings of the Royal Irish Academy, 60:* 79–134.

WATTS, W. A. & BRIGHT, R. C., 1968. Pollen, seed and mollusk analyses of a sediment core from Pickerel Lake, north eastern South Dakota. *Bulletin of the Geological Society of America, 79:* 855–876.

WATTS, W. A. & WINTER, T. C., 1966. Plant microfossils from Kirchner Marsh Minnesota—A paleoecological study. *Bulletin of the Geological Society of America, 77:* 1339–1360.

WATTS, W. A. & WRIGHT, H. E., 1966. Late-Wisconsin pollen and seed analysis from the Nebraska sandhills. *Ecology, 47:* 202–210.

WELCH, D., 1966. *Juncus squarrosus* L. Biological flora of the British Isles. *Journal of Ecology, 54:* 535–548.

WEST, R. G. & GAY-WILSON, D., 1968. Plant remains from the Corton Beds at Lowestoft, Suffolk. *Geological Magazine, 105:* 116–123.

WEYLAND, H., PFLUG, H. & MUELLER, H., 1960. Die Pflanzenreste der Pliozanen Braunkohle von Ptolemais in Nordgriechenland II. *Palaeontographica, B, 106:* 71–98.

WILSON, M. V. H., 1977. Paleoecology of Eocene lacustrine varves at Horsefly, British Colombia. *Canadian Journal of Earth Sciences, 14:* 953–962.

WILSON, M. V. H., 1980. Eocene lake environments: depth and distance-from-shore variation in fish, insect and plant assemblages. *Palaeogeography, Palaeoclimatology, Palaeoecology, 32:* 21–44.

WODEHOUSE, R. P., 1933. Tertiary pollen II The oil shales of the Eocene Green River Formation. *Bulletin of the Torrey Botanical Club, 60:* 479–524.

WOOD, R. D., 1952a. An analysis of ecological factors in the occurrence of Characeae in the Woods Hole region, Massachusetts. *Ecology, 33:* 104–109.

WOOD, R. D., 1952b. The Characeae. *Botanical Review, 18:* 317–353.

WOODWARD, H., 1877. On the occurrence of *Branchipus* (or *Chirocephalus*) in a fossil state in the upper part of the fluvio-marine series (Middle Eocene) at Gurnet and Thorness Bay, near Cowes, Isle of Wight. *Report of the British Association for the Advancement of Science, Transactions of the Geology subsection, 78.*

WOODWARD, H., 1879. On the occurrence of *Branchipus* (or *Chirocephalus*) in a fossil state, associated with Eosphraeroma and with numerous insect remains, in the Eocene freshwater (Bembridge) Limestone of Gurnet Bay, Isle of Wight. *Quarterly Journal of the Geological Society of London, 35:* 342–350.

YOUNT, J. L., 1963. South Atlantic States. In D. G. Frey (Ed.), *Limnology in North America:* 269–286. Madison: University of Wisconsin Press.

ZANGERL, R. & RICHARDSON, E. S., 1963. The palaeoecological history of two Pennsylvanian black shales. *Fieldiana: Geology Memoirs, 4:* 1–352.